The Book of Wilding

Isabella Tree & Charlie Burrell

The Book of
WILDING

A Practical Guide to Rewilding
Big and Small

BLOOMSBURY PUBLISHING
LONDON · OXFORD · NEW YORK · NEW DELHI · SYDNEY

The Book of Wilding

Contents

Introduction
The ripple effect of rewilding

In 2002, we began rewilding a 1,400-hectare estate in
West Sussex inherited from Charlie's grandparents.
The land is heavy Low Weald clay, which is notoriously
challenging for agriculture and generally considered
'marginal' by farmers.

Propped up by subsidies, however, Knepp Estate had been
intensively farmed for arable and dairy since the Second World
War. By the time we took over, in 1987, the farm was making heavy
losses, even after subsidies. Charlie – then in his early twenties, fresh
out of agricultural college and full of ideas about new crop varieties,
modern livestock breeds and the latest technology – was convinced
he could make a go of it. It took sixteen years and an overdraft of
£1.5 million to make him realise the clay had won.

   In long, gloomy farm meetings it seemed the writing was on the
wall. Selling off the estate was not an option for us – or, at least,
one we would resort to only at the very last ditch. Knepp has been in
Charlie's family for more than 200 years. The landholding itself harks
back to the time of King John in the early thirteenth century, when it
was a hunting 'forest' inhabited by fallow deer and wild boar and pre-
sided over by Old Knepp Castle, now a single ruined tower. We knew
we had to find a different way for the future of the estate, something
that would be viable for the long term.

It had to be something that worked with the land, rather than battling against it.

Our conversion to rewilding was a succession of epiphanies. In the 1990s rewilding was still a dirty word, dodged by conservationists, reviled by farmers and ignored by policymakers. To most people, if they'd heard it at all, it was about the reintroduction of wolves – unlikely to appeal to the denizens of England's southern counties. A trip to the Netherlands to meet the renowned ecologist Frans Vera lit a fuse for us. As he explained it, rewilding is – unlike conventional nature conservation – not about targeting certain species or groups of species with intensive human management that maximises conditions for their survival. Rather, it is about putting nature back in the driving seat, allowing habitats to evolve, to shift and change, to find their own way. Keeping an open mind about what species may return, and how the landscape might look, is key. That, in itself, was appealing to us. While micro-management in nature reserves has been vital for preserving endangered scraps of habitat and species that would other-wise have been lost, there were no such rarities at Knepp. We were starting from the lowest of baselines and had little to lose. We'd had enough of expensive, hands-on, intensive land management, and this looser approach seemed just what we were after.

But you can't simply close the farm gate, let the land go and expect miracles to happen. Land like ours, over decades of ploughing and chemical abuse, has undergone what ecologists call a 'catastrophic shift' to a depleted state of equilibrium: our soils were no longer func-tioning, and the dynamic natural processes that give rise to life were almost entirely missing. We would need some initial interventions to start the ball rolling, to put nature back in the driving seat.

Vera identifies large, free-roaming grazing animals as the primary drivers of nature restoration. Cattle, horses, bison, water buffalo, elk, reindeer, wild boar and beavers – animals that once ranged the con-tinent of Europe in vast numbers – are 'keystone species', capable of having a hugely beneficial impact on the environment. Their natural disturbance of the soil through trampling, rootling and puddling; their effect on vegetation through grazing and browsing; the trans-portation of seeds and nutrients from place to place in their dung, hooves and fur; their impact on watercourses through dam-building and trampling riverbanks; all these trigger habitat change and open up opportunities for other forms of life. Vera's version of rewilding is about using these animals – as many as are logistically available – as landscape engineers, allowing them the space and time to kick-start nature's own engine, to re-create dynamic, biodiverse ecosystems. The key for us, as human beings, is to stand back once the train has left the station, and try not to intervene.

Could Vera's approach, which was controversial among conservationists at the time, work its miracles at Knepp, we wondered? Devoting the estate to nature restoration began to seem desperately urgent. We had always been passionate about wildlife in an amateurish way, and had travelled the world to see it, but we'd never questioned why we had so little in our own backyard. We were now beginning to realise that our own intensive farming methods had been a major part of the problem. Thanks largely to the advance of industrial, chemical-based agriculture, landscapes the world over have undergone a devastating transformation. Pollution of land and sea is now pervasive. Fertile soil and natural water sources are disappearing. Deforestation is at galloping pace. Biodiversity is crashing. The life-support systems on which all species, including our own, depend are on the verge of collapse. Nowhere is this more evident than in the United Kingdom since the Second World War. Driven by farming subsidies, every available inch of land has been ploughed, loaded with artificial fertiliser and drenched with chemicals. According to the *State of Nature* report conducted in 2016 by more than fifty nature-preservation organisations, the UK has lost significantly more biodiversity than the world average. Ranked 29th lowest out of 218 countries, we are among the most nature-depleted countries in the world.

An experiment in nature restoration using low densities of free-roaming animals at Knepp seemed relatively straightforward and eminently suitable for our heavy clay soil. If we could secure transitional funding, this could be a relatively low-cost way of creating conditions for wildlife to thrive on our land again. We would be starting from scratch, and our ambitions were broad and holistic, in line with the open-ended, non-goal-orientated thinking of rewilding. If we could improve biodiversity at Knepp even slightly, we would be happy.

Bit by bit, as we secured agri-environment grants from the government in stages, and getting braver all the time, we took our fields out of arable production and allowed them to revert to nature. We removed internal fences and began introducing herds of free-roaming ponies, pigs and deer to mimic the actions of the lost herds in the landscapes of the past. We had no expectations of the outcome. We had absolutely no idea that, within twenty years, our depleted land at Knepp would become one of the most significant hotspots for wildlife in the country, and a magnet for some of the UK's rarest species.

As a kaleidoscope of new habitats began to spring up under the feet and noses of our free-roaming animals, extraordinary things began to happen. Nightingales, peregrine falcons, lesser spotted woodpeckers, purple emperor butterflies and numerous other nationally rare birds and insects found us. Some of them, having discovered at Knepp resources that are now scarce in the wider countryside, were showing behaviour very different from the way they're described in textbooks. We began to realise that modern science, which observes species in a depleted landscape, often has a skewed vision of what is 'natural'. The term 'shifting baselines' (the acceptance as normal, generation by generation, of increasingly degraded natural ecosystems) was becoming familiar to us. However, Knepp was now demonstrating a new 'normal', shifting the baseline in the other direction. The sheer number – the biomass – of more common species proliferating at Knepp was breathtaking, even to naturalists who had spent a lifetime in the field.

As wildlife poured in, people began coming, too. Teams from conservation NGOs and bird and butterfly enthusiasts at first; then, as word spread, members of the public who had read about us in the press. We started a low-impact eco-tourism business based on our experience of African safaris, including glamping, camping and guided tours. Rewilding had unearthed an income stream we could never have dreamed of, and the business now turns over £1 million a year with a 20 per cent margin that would have been unimaginable against the 1 per cent margin of our previous farming business. Converting farm buildings to office space, light industrial use and storage has produced another significant source of revenue that, together with meat sales from managing the numbers of our organic, free-roaming animals, makes us financially viable for the first time in many decades. We now have our own butchery and are about to open a café and farm shop. Our employment figure has risen from twenty-three full-time employees under farming to fifty under rewilding, and continues to rise every year. As we see it, rewilding is not about excluding people from the landscape, as some of its critics claim. Rather, it opens up opportunities for employment and much greater public engagement with the natural world.

In 2018, after countless visits to Knepp from government ministers, advisors and civil servants, the Department for Environment, Food & Rural Affairs (DEFRA) singled out Knepp in its 25 Year

Environment Plan as an outstanding example of 'landscape-scale restoration in recovering nature'. Now enshrined in the 2021 Environment Act, nature restoration has been declared a pivotal part of the UK government's efforts to combat climate change and bio-diversity loss. Rewilding is part of that strategy, considered a primary route to swift and cost-effective nature recovery. Gone are the days when government ministers would skirt around the 'R' word.

The government's ambition, now a legal commitment, is to protect 30 per cent of Britain's land and sea for nature by 2030. Part of the plan is for between ten and fifteen rewilding projects based on the Knepp model, each between 500 and 5,000 hectares. However, this is a goal that many conservation experts predict will be missed. It is one thing to declare high-minded policies, another to deliver them. And, of course, governments are notoriously fickle. In times of economic and political crisis, with funding under pressure, the environment invariably draws the short straw.

Whether committed to nature restoration or not, governments on their own are unlikely ever to have the money or the will to do it on the scale that is needed to avert environmental disaster. The United Nations estimates that at least $500 billion a year must be made available for 'redirecting, repurposing or eliminating incentives' that harm biodiversity – basically, to halt further decline.[1] Clearly, far, far more will be needed to push biodiversity back into recovery. The private sector is set to play a much more significant role in nature recovery in the years to come by funding rewilding through carbon and biodiversity credits, a move that is happening across the globe (see Chapter 9). This rapidly evolving market opens up opportunities for a whole new spectrum of rewilders, and could be the way we begin to turn the *Titanic*.

Explaining rewilding, how to do it and how to pay for it has become an increasingly important part of what we do at Knepp, and is the catalyst behind this book. Everything we've learned, the research we've done, the experts we've talked to, has fed into these pages. Hundreds of farmers and landowners, some of them from farm clusters and community groups, some from Europe and further afield, have come on Rewilding Workshops at Knepp. What we teach in these workshops forms the basis of the chapters about rewilding at landscape scale (hundreds or thousands of hectares) and on a smaller scale (areas of 10–100 hectares). The latter might include rewilding part of a farm, not just as nature for nature's sake, but as part of a strategy to boost productivity and sustainability. Our experience also informs the chapters on rewilding gardens and making space for nature in cities.

One question we are often asked is where food production sits with rewilding. Many consider rewilding as being at odds with our need to grow food, because it might compete directly for land. We see this very differently. Almost always unaddressed in these conversations is the shocking amount of food that is wasted. Globally, we produce enough food for over 10 billion people, 2 billion more than the current world population. But a third is wasted – 1.3 billion tonnes of edible food a year.[2]

The most pressing question concerning food security is not the quantity of food we're producing, but the way we produce it. Globally, we need to shift to regenerative agriculture (farming practices that enhance ecosystems), which can help to restore our soils and protect our water sources through no ploughing, using cover crops, rotating crops with livestock, and agroforestry. Most urgently, we need to move away from the entirely unsustainable and unethical system of industrialised livestock production that relies on growing vast quantities of grain for animal feed – and which will almost certainly mean eating less meat and dairy. And we need to produce more of our own crops in our own countries, rather than importing them from far-flung parts of the globe, thereby depleting soil and resources in places we will never see. The war in Ukraine has shown us how precarious food supply lines can be. But a recent study has found that if Europeans reduced their consumption of meat and dairy by just 15 per cent, this would knock out entirely the need for imports of grain from Ukraine and Russia.[3]

Even with increased food production in the UK, there is enough land for agriculture *and* for rewilding. Indeed, the two work hand in glove. Rewilding will provide the life-support system needed to increase yields and protect farmland from the impact of climate change. It can provide the pollinating insects for crops and the natural predators to control pests and outbreaks of disease. It will help to replenish water tables, purify polluted water courses and protect farmland from both flooding and drought. Swathes of natural vegetation, running like webbing through our agricultural landscapes, will act as buffers against extreme weather events as the effects of climate change take hold. Far from being the enemy of farming, rewilding is its natural ally, securing its long-term future.

While farmers on marginal land may be interested in shifting entirely to rewilding, those on better soils will clearly want to continue

producing food. What we're seeing now, though, is a deeper appreciation of the relationship between rewilding and agriculture. Many
farmers, even those on the most productive land, are now interested
in how aspects of rewilding, such as restoring natural water courses
and creating wildlife corridors, can increase space for nature, generate
additional income and improve the conditions for farming.

As the transformation of Knepp has begun to mature, the wider
public response to what is happening here has been overwhelming.
People who camp at Knepp for a night or two, go on a safari or
simply walk the footpaths tell us how moved they have been by their
experience. We receive reams of letters, emails and messages: the
elderly, inspired by hearing woodlarks and nightingales, recalling the
last time they'd heard them, long ago, as children; retired farmers and
agronomists expressing regret, realising the damage they, like us, have
unwittingly done in their lives; the young exclaiming in joy at hearing
their first turtle dove or cuckoo. Some have been moved to create
poems and paintings, even to compose music. What they all express is
the sense of hope that Knepp brings. Seeing how nature can bounce
back in such profusion and with such astonishing speed, especially on
land as unpromising as ours, is both profoundly reassuring and galvanising. In this age of eco-anxiety, when we can so easily feel utterly
powerless and overwhelmed by the challenges of climate change and
biodiversity loss, experiencing rewilding seems to restore a sense of
agency and ambition. We now receive requests for information and
advice from people who have been inspired by what they have seen,
heard and felt at Knepp, who want to know how to become part of
the movement for change: how to rewild their garden, allotment,
orchard or even window box; how to influence the management of
public green space, such as roadside verges, avenues, local parks,
towpaths, embankments, churchyards and cemeteries; or how to
participate in a rewilding project.

This book, we hope, will answer many of those questions.
Inevitably, it is focused on the UK in its details. Regulations, funding,
organisations, cultural habits and many other factors will vary in
other countries. The general principles, however, apply to aspiring
rewilders everywhere. To our minds, rewilding is a spectrum and
everyone is on it, with the capacity to move even small spaces to ever
wilder degrees by connecting directly with existing habitat or acting
as a stepping stone between other areas of nature. Rewilding is about
thinking holistically, seeing oneself as an integral part of the much

bigger picture. Even if the 'Rewilding Your Garden' chapter seems the only one of practical use to you – or, indeed, if you only have a window box – the principles of rewilding discussed in the earlier chapters, such as restoring vegetation and natural water systems, and using large herbivores as drivers of recovery, are still relevant. By beginning to understand how nature works at scale, in the wild, we can learn how to replicate some of those processes and maximise conditions for biodiversity in smaller, confined spaces. Where natural processes are missing, and have no room to perform, we, as human beings, can become the keystone species. We can take lessons from the beaver, wild boar and bison. We can act for the good of nature, rather than against it. There has never been a more concerning time to live on this earth but, equally, there has never been a more exciting one. In recognising the miraculous ability of nature to restore itself, we can realise our own capacity to contribute to the rewilding of this planet, our home.

Exmoor ponies roam free in the Southern Block of Knepp's 1,400-hectare rewilding project in West Sussex.

1

What is Rewilding?
The essential principles

Old English longhorns graze right up to the front
of Knepp Castle, in the heart of our 1,400-hectare
rewilding project in West Sussex. They act as proxies
for their extinct ancestor, the aurochs, which once
roamed Britain in huge numbers.

Introducing large, free-roaming herbivores is one of
the key interventions rewilders can make at large
scale to kickstart a dynamic ecosystem. Reintroductions of other missing species – such as the white
stork you can see nesting on our chimney – help
inspire a desire for change, to live in a world rich in
wildlife once again. These are the first white storks
breeding successfully in Britain for over 600 years.

Over the past few years, 'rewilding' has become a household word. What began as the buzzword for allowing nature back into our landscapes has become synonymous with making anything wilder and more dynamic, including much smaller areas, such as gardens, churchyards, playgrounds and roadside verges, right down to window boxes. Inspired by its association with radical change and 'letting go', we've even begun to talk about rewilding institutions, belief systems, urban infrastructure, social conventions and – perhaps the biggest challenge of all – ourselves. But what does rewilding actually mean?

The term 'rewilding' was coined in the 1980s by a group of conservationists led by Dave Foreman, founder of the Wildlands Network and the Rewilding Institute in the United States. It appeared in print for the first time in 1990.[1] In 1998 the American biologists Michael Soulé and Reed Noss refined the idea by focusing on three principles: cores, corridors and carnivores.[2] They envisaged nature restoration on a vast scale, centred mainly on existing wildernesses and national parks: the 'cores'. They emphasised the importance of connectivity: joining up these isolated biodiversity hotspots using wildlife corridors so that wild animals and plants could move more freely and natural processes could function on a significant scale again.

Soulé and Noss also championed the role of apex predators: the carnivores at the top of the food chain, such as wolves, bears and mountain lions. This is something the father of modern conservation – and arguably the first rewilder – the American author and ecologist Aldo Leopold, had identified half a century earlier. Yellowstone National Park has become a flagship example of the rewilding movement in the USA ever since the reintroduction of wolves in 1995 was seen to trigger a staggering domino-effect increase in biodiversity, a phenomenon that has come to be known as the 'apex predator trophic cascade'. The wolves, for example, harried the herds of elk, pushing them away from their preferred easy pickings on riverbanks. This enabled the regrowth of aspens and willows, providing new habitat for songbirds. The increased shade along the river cooled the water, boosting populations of fish. With riverside willows to eat, beavers returned to the river, building dams which provide protective nurseries for aquatic invertebrates and fish fry. The wolves also outcompete smaller carnivores such as coyotes, pushing them out of wolf territory. With the drop in coyote numbers, populations of small mammals – their prey – have rocketed, providing more food for eagles, hawks and osprey.

The Yellowstone to Yukon Conservation Initiative, or Y2Y, established in 1997, was the most ambitious wildlife corridor of all. It was inspired by the vast roaming of a radio-collared she-wolf.[3] It is 3,200 kilometres long and 480–800 kilometres wide, and covers an area of about 1.3 million square kilometres, along the spine of North America's Rocky Mountains from the Greater Yellowstone ecosystem in Wyoming to Canada's Yukon Territory – an area well over five times the size of the UK.

Rewilding in Europe

In Europe – a continent half the size of North America, densely populated, heavily industrialised and historically fragmented – the concept of rewilding has evolved differently. In contrast to the USA, there are few true wilderness areas left in mainland Europe. Even remote mountains and national parks contain people, and rewilding will always involve them. Therefore, European rewilders think in terms of 'novel ecosystems'; of kick-starting dynamic natural processes in nature-depleted areas, and accepting that the new forms of nature that emerge might be very different from the natural systems of the past.

In Europe, the areas where nature can be considered to be properly functioning are generally remote, lightly populated and/or mountainous: they include northern Scandinavia, Estonia, parts of Romania, Slovakia, Bulgaria, Spain and northern Greece; the Apennines, the Alps, the Pyrenees, the Danube Delta and the islands of Corsica and Sardinia. In the UK, nowhere, not even the mountains of Wales and Scotland, has 'ecological integrity'. This means that nature in these areas is not supported by a functioning ecosystem, and is subject to intense human pressure.

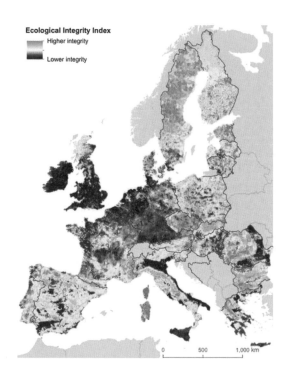

Ecological Integrity Index
Higher integrity
Lower integrity

The importance of connectivity is emphasised by Natura 2000, the network of nature protection areas in the European Union. This map from 2020 shows potential corridors that could link up fragmented areas of nature and reanimate the whole continent.[4]

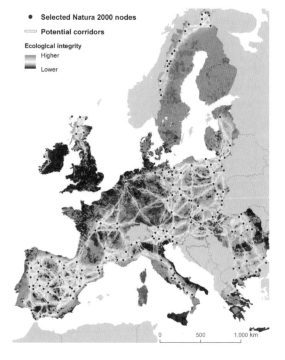

• Selected Natura 2000 nodes
▭ Potential corridors
Ecological integrity
Higher
Lower

The Netherlands is one of the most densely populated countries in the world, and it has been hugely influential in formulating ideas about natural processes. In the 1980s the Dutch government embarked on a radical new policy of 'nature development' based on reconnecting remaining areas of nature. The long-term aim was to get natural ecological systems functioning and evolving without the constant intensive management that is normally deployed in European nature reserves.

The Dutch have also focused on restoring natural water systems. Catastrophic floods in 1993 and 1995 led to the 'Room for the River' project (2006–15), which gave back hard-won reclaimed land (polder) to the rivers, cutting meanders back into the floodplains and restoring old marshes and wetlands. This radical step of 'rewilding' rivers in the Netherlands has reduced the predicted risk of extreme flooding from once every 100 years to once every 1,250 years, and advanced the idea that restoring natural hydrology benefits both the economy and biodiversity.

In the 1990s, a group of Dutch ecologists put forward another idea that has become a feature of European rewilding: the importance of grazing animals in natural systems. A key figure behind this idea is Dr Frans Vera, a biologist and conservationist whose work, which strives for more dynamic natural ecosystems, has influenced the ecological strategy for the Netherlands. Vera's book *Grazing Ecology and Forest History* (2000) explains how free-roaming herbivores can be the creators of habitats and drivers of dynamic natural processes. He argues that, historically, ecology has been too focused on botany. In our efforts to understand how the landscapes of the past would have looked, and hugely influenced by the study of pollen in the fossil record, we have concentrated on 'vegetation succession' (the growth stages of plants and trees) and almost entirely neglected zoology. We have forgotten about the impact that free-roaming herds of large animals, such as aurochs (the original wild ox), bison, tarpan (the European wild horse), water buffalo, elk, reindeer, red deer and wild boar would have had on the landscape before humans hunted them to extinction or drove them into inaccessible margins where they survive only in low numbers, if at all.

After the end of the last ice age (from about 12,000 years ago), great herds of herbivores would have wandered freely across Europe and everywhere else on the planet, much like the herds that still range parts of Africa. Vera argues that three broad characteristics – their disturbance of vegetation (trampling, wallowing, rootling, grazing

19

and browsing, uprooting shrubs and de-barking trees); the way they transport seeds in their gut, hooves and fur and replenish the soil with their dung, urine and decomposing carcasses; and their migrations that moved nutrients over great distances – had a fundamental impact on the environment and especially on vegetation cover. In temperate Europe, large herbivores would have prevented trees and woody shrubs from winning out, and created a much more open, complex landscape than the primal, closed-canopy forest that has come to prevail as Europe's origin myth. Large herbivores are all keystone species: animals that have a disproportionate effect on their environment, and that therefore boost biodiversity.

The radical experiment of the Oostvaardersplassen began in 1983, just 32 kilometres northeast of Amsterdam. It was the first trial on a significant scale of the introduction of large, free-roaming herbivores (intended as proxies for some of those now extinct in Europe), to see if they would generate habitats and increase biodiversity. The results shifted ecological understanding profoundly.

'Wilded' herds of Heck cattle, Konik ponies and red deer were released into 60 square kilometres of polder that was being colonised rapidly by willow saplings and on its way to becoming closed-canopy woodland. In just a few years, these animals transformed the area into a Serengeti-style savannah teeming with small mammals and birds including bitterns, spoonbills, lapwings, skylarks, white-tailed eagles and thousands of geese.[5] The soil, once seabed, became a functioning, living ecosystem replete with soil biota including earthworms and dung beetles – a fundamental resource for the food chain. It is unsurprising that biodiversity there is far greater than in surrounding farmland. But it is also greater than areas of a similar size and character that are managed as nature reserves using conventional methods. Perhaps most interestingly, boom-and-bust cycles in the Oostvaardersplassen produce a spillover effect that radiates species into other areas. Bearded tits, great white egrets and spoonbills from the reserve, for example, have colonised the UK.

While the emphasis in the USA has been primarily on the domino effect of reintroducing apex predators into wilderness areas, in Europe the introduction of free-roaming herbivores has come to be seen as a fundamental step in kick-starting natural processes, particularly in areas devoid of wildlife or that have been depleted through centuries of human exploitation. This is something that

apex predators on their own cannot do. The famous trophic cascade of Yellowstone happened because wolves and mountain lions were returned to an almost fully functioning ecosystem, replete with prey. Put wolves or lynx into species-poor, closed-canopy woodland, forestry plantations, floodplains or degraded grassland, however, and not much will change, even if the predators manage to survive. But as soon as bison, beavers and wild boar are involved, change is dramatic; a complex mosaic of habitats begins to emerge and miracles happen. Wildlife pours in. The diversity and abundance of species rockets, and even rare species start to appear. As Europeans see it, large predators are the icing on the cake. Prey species and habitat must be present on a significant scale before apex predators can thrive and contribute in a positive way.

This imagined river delta shows how nature restoration can work. In Europe, animals such as red deer, elk, ponies, bison and beavers regenerate wetland habitats, and natural water systems allow passage for mass migrations of fish. In the Danube delta, Asian water buffalo have been introduced as a proxy for the extinct European water buffalo to help manage the reeds.

Missing animals

In the 1990s the idea that large herbivores could drive ecosystems and biodiversity began to resonate on both sides of the Atlantic.[6] Soon, ecologists were taking the idea further. In his book *Twilight of the Mammoths: Ice Age Extinctions and the Rewilding of America* (2005), palaeoecologist Paul Martin argues that simply conserving ecosystems as they exist in the present day is short-sighted and insufficient. Even our precious, protected wilderness areas, he says, are impoverished, their communities of wildlife incomplete – a mere shade of their true potential. To understand how simplified ecosystems across the planet have become, we must look further back, to the Pleistocene – before the last ice age – and the time of the great extinctions of megafauna (large or giant animals), which began about 45,000 years ago and accelerated dramatically about 22,000 years ago.[7]

The Pleistocene–Holocene transition, coinciding with the spread of humans around the globe, saw great extinctions of megafauna. The black silhouettes show the proportion of losses for each continent, with the greatest (58 extinctions) in South America.[8]

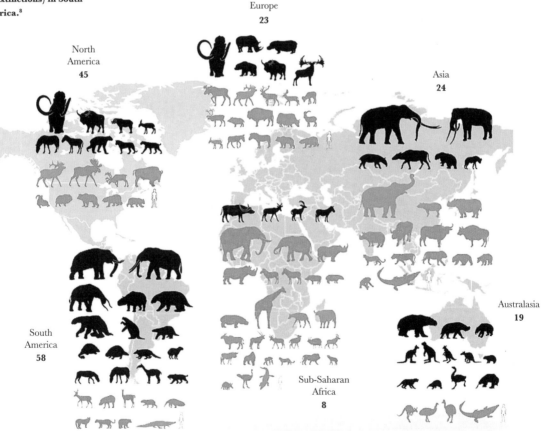

Europe
23

North
America
45

Asia
24

South
America
58

Australasia
19

Sub-Saharan
Africa
8

Conventional conservationists find it hard to consider how ecosystems worked before human impact because they are hampered by Western bias and a focus on the wrong baselines. In North America, the benchmark for nature, and the level at which most conservation efforts are focused, is how things were when Christopher Columbus arrived in 1492. In many other parts of the world, ecological ambitions are often aimed at recovering ecology as it was at the time of European colonisation, when Western naturalists began avidly collecting and naming all the new animals and plants they encountered. For Australia, it's the arrival of Captain James Cook in 1770. In Europe, the baseline tends to be the bucolic rural eighteenth century.

Martin argues that these baselines do not take into account the devastating environmental impact that humans had already had on the planet. The continents encountered by naturalists during the seafaring age of Europe's empires had already undergone what is known in ecological terms as a 'catastrophic shift'. Biodiversity had plummeted and ecosystems had become far less complex and dynamic – and this was because, about 40,000 years ago, the world's megafauna began to go extinct. The timing of the extinctions coincides with the spread of *Homo sapiens* around the planet, and a link between the two events is now almost universally accepted.[9]

It is much easier than one might expect to drive a large creature to extinction. Uncomfortable though it is, the theory of 'overkill' accounts for what happened to almost all super-megafauna (animals weighing over 2,000 kilograms) on the planet. Our only remaining terrestrial super-megafauna today are three species of elephant and five species of rhino (there have been as many as 250 species of rhino in the past, although not all at the same time). Some of these are still recovering from or experiencing overkill, and some are still bordering on extinction. Like many predators, including leopards and wolves, the instincts of our hunting ancestors compelled them to take advantage of easy pickings when they could. Wielding weapons, and with sophisticated language and social cooperation to their advantage, humans became the most organised, efficient and widespread predator the world has ever known, especially when they dispersed into areas where animals were naïve to the ways of humans.[10]

It was particularly easy and rewarding, albeit manifestly wasteful, for spear-throwing hunters to kill herds of large herbivores en masse, trapping them in cul-de-sacs or driving them over cliffs. The tipping point at which a very large mammal can no longer reproduce quickly enough to prevent decline towards extinction is reached relatively swiftly because of its low and slow rate of reproduction. The larger an animal, the more vulnerable it tends to be.[11]

Forty-five species of megafauna disappeared in North America after humans entered the continent.[12] Until recently, the date of *Homo sapiens'* arrival in North America was thought to be around 14,000 years ago, but recent studies have moved it to 25,000–27,000 years before the present, and perhaps even as far back as 50,000 years.[13] The period for human extinction of megafauna is consequently far longer than was originally thought. The animals that succumbed included giant sloths, peccaries and armadillos, four species of mammoth, mastodons, horses, gigantic beavers and several species of native camel.

Extinctions have been fewer in sub-Saharan Africa (just eight out of fifty species of megafauna), probably because humans evolved there and our prey had time to learn how to fear and evade us. Likewise, in Europe, the relatively slow spread of humans radiating out of Africa gave animals a chance to learn wariness. The first hominids colonising Europe from Africa had poorer tools and were, perhaps, less efficient killers than the organised bands of *Homo sapiens* that entered North America much later. The impact was particularly dramatic in Australia. More than two-thirds of the gigantic marsupial species went extinct shortly after first contact with human beings, around 45,000 years ago. On islands, the pattern holds true, with the largest animals and flightless birds generally succumbing first.

Before the continental disappearance of megafauna, the planet was at peak biodiversity. It was also, largely, a planet of grasslands and wood pasture.[14] The extinctions affected vegetation cover and food webs, diminishing and sometimes extinguishing species further down the food chain that were intimately connected to the larger animals through complex trophic cascades.[15] They also had a profound impact on natural processes, such as flows of energy, water, gases, organisms and nutrients.[16] Large animals once provided a vital service to ecosystems by transferring nutrients from nutrient-rich to nutrient-poor areas. Their digestive tracts are more efficient at breaking down plant and animal material than most other types of decomposition such as fungi and soil bacteria, which work relatively slowly. An animal, particularly a very large one, may therefore travel a considerable distance – perhaps tens of kilometres – before defecating and returning nutrients to the soil.[17]

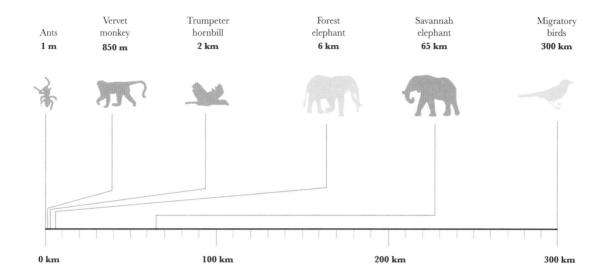

Ants	Vervet monkey	Trumpeter hornbill	Forest elephant	Savannah elephant	Migratory birds
1 m	**850 m**	**2 km**	**6 km**	**65 km**	**300 km**

0 km   100 km   200 km   300 km

Without free-roaming megafauna we have lost one of the planet's most effective systems of nutrient transfer.[18] The movement of phosphorus (the key nutrient required by plants and trees for growth) from the alluvial floodplains of the Amazon and its tributaries into the forest interior has declined by 98 per cent since the extinction of megafauna such as giant ground sloths. Before commercial whaling, whales would have moved 375 million kilograms of phosphorus to the surface of the sea; today, it's 82.5 million kilograms. When the herds of mammoths, woolly rhinoceroses, bison, horses, musk oxen, elk, saiga and yaks were hunted to extinction in Alaska and the Yukon, woody vegetation and mosses took over the steppe, absorbing heat from the sun and causing an estimated 0.2–1 °C warming in Siberia and Beringia. This may well have been the first human-induced global warming.[19]

In Australia, the loss of giant marsupial browsers ushered in a new ecosystem that was prone to fires. The temperate, complex, self-regulating forests of Australia's past are no more. A new league of fire-tolerant trees such as eucalypts, as well as trees and shrubs that actually require burning to reproduce, have taken over.[20] Other fire-prone regions of the world today, such as California and semi-arid savannahs in Africa, may be suffering similarly because of the loss of large herbivores.[21] The environmental organisation Rewilding Europe suggests the introduction of herbivores as a way of building fire resilience in Mediterranean countries.

The distances over which different animals can transfer nutrients. In today's world, birds are the most significicant dispersers of seeds. But in the past, at the height of megafaunal diversity, numerous large mammals would have transported much larger seeds and nutrients over great distances. The savannah elephant is the last remaining 'super-herbivore' on earth performing this role.

Many indigenous communities, for example those in tropical rainforests, the Arctic tundra, Australia and the Kalahari, ultimately came to realise the folly of overkill and have learned to live sustainably with their remaining large mammals. But generally, people across the world have continued hunting to extinction right up to the present day. In Europe, the last aurochs was hunted in Poland in 1627; the last wild tarpan is believed to have died in 1887, though the pure breed is likely to have disappeared much earlier; and the Eurasian beaver, once numbering hundreds of millions, was down to 1,200 individuals by 1900.

The catastrophic impact the loss of megafauna has had on global ecosystems and climate has led some people in the rewilding movement to argue for an earlier start date for the Anthropocene (the period during which human activity has been the dominant influence on the planet). This epoch is generally considered to start at the beginning of the Industrial Revolution, the moment when the human signature arguably becomes visible in the rock record. But some now propose another start date, one that isn't written in rock, but which begins in the late Pleistocene period of overkill and megafauna extinctions, the reverberations of which are still being felt.

Reintroductions

The catastrophic impact of megafauna extinctions so long ago may sound like just another disaster. But it has also opened up exciting new solutions to the problem of collapsing nature in the present day: the return of missing species.[22]

One of the earliest and most dramatic examples of introductions of missing megafauna acting as a catalyst for restoring natural systems occurred in the early 2000s on two small islands that were once home to the dodo, Round Island (169 hectares) and Île aux Aigrettes (25 hectares), off Mauritius in the Indian Ocean.[23] Conservation efforts on these islands in the 1980s focused on eradicating non-native rats, feral cats, rabbits and goats, all introduced by seafarers, that threatened critically endangered native species such as the Mauritius kestrel, the echo parakeet, the Mauritius fody (a rare bird in the weaver family), the golden bat and the pink pigeon. The successful elimination of these predators removed one significant risk for the endangered species, but also – unexpectedly – imposed another. Left with no grazers at all, the islands' flora took a beating. Tall plants and thuggish grasses took over, outcompeting other native grasses and endangering endemic plant species, such as the Mascarene amaranth, dramatically reducing floral diversity and smothering the natural tussocky grasslands of the island.

Ecologists realised that two extinct species of Mauritian giant tortoise must have been vital to the functioning of these islands' eco-systems. Giant tortoises – popular with seafarers because they could stay alive as food in a ship's hold for months – were probably extinct on Mauritius by 1700, and in its outlying islands by around 1735. A close living relative is itself a protected species: the giant tortoise of Aldabra, one of the remoter islands in the Seychelles archipelago, 1,600 kilometres away in the Indian Ocean.

Releasing an endangered species onto tiny, vulnerable islands where it had never lived and where it could be considered invasive was – and still is – controversial. The architects of the idea won the day largely because they argued that it would be easy to remove the giant tortoises if their impact was not proving beneficial.[24] In 2000 four Aldabra tortoises were released on to Île aux Aigrettes, where there is now a population of about twenty-five adults and several dozen juveniles. Twelve, harvested from there, were introduced on to Round Island, where, following further introductions, more than 700 now roam. Their impact, by creating a mosaic of vegetation types, has been astonishing and in many ways unexpected.[25] Research suggests, however, that the full benefit will be seen only when the tortoises reach densities of about 1,200 individuals per square kilometre.

Proxies and non-natives

Herbivory is now recognised as a fundamental ecological process, and the reintroduction of missing large herbivores is a key aspect of rewilding. But while surviving species of large herbivore can, at least in theory, be returned to areas where they were once present – such as bison on the great plains of North America or reindeer in the Arctic tundra – there's obviously a problem where species are extinct.

The Round Island and Île aux Aigrettes project demonstrated the efficacy of using a close relative as a substitute for an extinct species. The behaviour, disturbance, grazing and browsing preferences of the Aldabra giant tortoise are close enough – perhaps even identical – to those of its extinct cousins to generate the same, or very similar, outcomes. Non-native giant tortoises are now being introduced as 'vegetation management tools' in other places where giant tortoise species have become extinct, such as in the Galápagos, Madagascar and other parts of the Seychelles (where they benefit the Seychelles magpie robin – there were just sixteen individuals in 1960, but they have now recovered to more than 200 – by disturbing the ground).[26]

This idea is now one of the main tenets of rewilding in Europe. The bison, also known as the wisent, was once one of the most significant animals on the continent in terms both of numbers and of impact. It almost disappeared for ever. Of the various species and subspecies of European bison, just fifty-four individuals of one sub-species, *Bison bonasus bonasus*, survived into the twentieth century.

There are now around 7,000 free-roaming individuals of this sub-species, now known simply as European bison, in rewilding projects in fifteen European countries.[27] We will never know the differences in impact, subtle or otherwise, between *B. bonasus bonasus* and its extinct cousins. But in the absence of the latter, the former is proving an effective stimulus for increasing biodiversity. In rewilding terms, it's a case of using the tools at our disposal – the species that are left to us – and focusing on the restoration of dynamic natural processes, rather than getting hung up on the impossibility of restoring the ecosystems of the past. The small genetic pool of *B. bonasus bonasus* is a worry, though, and some ecologists argue for the benefits of crossing it with its close cousin the American bison (*B. bison*) to improve its genetic prospects in the long term – another example of the pragmatic approach of rewilding.

Old breeds of domesticated livestock can also be used as proxies for extinct species, as the pioneering work of the Oostvaardersplassen has demonstrated. In Europe, by the seventeenth century, tarpan (Europe's original wild horse) and aurochs (the original wild ox) had been hunted to extinction. But for thousands of years humans had been breeding domesticated strains of tarpan and aurochs for use as draught animals, for riding, and for meat, hides and milk. Domesticated horses and cattle were periodically turned loose to breed with their wild cousins, in order to restore hybrid vigour. Throughout the Middle Ages, herds of domesticated cattle, horses and pigs, released into the forests of Europe for browse and pannage, and driven from winter to summer grazing in the mountains by pastoral herders, maintained wood pasture landscapes and herd migrations similar to those that would originally have been generated by their wild ancestors. Herds of old-breed cattle and horses are now roaming free in rewilding projects across Europe, including at Knepp, proving their worth as catalysts for the restoration of dynamic ecosystems. In the same way, in areas where wild boar no longer survive, old pig breeds can be used as proxies for their wild relative.

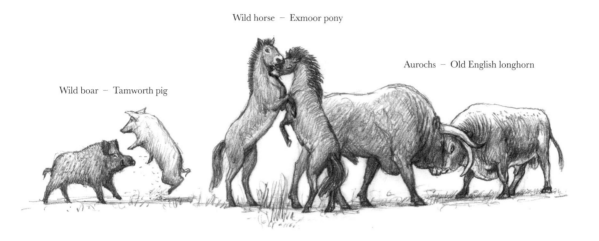

Wild horse – Exmoor pony

Aurochs – Old English longhorn

Wild boar – Tamworth pig

More controversial, perhaps, is the idea that surviving species may be useful proxies for quite different extinct species. Feral dromedaries in Australia browse in the harsh desert interior, effectively replacing the many extinct species of giant marsupial browser (extant species of kangaroos are predominantly grazers and smaller than their lost cousins).[28] Escaped hippos from the drug lord Pablo Escobar's private zoo graze the banks of the river Magdalena in Colombia, and so act as proxies for extinct megafauna in South America.[30]

Old breeds of domesticated livestock can act as proxies of their extinct ancestors, or when wild animals such as boar are unavailable.[29]

Pleistocene rewilding

Retracing our steps into the Pleistocene epoch reveals some surprises about the origin of species and throws up interesting challenges for rewilding and the future of conservation. Horses, for example, evolved in North America and became extinct there about 8,000 years ago, but only after they had dispersed into the rest of the world. Now considered 'non-native' in the continent of their origin, feral horses and donkeys (or 'burros'), brought to the Americas in the sixteenth century by the Spanish, roam wild in some areas of the States. They are broadly considered to be a problem despite, or perhaps because of, being protected under the 1971 Wild Free-Roaming Horses and Burros Act. That they are stepping back into their ecological niche is hard for most people – perhaps especially conservationists – to accept, given our reluctance to explore what ecosystems looked like behind the 'Columbian curtain' (before European conquest). As always, it's largely a perception of numbers of animals, and a prevailing view of how the landscape should look. What is 'too many' wild horses; when is their impact too great? It is hard to imagine the vast herds of the Pleistocene, let alone countenance them in the landscapes of today.

The reproductive strategies of plants often point to missing species. Until recently, ecologists puzzled over certain trees, particularly those in the tropics that have evolved excessively large fruit and/or rock-hard seeds, that appeared to have no means of natural dispersal. Some of these trees – those with edible fruit, such as avocado, guava and jackfruit, or scented trees, such as the honey locust – have survived only because people liked and cultivated them. The extinct megafauna with which these plants evolved explain the mystery. Native American mastodons and mammoths once fed on the pods of the honey locust, dispersing the seeds in their dung, just as today's elephants disperse baobab and marula seeds in Africa, the passage through their gut aiding germination. Likewise, fruit-eating giant ground sloths, giant armadillos and elephant-like mammals known as gomphotheres in South America helped to disperse tropical palms such as *Scheelea rostrata*, a species that is now only just clinging on in isolated remnants. Many of our plant species still yearn for the mouths that once fed on them, and may even be lost without them.[31]

Expanding our thinking to encompass these long-lost herbivore–plant relationships opens up the possibilities for recovering natural systems using reintroductions – or even introductions – of megafauna. The idea of releasing camels in North America, African elephants in the Amazon, and African forest elephants (*Loxodonta cyclotis*, a very close relative of the extinct European straight-tusked elephant) into Europe and the UK might sound unnervingly radical.[32] But it's an important ambition for the horizons of rewilding. At the very least it encourages us to conjure up the ghosts of missing megafauna in our minds, to work out the beneficial impact they might have had on the environment and, in the absence of an alternative, try to mimic those effects ourselves. Most importantly of all, it shifts the baseline for

A straight-tusked elephant as it would have looked wandering through the British landscape around 115,000 years ago. It would have pushed over trees, ripped off branches and smashed through scrub, just as its cousins do in Africa.[33] Our trees and shrubs evolved to survive their assaults, bouncing back with new shoots – which is why they respond so well to coppicing, pollarding and hedge-laying.[34]

human impact. To understand dynamic, complex systems of nature we should look to an era much earlier than the seventeenth or eighteenth century. Currently, the International Union for Conservation of Nature, the global authority on the status of the natural world, allows the word 'reintroduction' only for species that became extinct less than 200–300 years ago.[35] Rewilding is much longer-sighted than that.

One audacious rewilding experiment in northeastern Siberia, known as the Pleistocene Park and begun in 1996, aims to restore the mammoth steppe and, perhaps, reveal a way to combat climate change.[36] Its founder, the Russian geophysicist Sergey Zimov, has fenced 1,500 hectares and introduced musk ox, bison, elk and hardy Yakutian horses to see if they can indeed return the habitat of moss and trees – the *tundra* and *taiga*, usually considered to be the 'natural' environment of the Arctic – to grassland steppe.

Today, with warming temperatures, the permafrost – the accumulation of frozen, decomposed vegetation nearly 1.6 kilometres thick in places and covering 23.3 million square kilometres at the top of the planet – is melting, releasing vast quantities of stored carbon into the atmosphere.[37] Globally, permafrost holds up to 1,600 gigatons of carbon, nearly twice the amount of carbon dioxide in the atmosphere. For every 1°C rise in the Earth's temperature, scientists calculate the permafrost will release the equivalent of between four and six years' worth of global coal, oil and natural gas emissions.[38] Already the permafrost is thawing faster than all previous predictions. Zimov hopes that returning large herbivores to this area could, by reducing the growth of woody vegetation on the surface, lower the temperature of the soil, which would protect the permafrost and dramatically reduce carbon emissions. Evidence from the park, so far, is promising. Even with only about a hundred grazing animals, the grasslands are staying cooler than ground in the surrounding area.

The Pleistocene Park may provide evidence for the principle of megafaunal geoengineering. However, the logistics of scaling up the experiment to recover the Arctic steppe in its entirety (including sourcing, transporting and acclimatising enough bison, horses, musk ox, reindeer and other animals to do the job) seem nearly insurmountable, even should the political will be there.

De-extinction

Advances in DNA extraction, gene editing and cloning have opened the doors to a new and controversial realm of species conservation. In 2002, cryopreserved cells from 'Celia', the world's last Pyrenean ibex or *bucardo* (a wild mountain goat from northern Spain), gave life – for a few minutes – to a cloned *bucardo* kid born to a surrogate domestic goat two years after Celia herself had died. The *bucardo* – uniquely – became extinct for a second time, but the event captured the imagination of many in the conservation world. The organisation Revive & Restore, founded in the USA in 2012 by Ryan Phelan and Stewart Brand, funds projects to bring back species from extinction, from the passenger pigeon and heath hen (a subspecies of the greater prairie chicken) to the woolly mammoth.[39]

It is now possible to construct whole genomes from degraded ancient DNA extracted from a fossilised bone. A team at Harvard led by George Church is working to identify the cold-climate-adapted alleles (gene forms) of the mammoth genome so that they can edit them into the living cells of the mammoth's close surviving cousin, the Asian elephant, and produce a 'mammophant'. What seemed the stuff of science fiction only a decade or so ago is creeping closer to the realms of possibility, although, of course, serious ethical questions and challenges remain. It is one thing to create an individual with the physical characteristics of an extinct species, for example, but quite another to resuscitate the acquired knowledge, wisdom and social behaviour of a herd animal. Would the mammophant know how to *be* a mammophant?

Critics argue that the millions of dollars raised for these 'Frankenstein projects' would be better spent on habitat restoration and the protection of the endangered species that are clinging on to life today. Yet, as with Formula One or missions to Mars, the advances in science gained through ambitious vision almost invariably have applications at a more mundane level. The de-extinction movement is developing techniques to increase the genetic diversity of endangered species.[40] One of Revive & Restore's projects is the recovery of the black-footed ferret, reduced to just eighteen individuals by 1981. Genetic modification might improve resistance to disease in species such as the endangered Tasmanian devil, which is currently beset by a contagious cancer, and the endemic birds in Hawaii affected by a form of avian malaria. It might also advance species' adaptations to cope with the rapidly changing climate. Migrating fish such as salmon, for example, might be helped to survive in warmer rivers.

Less controversial, perhaps, is the Tauros Programme, based in the Netherlands.[41] It has produced a modern equivalent of the extinct aurochs, primarily as a driver for rewilding, through the more conventional practice of animal breeding. The founder, Ronald Goderie, realised that aurochs DNA survived in Europe's domesticated cattle, particularly ancient breeds. He concluded that if we compare an aurochs' gene sequence (the first whole genome sequence was published in 2015 from DNA extracted from a 6,750-year-old British aurochs bone) with that of existing cattle, a programme of cross-breeding could produce animals with genomes increasingly approaching that of their extinct ancestor.[42] Herds of free-roaming Tauros cattle, now in their fifth breed-back generation and numbering 700 individuals, have been introduced into rewilding projects in Germany, Denmark, Hungary, Latvia and the Netherlands.

Can we spare land for rewilding?

In his book *Half-Earth: Our Planet's Fight for Life* (2016), the American biologist E. O. Wilson says that if we are to preserve biodiversity and the systems on which all species – including our own – depend, we must dedicate half the Earth's lands to nature. Our existing wilderness areas, national parks and nature reserves are nowhere near enough. We have to get functioning nature back on to every available piece of land, including our own backyards.

But how are we going to produce food for our 8 billion people, expected by the United Nations Population Division to level out at 11 billion by 2100? Can we spare half the Earth's land mass for nature? Surprisingly, food itself is, globally, not scarce, although the mighty food and farming industries have vested interests in perpetuating that myth. Distressing images of famine in war-torn and politically unstable regions of the world daily reinforce the idea that there is not enough food to go around. The message being driven by food producers and retailers, agribusiness and farmers' unions is that global food production must increase by 70–100 per cent. Yet across the planet we are already producing enough food to feed 10 billion people, 2 billion more than are alive today.[43] The shocking reality, according to the UN Food and Agriculture Organisation, is that we waste at least a third of it.[44] Not only are we over-producing (and now growing grain to feed cattle and maize to fuel cars), but also we are using less land than ever before to grow crops. In the UK, yields have continued to rise, year on year, while the acreage devoted to wheat and barley in Britain has fallen by 25 per cent since the 1980s.[46]

Globally, the area of arable land needed to produce the same quantity of crops has fallen dramatically since the 1960s. Greater efficiencies, new crop varieties, better distribution and new technology have all contributed to this decrease. We are using less land than ever to produce the same amount of food.[45]

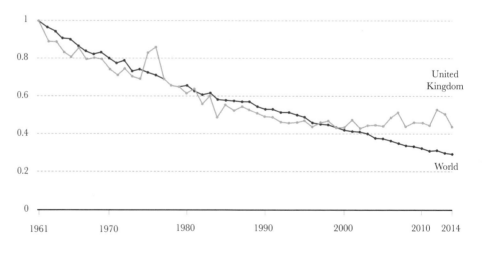

Arable land needed to produce a fixed quantity of crops (1961 = 1), 1961–2014

Globalisation – including importing cheaper food from abroad – has played a part in shifting the distribution of agriculture across the world. But precision machinery, improved storage and high-yielding varieties of crops have also contributed to greater efficiency. Increased productivity has, inevitably, led to a collapse in commodity prices – one of the factors underpinning our decision to give up farming at Knepp. Even though the UK population has increased by nearly 20 million since 1939, there is now the smallest area of land (6 million hectares) devoted to arable since before the Second World War.[47]

Across Europe, the fundamental shift in land use – and particularly the demise of farming on marginal land – is even more dramatic. Remote villages are emptying, with only a handful of the oldest occupants remaining. It is predicted that by 2030 more than 30 million hectares – an area the size of Italy – will no longer be in agricultural production.[48] Reform of Europe's farming subsidies (the Common Agricultural Policy) is likely to accelerate land abandonment even further. For years, conservationists and economists have been calling for an end to farming subsidies, which at present total around 31 per cent of the total EU budget and which have, for the past half-century, incentivised farmers to grow mainly arable crops in intensive, chemically dependent systems, irrespective of the suitability of the land and the damage done to the soil.[49] Soil degradation, the decline in pollinating insects, chemical pollution and the loss of water are now so serious as to jeopardise our ability to grow food effectively in decades to come.

Currently half the world's land mass is under agriculture, 77 per cent of which is dedicated to (mostly) intensive, grain-fed meat and dairy production, and responsible for only a small percentage of global calories. Reducing meat and dairy consumption and shifting to regenerative agriculture and alternative proteins such as pulses, insects, algae and precision fermentation will ensure a sustainable global supply of food and potentially reduce the amount of land needed for agriculture.[50]

Global land use for food production

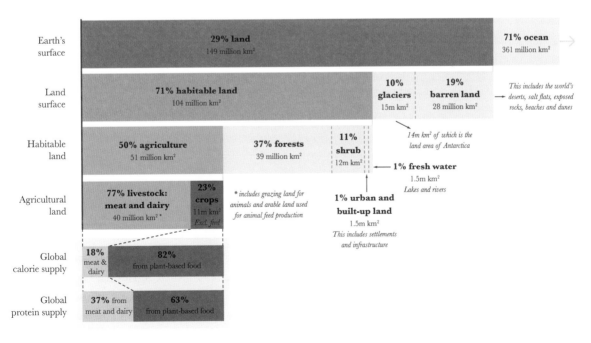

35

While countries must, wherever possible (depending on their size, of course), make every effort to home-produce in order to decrease carbon costs and shoulder a fairer burden in the use of natural resources, the availability of land itself is not a concern in terms of food production.

But we have to switch to more sustainable ways of farming and producing food. Terminating our love affair with industrialised grain-fed meat production will be critical, and shifting the emphasis from arable crops to more calorific and nutritious legumes and pulses could have a huge impact. New systems of food production, including sourcing protein from kelp and insects, will undoubtedly be part of the mix. Other hi-tech innovations have received media attention, including lab-created proteins from animal cells and bacteria, and precision fermentation. We must take care, though, that new technology doesn't simply lead us down a different route to unsustainability. As a rule, if a food production process doesn't sequester carbon, it will not be sustainable in the long term. Being carbon *neutral* is probably not good enough.

The biggest change in food production – and the one that is likely to have the most positive effect on food security, soil, water, biodiversity, carbon sequestration and food nutrition for the future – is regenerative agriculture, also known as 'agroecological farming' and 'conservation agriculture'. This revolutionary approach to farming involves planting seeds straight into the field ('direct drilling'), so avoiding ploughing entirely; using a wide variety of 'cover crops' to protect and nourish the soil; rotating crops to reduce soil exhaustion and the build-up of pests; and using livestock to fertilise the soil.

Incentivising farmers to move from intensive, chemical systems will almost certainly become EU policy, and we hope that this will happen sooner rather than later.[51] The UK is already shifting in this direction with its post-Brexit Agricultural Bill. There, farms on unproductive marginal land, which currently depend on subsidies, are likely to struggle.

Concerns that land abandonment and switching productive farmland to regenerative practices pose a threat to food security are largely unfounded. Farmers themselves – among them Gabe Brown, Joel Salatin and Mark Shepard in the USA, and John and Paul Cherry of Groundswell in the UK – are challenging the industrial mindset.[52] They have proven that organic, no-till farming, sustainable horticulture and agroforestry can produce just as much food as chemical farming, and result in significantly increased profits (because of the low inputs) for the producer. At the same time, regenerative agriculture restores soil function, water resources and pollinating insects, and sequesters carbon.[53]

The combined effect of shifting productive farmland to regenerative agriculture and rewilding much of the remaining land could provide the 'half-Earth' for nature and restore functioning systems across entire landscapes. The two would work hand in hand, with the webbing of rewilding (wildlife corridors, green belts, field margins for nature, natural hydrology) supporting farming systems and increasing food security and productivity. Rather than being seen as a threat to agriculture – as it's often depicted in the press and by some farming lobbyists – rewilding could be its greatest ally. It has the potential to safeguard farmland from flooding, restore water tables, provide microclimates and windbreaks to mitigate climate change and extreme weather events, and increase numbers of pollinating insects and predators for pest control, as well as promoting natural systems of nutrient transfer to replenish agricultural land.

Rewilding in cultural landscapes

Given the unsettling predicament of land abandonment in the UK and Europe, some people find it a challenge to embrace rewilding. They fear it as a signal of the end of the cherished 'cultural' landscapes of the past. Certainly, landscapes under rewilding can look very different from the heavily worked countryside we are used to, but rewilding may also provide solutions for rural life and traditions currently affected by land abandonment. Old pastoral systems of grazing, such as those practised in the *dehesas* of Spain and Portugal and the wood pastures of Romania, can, if carefully orchestrated, create the kind of dynamic conditions required to restore ecosystems by imitating the wild herd migrations of the past. There may be opportunities for managing livestock in new areas, such as water buffalo, which have been introduced to control reeds in the Danube Delta in Romania and Lake Prespa in northern Greece. Wildlife tourism – almost invariably a consequence of rewilding – can help to keep ancient villages and traditions alive. And, in the future, government payments for services essential for public well-being and prosperity could underpin the survival of rural smallholders and land

managers. These vital services include soil restoration, carbon sequestration, water and air purification, flood mitigation, restoration of
biodiversity, extensive food production and foraging, and the provision of wild areas to promote human health and allow recreation.

We can also value features of the cultural landscape as homes for
wildlife and for the pleasure humans take in beauty. Old buildings,
dry-stone walls, hedges, solitary trees and avenues, ancient coppice
woodlands, hay meadows and dew ponds can all be considered a
'public good'. If they are recognised as such, we should be willing to
subsidise them. In the debate about nature conservation it is often
tempting to exclude people – the destroyers of nature – as participants in the scene. But humans can be a keystone species, too. If we
act in tune with nature we can be a tremendous positive force for
biodiversity. To write ourselves out of the picture not only under-
mines our potential for good, but also denies our identity as a species,
as part of nature, with unfathomably complex interrelationships with
all other forms of life.

What size is right for rewilding?

It has been shown conclusively that rewilding – restoring water
systems and vegetation, introducing keystone species, and allowing
natural processes the freedom to function – brings many benefits for
people and biodiversity. It can even solve our most pressing environmental concerns, such as climate change and soil degradation. But
ecologists and conservationists still disagree about the form that rewilding should take. Scientists would like to pin the concept within fixed
boundaries, a framework they can work to, and often this framework
is defined more than anything by a person's training, world view and
even temperament. Rewilding is a minefield of conflicting ideals and
emotions. In the two decades since the word has become common
currency, papers have been written, debates waged and hairs split, but
we still cannot agree on a definition.[54]

Knepp, for example, is often described as a 'rewilding' project.
But, especially in the early days, some academics argued that only
areas left entirely without human management, and preferably
encompassing free-roaming apex predators, could properly be considered rewilding. Certainly, compared with the great rewilding projects
underway in Europe, Knepp – at 1,400 hectares – is small fry. One

day we hope to introduce bison, wild boar and even European elk on to our land. But the prospect of seeing wolves, bears and lynx in the southeast of England, even in the distant future, is remote. At Knepp, too, we manage the numbers of herbivores by culling, which rewilding purists consider to be interventionist.

Intrigued by the debate over the word 'rewilding', we asked our Advisory Board of thirty-two ecologists and conservationists to come up with their own definition of our experiment at Knepp. What could we use instead of the controversial 'R' word? After much scratching of heads, they declared Knepp to be a 'long-term, minimum intervention, natural process-led area'. Not exactly catchy. In characteristic academic fashion, there was still much wrangling about specifics: our eminent advisors could not agree on how long was 'long-term'; what was meant by 'minimum intervention'; and how exactly to define 'natural' and 'process-led'.

The word 'area' proved particularly contentious. How big must a project be for it to manifest natural processes, to be considered dynamic and 'wild'? For example, Kraansvlak in the Netherlands is just 330 hectares, less than a quarter of the size of Knepp. This small nature reserve has shown dramatic increases in biodiversity since twenty bison were introduced there in 2007, and a small herd of Konik ponies in 2009. A sensitive sand-dune ecosystem has sprung back to life. Most would argue that what is happening in Kraansvlak is rewilding. Perhaps the term 'rewilding' can apply to even smaller areas.

One of our board members at Knepp, the eminent biologist John Lawton, came up with a simple 'cartoon' (as he called it), on the next page. It demonstrates how the amount of management needed to encourage and sustain dynamic natural processes relates directly to the amount of land involved. Very broadly, the larger the protected area, the less humans need to manage it. Wilderness areas such as Yellowstone National Park in the USA and the Okavango Delta in Botswana require no management at all. The smaller the area, the more intervention is required to sustain a specific habitat. The single-hectare nature reserve of Badgeworth, near Cheltenham in Gloucestershire, for example (designated a Site of Special Scientific Interest, or SSSI, by Natural England), is micro-managed to protect the rare adder's tongue spearwort (*Ranunculus ophioglossifolius*), known as the 'Badgeworth buttercup'.

The rewilding spectrum

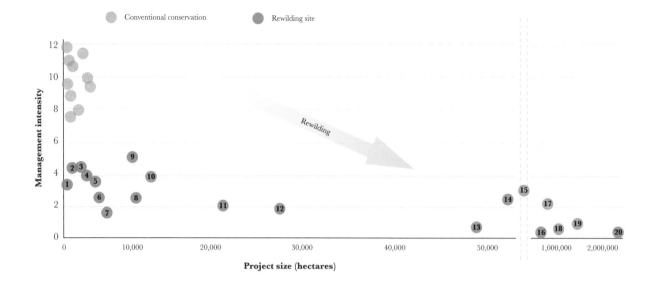

A conceptual graph, inspired by John Lawton, showing how the intensity of human management diminishes as the area under rewilding gets bigger. Dynamic, functioning ecosystems such as Yellowstone National Park need little or no human involvement. But in smaller areas, where natural processes such as those caused by large animal disturbance cannot happen, interventions are needed to stimulate dynamism and flux. Humans become the keystone species. But even at this end of the spectrum, rewilding tends to be more 'hands-off' than conventional conservation, which tends to focus on small areas where species and/or habitats are critically endangered.

Graph no.	Project	Region	Size (ha)
1	Kraansvlak	Netherlands	320
2	Doddington Estate	England	400
3	Wild Ken Hill	England	800
4	Knepp	England	1,400
5	Langholm Initiative	Scotland	4,000
6	Wild Ennerdale	England	4,800
7	Oostvaardersplassen	Netherlands	5,550
8	Alladale Wilderness Reserve	Scotland	10,000
9	The Yearn Stane Project	Scotland	10,000
10	Lowther Estate	England	12,000
11	Velebit Mountains	Croatia	22,500
12	Mar Lodge Estate	Scotland	28,000
13	Foundation Conservation Carpathia	Romania	50,000
14	Cairngorms Connect	Scotland	60,000
15	Greater Côa Valley	Portugal	120,000
16	Pumalín National Park	Chile	405,000
17	Danube Delta	Romania	580,000
18	Yellowstone National Park	USA	900,000
19	American Prairie Reserve	USA	1,300,000
20	Okavango Delta	Botswana	2,000,000

For us, Lawton's graph sheds valuable light on how conflicting views about rewilding arise, and on the degree to which people can be involved. Rewilding has no single, fixed definition, nor is it a practice with set parameters. It is a progression with degrees of management on a sliding scale. This graph has, in a sense, 'rewilded' rewilding. One's views on rewilding depend, to a large extent, on where one stands on the spectrum. At one end – the wilderness end – are the rewilding purists and adventurers. They would like to see the creation of vast wild areas with as little ongoing human involvement as possible. They espouse the reintroduction of wolves and bears; they might even consider reintroducing Pleistocene megafauna such as elephants and rhinoceroses into Britain, and they are excited by the idea of conjuring up mammoths from fossil DNA. At the other end are the friends – typically regional conservationists and nature wardens – of the UK's many tiny Noah's arks of protected natural habitat. Mindful of the legacy of conservation history, and especially all the recent battles for nature won and lost, they tend to regard human management as vital to environmental protection. Inevitably, perhaps, they are more cautious about surrendering control and launching into species introductions and other rewilding techniques, especially in designated nature reserves. They need reassurance and feasibility studies.

As Alastair Driver, another of our board members and director of Rewilding Britain, puts it, the present participle is a guiding light: 'rewilding' is a journey towards becoming 'rewilded'. Even at the smallest scale, the aim should always be to become wilder, more diverse, more dynamic, self-sustaining. The larger the area of land under rewilding, the closer it comes to the goal of being truly 'self-willed'. The full rewilding goal is reached at the largest end of the spectrum, on the scale of Yellowstone, the Okavango Delta and the Serengeti, perhaps. There, no human assistance or intervention is desired or necessary. The full, or nearly full, complement of species are present and thriving, and nature is playing out its full potential, dancing unfettered to the dynamic rhythm of boom-and-bust cycles. Only then can land be described as 'rewilded'. And, even then, laws must be written and guards must roam to ensure that rewilded land remains protected.

The possibility of reaching the Yellowstone end of the spectrum are, in Britain at least, essentially nil. On these densely populated islands there are few, if any, opportunities for wilderness areas on this scale, although areas of nature could be connected in the mountains.

In rewilding projects where
large herbivores are the main
drivers of the system, the
level of human management
varies enormously, depending
on the area of land. Predators
reign in areas above 100,000
hectares. Without them, in
areas above 10,000 hectares,
large herbivores may need
to be controlled. At Knepp
scale (1,400 hectares), where
we use domestic animals
as proxies, controlling the
numbers is still virtually
the only intervention we do.
At around 250 hectares the
approach shifts from minimal
intervention to hands-on
management. Below 100
hectares is unlikely to sustain
free-roaming animals all year
round; they may need to be
brought in periodically. Below
1 hectare, human tools and
effort are needed to mimic
large animal disturbance.

It is the journey, the intention, the striving for an increase in wildness
and natural processes and a progressive release of human management that are important in Britain.

Increasing management			Increasing naturalness
Non intervention	Very large-region – predators	100,000 ha plus	
Predator management/ mimicking	Large – wild grazers	10,000 ha plus	
Domestic proxies	Quite large	c.1000 ha plus	
	Larger		
	Smaller		
Managed domestic proxies	Medium	100 ha plus	
Managed domestic animals	Small	10 ha plus	
Managed vegetation	Very small	1 ha plus	

The table above gives a very general idea of the kind of management
that might be expected for a given amount of land under rewilding
when large herbivores are used as the main drivers of the system.
Obviously, variables such as soil type, vegetation, topography, altitude
and climate significantly affect the biological productivity of the land.
This dictates how many and what kind of herbivores the land can
support, and that in turn dictates the amount of human management
that will be needed. Other factors will affect the vegetation, too – such
as hydrology and soil erosion/accretion (which are generally dictated by a combination of topography and climate) – and these vary
hugely from site to site. The table does not account, either, for the
effect on vegetation of episodic natural disturbance caused by disease,
fire, flood and wind. But it gives a useful scale from which to gauge
the levels of human intervention that may be necessary to sustain
dynamic natural processes.

Around the 250-hectare mark – the dotted line – is where most
rewilders, including Knepp and Rewilding Britain, reckon there is
a marked shift in approach from minimal intervention (above it) to
hands-on management (below it). Below the 100-hectare mark is

a tipping point where the area is no longer big enough to sustain free-roaming animals in significant numbers all year round. The amount of human intervention required to stimulate dynamic natural processes therefore rises exponentially until, at the very smallest end of the scale, the options are almost entirely dependent on human management.

The importance of connectivity

Above all, Lawton's sliding scale underlines the importance of connectivity. A small area of nature can move along the rewilding axis towards self-willed processes and sustainability – the wilderness end – only if it increases in size. As soon as a tiny 10-hectare SSSI or SPA (Special Protection Area) can connect with other areas of nature – through wildlife corridors, stepping stones or rewilded lands, such as Knepp – natural processes and biodiversity rise exponentially. Suddenly isolated populations of species can interact with other populations, increase their gene pool, and move in reaction to climate change, pollution and extreme weather. They are likely to have more places in which to shelter and a greater chance of finding food and meeting the needs of complex life cycles, if they have them. They become far less vulnerable to extinction, far more elastic and dynamic. Even at the wildest and largest end of the scale, linking up existing wilderness areas, such as Yellowstone and Yukon in the Y2Y initiative in the USA, has an exponential effect on natural processes and the survival of wildlife. Perhaps, in the end, a truly rewilded state will only ever be reached with continental, or even intercontinental, connectivity – something approaching Wilson's vision of 'half-Earth' returning to nature.

Emergent properties

When the arteries of life are reconnected, the phenomenon of 'emergent properties' comes into play: the miracle of reactivation, when previously missing or dormant components come together with extraordinary and unexpected results. We often think we know what a given species needs in order to thrive. But the circumstances, relationships and resources involved in that species' existence may be far more complex than we appreciate, and this is one of the reasons conventional, human-directed conservation so often fails. For instance, small populations of the purple emperor butterfly may survive in fragments of ancient oak woodland, and that is why it is known as a woodland species. But where it really takes off in numbers, as we discovered at Knepp, is in prolific, sallow scrub, a habitat that is rarely tolerated in modern farming landscapes. But sallow (naturally hybridising native willow) itself may not be the only determining factor. Who knows what other components – such as territorial display sites, microclimates, moisture, the right variety of leaves, mineral licks, fox scat and sap runs – come together to provide the purple emperor with the optimal conditions for life? Nightjars, meanwhile, are not exclusively 'heathland' birds, as the guidebooks describe them. That is simply where they have ended up in today's world, having lost their other options, such as quiet, sandy lanes and low scrub.

Taking our hands off the steering wheel and allowing nature the time and space to express itself is one of the fundamentals of rewilding. It teaches us new things about even those species we think we know intimately. Connectivity – allowing habitats to merge and morph into a kaleidoscope of shifting, dynamic scenarios – provides species (particularly those with complex requirements or different stages in their life cycle) with an infinitely greater array of niches and opportunities. Suddenly we find species behaving differently, defying the 'scientific' envelopes we've put them in. At Knepp we have peregrine falcons – a species that is renowned for living on cliffs or in buildings, like cathedrals, that resemble them – nesting in an oak tree. We forget that our modern landscape is so changed, so desperately impoverished, that we may be observing species not in their preferred habitats at all, but pushed to the very limit of their range, where they cling to existence. Rewilding opens up the possibilities of rediscovery by giving animals and plants freedom.

Novel ecosystems

One of the criticisms academics most often level at rewilding, perhaps because of its associations with reintroducing species, is that it aims to replicate the environment as it existed before humans had an impact – something that is clearly impossible to do. The stubborn little 're-' prefix in rewilding (a prefix that usually means doing something again) is, in some part, responsible for this; *re*-wilding suggests returning to, or recovering, the past. As a reminder that this is not our aim, we have chosen to call this book *The Book of Wilding*. Of course, 'rewilding' remains the term that most people know and will no doubt adopt. We hope that, as people become more familiar with the concept, the term will lose these unhelpful connotations.

In reality, rewilding is much more open. Even those who espouse the resuscitation of mammoths and 'Pleistocene rewilding' appreciate that recovering ecosystems as they once were is impossible. Humans have altered the planet profoundly, not just for centuries but for millennia, exterminating species and habitats, even irreversibly changing soils, the climate and the oceans.

What rewilding can do, however, is use our knowledge of the past – of the way ecosystems functioned before humans took over – to inform our ideas about conservation in the future. Using the tools remaining to us today, we can encourage the return of functioning, dynamic natural processes wherever possible. Almost invariably, these systems – what we now call 'novel ecosystems' – will never have occurred on the planet before. Communities of animals, insects, plants and micro-organisms that may never have previously interacted are now coming together and associating in unprecedented environmental and climatic conditions. The globalisation of transport and the effects of climate change are allowing species to move around the globe and colonise new areas at an extraordinary pace, and accordingly the media revels in alarmist headlines of 'invasive' and 'alien' species.

The reintroduction, or introduction, of keystone species often worries people, especially when those species have been absent for many hundreds of years. The idea of novel ecosystems can feel a little alarming, too, until one realises that nature is constantly in flux, constantly evolving and constantly recombining into new communities of plants and animals. Nothing ever stands still. In a sense, every ecosystem that exists or has ever existed on Earth is a novel ecosystem. Far more important than the consideration of whether a particular species should be present is ecological function, and the most obvious measures of that are abundance of diversity and life.

Stepping back

Another fundamental principle of rewilding is to recognise the ability of nature to accommodate change, to respond, adapt and modulate in infinitely complex and subtle ways. Even with the best will in the world, and any amount of funding, humans cannot emulate that. The idea that nature knows best is both reassuring and challenging. It releases mankind from the burden of intensive, expensive, time-consuming micro-management and the godlike responsibilities many societies (particularly industrialised ones) have taken upon themselves when it comes to conserving nature. It forgives us many of our sins. We know that bacteria and mycorrhizal fungi are already evolving to be able to consume plastics and neutralise chemical pollutants as toxic as the pesticide DDT and PCBs (polychlorinated biphenyls) used in manufacturing; that viruses may provide the next generation of anti-biotics; and that urban foxes are evolving wider and shorter muzzles, faces that look less threatening, attracting people to feed them.[55] We know that healthy soils, vegetation, peatlands and sea grass are powerful carbon sinks; that forests and vegetation can stimulate rain; that healthy topsoil can increase in depth; and that desertification can be reversed under the pulsing, intermittent influence of herbivores, or 'mob grazing' (see page 71).[56]

Allowing, encouraging and even accelerating natural processes is the most effective and, arguably, the only way we can solve the planet-wide environmental crises we have caused. But it also exposes us to uncertainty and surprise, and demands of us the unaccustomed but ancient virtues of patience and humility. It is a hard lesson to teach ourselves, control freaks that we are. 'Don't just do something, stand there' is a hard order to obey. But doing nothing, once the right conditions are in place, can be the best option. Our dependence on chemical pesticides and herbicides over the past fifty years, in particu-lar, has masked the ability of nature to regulate itself.

The principle applies, on a smaller scale, to garden 'pests'. When the caterpillars of cabbage white butterflies start munching through the vegetable patch, a little patience gives natural predators, such as blackbirds, robins, blue tits and wasps, a chance to respond. Why reach for pesticides the moment the caterpillars appear? Why not sit it out? Perhaps accepting a ravaged crop for a season will allow predators to discover the new food source and rally their forces for the subsequent year. Providing the right plants for a wide range of insect

predators – not just pollinators – is also key. Scientists are beginning
to appreciate the fact that trophic cascades can work at micro, as
well as macro, levels. A garden, after all, is a mini-Serengeti. There
may not be wolves, but there are wolf spiders, hunting down their
prey and leaping on it. And trophic cascades work not just in vertical
hierarchies – such as the wolf in Yellowstone, or starfish predating on
mussels in rocky shore communities – but also horizontally, between
species of similar size and within the same genera.[57]

Rewilding: the spectrum

The graph on page 40 convinces us that rewilding applies across the
spectrum of human management and space. All approaches and all
definitions of rewilding count. In this debate we are all right; all rewilders, whatever the scale of their actions, positively affect biodiversity
and ecosystems. Indeed, a diversity of approach supports a diversity of
outcomes. We can all play a part, no matter how small the land under
our control, even down to a window box or a few pots on a balcony.
The area of private gardens in the UK – over 4 million hectares –
exceeds the area of the country's nature reserves combined.[58] There
are 330,000 allotments in England (and, with an estimated 90,000
people on waiting lists for them, it is to be hoped, many more to
come).[59] These are places where humans are a keystone species,
selecting and managing for maximum productivity, while welcoming
biodiversity in even the smallest of spaces. If all Britain's gardens and
allotments were managed organically, without poisonous pesticides and
herbicides, and encouraged rather than eliminated wildlife, biodiversity
in the UK would skyrocket. Imagine, too, if we also gently and organically looked after churchyards, cemeteries, orchards and school playgrounds. Imagine us weaving green corridors across our towns and
cities, connecting parks, riversides, embankments and towpaths. From
there, we could reach out into the wider countryside. Suddenly our
isolated pockets of nature become part of a pulsing, functioning whole,
a web of life that could revive our entire landscape, and so also revive us.

In the end, academics squabbling over definitions of rewilding
may be missing the point. Attempts to pin down this fluid, shape-
shifting word seem doomed to fail. Perhaps it is in the very nature of
rewilding to be indefinable, to defy parameters and boundaries. To us
at Knepp, rewilding is as much about a state of mind as anything. It

means a willingness to question preconceived ideas, particularly about how landscapes should look, whether in the countryside, villages, cities or our own backyards.

Shifting baselines

One of the greatest barriers to rewilding may be something as intangible and intransigent as our sense of what is beautiful. Humans tend to carry around an ideal image of what the countryside should look like. England's 'green and pleasant land' with its heavily managed agricultural fields, tight hedgerows, linear edges, isolated pockets of woodland, canalised streams and rivers, and barren hillsides, crisscrossed by roads, seems natural and normal, even beautiful, to most people who are familiar with it. It inspires a sense of security, productivity and historical continuity, and the tendency is to assume it has been this way for ever.

But familiarity blinds us to the creeping processes of degradation that often happen imperceptibly. Rarely do we compare a photograph of our local landscape today with one from thirty years, let alone a century, ago. Declines in levels of sound – insects and birdsong – are even harder to gauge. We literally don't know what we are missing. Compared to now, the noise of wildlife seventy years ago would have been deafening. There were 44 million more farmland birds in the UK's skies in the 1960s than there are today.[60]

If we do sense some sort of ecological loss and change, we tend to go only as far back as our own childhood. Those born around the middle of the twentieth century in the UK have memories of summer cuckoos and hedgehogs that are now almost gone. Or, perhaps, our parents or grandparents told us about skies filled with lapwing, skylarks by the dozen, and fields red with poppies and blue with cornflowers. We are blind to the fact that our grandparents' grandparents would have heard corncrakes and bitterns, black grouse and curlew, and nightjars down every country lane. Or that their grandparents would have known spotted flycatchers in every orchard, meadow pipits from the salt flats to the crowns of mountains, and banks of giant cod and migrating tuna in a North Sea as clear as gin, filtered by vast beds of oysters.[61] And *their* grandparents, living at the time of the last beaver in Britain (in the eighteenth century), would have known great bustards, and grey whales feeding in the estuaries, and would have watched shoals of herring 8 kilometres long and 5 kilometres broad migrating within sight of the shore, chased by dolphins and sperm whales and the occasional great white shark.[62]

Our continuous lowering of standards and accepting degraded nature as normal is known as 'shifting baseline syndrome'. It's a problem identified by the landscape architect Ian McHarg in his book *Design With Nature* (1969) and adopted in 1995 by the fisheries scientist Daniel Pauly.[63] Hundreds of years ago an area of sea would have been heaving with fish, Pauly noted. But the scientists' reference point for 'natural' populations is invariably pinned to levels dating back to within their living memory, no more than a few decades from the present. Each generation, Pauly realised, redefines what is 'natural'. It's not just the scientists who are affected by ecological amnesia. The sports fishermen and women holding up their catch on the same Florida Cay over a period of fifty years look just as thrilled with their haul in every celebratory picture. Yet the size of the fish diminishes dramatically over a relatively short period. The species, too, are different; by the 1980s there are no large pelagic (open-sea) species – the top of the food chain – left at all, and by the 2000s the catch is tiny both in size and in number. Only the smiling faces of the humans have stayed the same.

The changes in catch from charter boats in Key West, Florida over fifty years is a dramatic example of shifting baseline syndrome. In 1957 (top left) Goliath-size groupers dominate. By 1965 (top right) the biggest fish are no longer bigger than the fishermen. By 1980 (bottom left) the giant groupers are gone, while snapper and smaller fish abound. In 2007 (bottom right) the average fish caught is around 30cm long.

We can see shifting baseline syndrome at work in the public response to reintroductions and species recovery. In recent years, extinct or near-extinct species, such as white-tailed eagles, red kites, ospreys, beavers and polecats, have been reintroduced at a few sites in the UK. Populations of species such as ravens, buzzards, badgers, foxes, polecats, stoats and weasels have also risen considerably in some areas. A more visible presence, however, is often accompanied by public anxiety, even among conservationists. We're simply not used to seeing these species in numbers, and the question on everyone's lips is 'How many is natural?' We're already thinking about how to control them. But we don't have to go back far in the records to find a time when these species existed in the UK in almost unimaginable numbers. Gamekeeper kill figures at the Glengarry Estate in Scotland, for example, show that 198 wildcats and 15 golden eagles were killed in just three years between 1837–40.[64]

Something similar happened when the British Trust for Ornithology (BTO) set 1970 as its baseline year for monitoring British bird populations. Of course, a baseline has to be set somewhere, and the declines since then, meticulously recorded, have been dramatic. But the baseline itself begins to encourage pre-baseline blindness. We forget that once there was much, much more. Rewilding offers a chance to reset our baselines. At Knepp, in the scrubland of the Southern Block in spring, the sound of birdsong bowls you over. You can feel the vibrations from all around reverberating in your lungs and stomach. Ecologists have been amazed at the sheer quantity of birds – the biomass – able to co-exist in that space. The numbers are beyond what any conservationist would have dreamed of. In 2019 a BTO survey conducted in the Southern Block indicated that we may have a higher density of breeding songbirds than anywhere else in Britain. Rewilding without setting targets for the size of the population of a given species expands the possibilities and shows how prolific that species could naturally become. It is a way for us to see what we could – and should – be aiming for.

Our wilder future

The charity Rewilding Britain was set up in 2015, with Charlie as its chair.[65] Amazingly, its goals no longer seem as challenging to achieve as they did at the start. It aims to establish core rewilding areas across at least 5 per cent – approximately 1 million hectares – of Britain by 2030, with a rich mosaic of good habitat for nature across another 25 per cent, or 5 million hectares. The charity hopes that one day lost creatures, such as lynx, burbot, eagle owl and Dalmatian pelican, and perhaps even elk and wolf in the remotest places, will be living in the UK again.

Of course, rewilding will only ever be a part – and most likely a modest part – of the land-use mix of the UK. We will always need land for farming, housing and industry. But rewilding can provide us with new biodiversity hotspots, like Knepp, and the wildlife corridors to connect them. This web of rewilding will provide a life-support system for nature, agriculture and humans. It will give us all access to living green space – something that we know, especially after COVID-19, is vital for human health and well-being. As farmland, industry, rivers, towns and cities make the environmental improvements necessary for a sustainable future, all our land will become more permeable for wildlife. Otters and ospreys, rare songbirds, small mammals and butterflies, which will thrive in rewilded areas, will inevitably spill out into the wider countryside. Who knows, nightingales may even sing in London again, as they do in Berlin, a city that also boasts goshawks, beavers and wild boar.[66]

The vision of a wilder landscape now seems within reach. Since the turn of the millennium the UK government's thinking has come on in leaps and bounds in relation to rewilding and species reintroductions. A review of England's national parks, chaired by the journalist Julian Glover, has resulted in improvements to habitats and wildlife in our biggest, woefully neglected nature reserves. Beavers are now back in Scotland, with legal protection. In 2020, after a five-year study trial, the Department for Environment, Food and Rural Affairs (DEFRA) granted a licence to remain to the wild population of beavers on the river Otter in Devon. Between 2017 and 2022 it has granted beaver release licences to thirteen sites in England, including Knepp. Its announcement in 2022 to grant beavers legal protection will, we hope, pave the way – soon – for their return to freedom in England. White-tailed eagles have been introduced on the Isle of Wight, and white storks (introduced in 2017 to three release sites in the southeast of England, including Knepp) are now breeding in the wild for the first time in 600 years.

Rewilding is gathering pace. In Scotland, the community-based conservation charity John Muir Trust – an early ambassador for wild landscapes, founded in 1983 – aims to let nature repair itself over an area of 25,000 hectares, including the summit of Ben Nevis. The Great Trossachs Forest, a project partnered by Forestry and Land Scotland, Royal Society for the Protection of Birds (RSPB) Scotland and the Woodland Trust, has 14,000 hectares under ecological restoration, and the National Trust for Scotland's Mar Lodge Estate in Aberdeenshire more than twice that. Cairngorms Connect, a partnership of land managers funded by the Endangered Landscapes Programme, has even bigger dreams; it is restoring habitat and natural processes over 60,000 hectares within the Cairngorms National Park. Private estates such as Glenfeshie, Corrour and Alladale Wilderness Reserve are also rewilding. Communities, too, are buying land to rewild. By 2022, the Langholm Initiative in southern Scotland had raised £2.2 million, partly by crowdfunding, to rewild 4,000 hectares of a former grouse moor.[67]

In England, private estates and farms, such as Ken Hill in Norfolk, Doddington in Lincolnshire and Cabilla Manor Farm on Bodmin Moor in Cornwall, are rewilding. There are public-private partnerships, such as the National Trust, Dorset Wildlife Trust and the Ministry of Defence partnership on the Isle of Purbeck; the Lowther Estate with the RSPB and United Utilities in Cumbria; and Forestry England, the National Trust and United Utilities at Wild Ennerdale in the Lake District. Rewilding Britain now has seventy-two member projects with over 110,000 hectares and 305 square kilometres of sea in rewilding. We are still waiting for the British Army, the Anglican Church and the Crown Estates to follow suit. But when they do, it will be transformative.

Waking up every morning to news of environmental collapse can seem overwhelming, and feeling impotent and despairing can numb us into resignation or denial. It is little surprise that we're beginning to suffer from eco-anxiety. But there is a way to lift the clouds. It helps to know that improvements are happening. Taking action and rewilding a patch of land, whether one's own garden, an allotment, the local churchyard or a stretch of roadside verge, can turn feelings of uselessness into a sense of agency and hope. Watching the arrival of butterflies, the performance of dung beetles, and the emergence of wildflowers and bees is both balm for the soul and a window on to the recovery of our planet.

How conventional nature conservation compares with rewilding

	Conventional conservation	Rewilding
Scientific focus	Ecology	Ecology, mainly, but also multidisciplinary – including, for example, geography, chemistry, history, archaeology and anthropology
Perspective	Linear progression; stability and predictability	Systems thinking; holistic and cyclical; dynamism and unpredictability.
Restoration baseline	Pre-colonial, pre-industrial	Many, since late Pleistocene; potential for novel ecosystems
Restoration aim	Careful stewardship of specific habitats, species and populations; often small, isolated sites with high nature value in need of protection	Ecosystem function and natural processes; areas of low nature importance, such as marginal farmland, post-industrial sites and drained floodplains
Restoration outcome	Specified and managed; expensive	Uncertain, random and evolving; 'hands-off' management; inexpensive
Species focus	Priority and endangered plants and animals	Keystone species, such as beavers, earthworms, wood ants; megafauna and predators
Ecosystem focus	Physical components of an ecosystem and assemblies of species	Interactions and dynamics involving both biotic (living organisms, nutrient flux) and abiotic (light, wind, water, temperature, salinity and ocean currents) influences
Landscape focus	Connectivity between, and expansion of, existing areas of high biodiversity value	Connectivity, but also an interest in natural recolonisation of species and dynamic, large-scale natural disturbances
Level of intervention	Close, continuous management	Create conditions for 'self-willed' ecosystems
Monitoring focus	Species composition, status (rarity, etc.) and distribution	Trophic complexity, natural disturbance and connectivity
Ethos	Defensive, protectionist	Proactive, pragmatic, innovative
Nature and people	People generally considered separate from nature, at the top of the natural hierarchy, with top-down command.	People considered to be indivisible from nature, with obvious and unknown influences, but also potential to act positively as a keystone species

Conventional conservation and rewilding are not binary things. Elements of both approaches can be combined in different places, depending on the context. In a way, this table represents points on the spectrum, with a range of approaches in between.[68]

Paul Lister's 9,300-hectare Alladale Wilderness Reserve in the heart of the Scottish Highlands was once a conventional stalking estate denuded of vegetation. Native woodland is now regenerating through Gleann Mòr (the Great Glen), shown here.

Often, nature reserves exist because they are the last bastions for rare species or habitats. Preserving optimal conditions for particular species or habitats generally requires holding other natural influences – anything that might threaten their survival – at bay. Nature is halted in its tracks. This demands enormous effort in terms of hands-on management, and can only usually be achieved at small scale. The efforts of conventional conservation over the past century or so have been nothing short of heroic; conservation has created natural Noah's Arks and saved numerous species and habitats from extinction.

But it is clearly not enough. Across the board, biodiversity continues to decline as larger ecosystems continue to unravel. The miracle of rewilding is its ability to restore nature in severely depleted areas, places that conventional conservation might consider beyond hope. Because it is not driven by targets, and allows nature to resume the driving seat, rewilding demands the minimum amount of management. Time, effort and costs are therefore hugely reduced. Once the right conditions are in place and nature is functioning again, human management can take a back seat. In philosophical terms, this places humankind within, rather than above, nature.

Smaller-scale rewilding may be similar in many ways to conventional conservation in that it requires more interventions to keep natural processes at play. But here, humans are acting as a keystone species, rather than overall managers. And the aim of rewilding is always for greater connection, which inevitably reduces the need for intervention.

We cannot give up on intensively managed nature reserves while biodiversity is in crisis. There may be some rewilding approaches a nature reserve could adopt to great benefit, such as using free-roaming animals to control tree encroachment on heathland, or to manage scrub or keep waterways open. Indeed, in recent years many nature reserve managers have taken a wilder approach. But rewilding's main contribution is to restore depleted land between existing areas of nature, creating functioning ecosystems again, along with green corridors and new biodiversity hotspots. One day, once we have healed our broken land, once nature is no longer confined to tiny, unviable oases and we have, instead, an integrated network of dynamic, functioning ecosystems providing all that wildlife needs, the role of conventional conservation as caretaker of the last-chance saloon will – we hope – become redundant.

2

Rewilding in the UK
Healing our broken land

Rewilding involves rethinking what our landscapes should look like. Here, at Knepp, emerging thorny scrub provides habitat of astonishing richness and diversity, attracting nightingales, turtle doves and other rare species. Yet traditionally, scrubland is considered undesirable and is rigorously cleared wherever it appears.

Questioning conventions about land management – and what we tell ourselves is 'natural' – is key to embracing a wilder countryside.

It's impossible nowadays to envisage vast numbers of aurochs, tarpan, reindeer and elk wandering the British countryside, let alone being harried by packs of wolves. Within the restricted confines of the British Isles, humans hunted megafauna to extinction centuries earlier than in mainland Europe. Beavers probably died out in England 500 years ago. Elk, reindeer and wild boar – which are still relatively populous in parts of Europe – disappeared from Britain entirely, and the rugged highlands of Scotland and Wales were the last refuge for bears, lynx and (the last to go) wolves. A few wildcats cling on in Scotland today in numbers well below any possible future survival.

In the 500 years since the last lynx wandered Britain the landscape has been utterly transformed.[1] The area of land that is built on today, however, may sound surprisingly low (just 8 per cent) for a population of 68 million.[2] It's not housing development in itself that has had the catastrophic effect on wildlife (although green spaces in some heavily populated regions, such as the southeast, are under increasing pressure). It is what's happened to the land in between. The Industrial Revolution brought wholesale drainage and land 'improvement' for farming, as a result of which 90 per cent of English wetlands and

80 per cent of UK lowland heathland were lost. The Dig for Victory campaign during the Second World War and the advanced technology of the Green Revolution in the 1960s accelerated the drive to put every available bit of land under the plough. Agricultural subsidies, first under the British government and then under the EU, incentivised farmers to increase the acreage of their farms. Tens of thousands of hectares of ancient woods were lost in the forty years after the war – more than in the previous four hundred.[3] Some 113,000 kilometres of hedgerow were pulled up, and countless – because they literally weren't counted – mature trees in fields were felled to make life easier for big machinery. Of wildflower meadows, 97 per cent were lost.

Britain's lowland landscapes are now dominated by rye grass, oilseed rape and cereal crops, and the topsoil has been killed by continual ploughing and agricultural chemicals, which run off into the rivers, polluting the water. Uplands are denuded by historic deforestation and overgrazing, and by skyrocketing deer populations in the Scottish Highlands. Their soils are eroding, and these bare hills cause widespread flooding. Roads and railways cut through every pocket of the country. The total road length in the UK is estimated to be around 400,000 kilometres, and although physically covering only 0.9 per cent of the landmass they have an ecological impact on approximately 20 per cent of the land.[4] Roads kill wildlife that tries to cross them, and break up and isolate habitats. The pollution of roadside edges and surrounding areas caused by particulates from vehicle exhausts and tyres, litter, and salt and grit put down in freezing weather also destroys life.[5] And roads, inevitably, breed more roads. The only remaining refuges for many species of wildflower, insect and songbird are scattered fragments of undermanaged woods and remnant hedgerows. The *State of Nature* report of 2016 ranked the UK 189th out of 218 countries on the Biodiversity Intactness Index (which measures how much of a country's original biodiversity remains in the face of human land use and related pressures).[6]

As John Lawton stated in his DEFRA report *Making Space for Nature: A Review of England's Wildlife Sites and Ecological Network* (2010), the UK needs 'more, better, bigger and joined up' areas of nature. So, how can that be done? How can we attempt to restore and rewild the fragmented and degraded landscapes? How do we restart natural processes, and at scale? Can we re-create complex, shifting habitats in this densely populated, historically modified countryside? Can we accommodate free-roaming animals – and perhaps, one day, even apex predators?

How rewilding can work in lowland Britain

To show how rewilding can work in the British landscape, and particularly how it can integrate with agriculture and infrastructure, we took a typical scene of lowland countryside in England and overlaid it with an artistic interpretation of how it could look when the key elements of rewilding are introduced. It gives an idea of how landowners and land managers can participate in reconstructing a connected landscape, with a large-scale rewilding project, or a smaller plot of land to connect with a larger project or form part of a wildlife corridor, or simply by improving woodland, hedges and wetlands for nature.

This typical scene of lowland rural England today – where the M6 crosses the Lancaster Canal near Farleton Knott in Cumbria – would be considered beautiful and timeless by many, a picture-postcard view of the British countryside. But a closer look shows a scene of loss, pollution and instability. Fragmented by roads, the intensive, chemically farmed fields and canalised river leave barely any space for nature.

Changing the aesthetic

To many eyes, this rural scene might seem an idyll of lowland British countryside, the personification of our 'green and pleasant land', a patchwork quilt of orderliness and productivity. We often assume that the land has looked like this for centuries, and that is reassuring. We sense historical continuity and security in this 'cultural landscape'.

But look more closely and you can see how tightly controlled and simplified it is, and how little wildlife habitat it contains. Much of this has happened since 1950. The fields are intensively farmed with monocrops; there are no surrounding margins of wildflowers for pollinating insects; hedges are thin and poorly maintained. Once there would have been wetlands. Now the river is canalised, and the fields around it drained. The copse next to the canal, perhaps a remnant of an ancient wood and already bisected by the road, is, in effect, an island, marooning any non-flying creatures living there, diminishing their ability to colonise, connect and breed. In the bottom left of the picture, the block of trees is a single-generation closed-canopy plantation offering very poor habitat.

Far from embodying efficiency and security, this landscape is supremely unstable – and costly. The soil, depleted by decades of ploughing and chemicals, has died. It has become little more than infertile 'dirt', a medium in which crops can be grown only with the input of expensive artificial fertilisers.[7] Compacted and eroded, with no organic matter to hold it together, topsoil washes off into watercourses when it rains, and blows away in the wind when it is dry. Soil degradation costs the UK £1.2 billion every year.[8] The depletion of topsoil is so severe in the UK that in 2014 *Farmers Weekly* announced that there are only a hundred harvests left in the country.[9]

The canalising of rivers (which was largely done during the Industrial Revolution) was originally intended to help drainage by getting water off the land as quickly as possible. For centuries it has been assumed that this is the surest way to prevent flooding, but the opposite has, in fact, proved to be the case. In big rains, water pouring off the compacted fields can quickly overwhelm canals and even the roads that have been built over them. With no meanders or functioning floodplains to soak up excess water and 'slow the flow', the canalised river speeds the travel of water and increases the likelihood of flooding downstream. Flood damage currently costs the UK £1.3 billion each year.[10]

With climate change, heavy rains are becoming more common. According to the UK Meteorological Office, the decade 2009–18 was on average 1 per cent wetter than 1981–2010 and 5 per cent wetter than 1961–90 in the UK overall. More importantly, the amount of rain from extremely wet days has increased by 17 per cent compared with the same periods. By the end of the twenty-first century severe flash flooding will be five times more likely than it is now. If, as is almost inevitable, global temperatures rise by 2°C, the number of British people at significant risk of flooding is expected to be 2.6 million. Under a 4°C scenario, 3.3 million people will be vulnerable to flooding. During the 2020s some 35,000 hectares of high-quality horticultural and arable land are likely to be flooded at least once every three years. By the 2080s, it will be 130,000 hectares.

The river in our picture is also likely to be contaminated by chemical run-off from the intensive agriculture around it. Between 2004–5 and 2008–9, water companies in England alone spent £189 million removing nitrates and £92 million removing pesticides from water supplies in order to meet drinking water standards.[11] Since 1975 some 146 groundwater sources used for public supply have been closed because of problems with quality.[12] The cost of water pollution in England and Wales is estimated to be £1.3 billion a year and rising – a cost that is, inevitably, reflected in water bills.[13]

It may seem that any tree cover is good. But a single-species, single-generation, closed-canopy, short-rotation commercial plantation such as the one in the bottom left of our picture is not only poor habitat for wildlife; it also puts the trees themselves at risk. They will have been planted as saplings from a single genetic source, propagated in a commercial nursery, possibly in continental Europe. They will have none of the natural immunity, genetic diversity or regional adaptations of trees that have naturally regenerated or been grown from local seed sources. Importing saplings runs the risk of introducing disease. This is how the fungus that causes ash dieback entered the UK. It also comes at considerable carbon cost, as does the whole process of propagation, distribution and planting by hand – something that is not always fully factored into the carbon sequestration calculations of commercial forestry. A plantation like this is also more vulnerable to storms, flooding and drought. It will be clear-felled (all the trees removed) at harvest, between thirty and sixty years before the trees have reached their full potential for carbon sequestration, and then it will simply be replanted in the same way.

Even if we do appreciate that a local landscape is depleted of wildlife and, in many ways, dysfunctional, we may still find ourselves attached to it through a sense of familiarity. We may have grown up there, or in a place like it, or spent holidays there; we may be nostalgically bound to what we are fond of, or to notions imposed by education or cultural background. Barren hillsides and sharp edges have a certain appeal to humans' orderly minds, and it is hard to disentangle what makes us believe something is beautiful and worth having. But function has to be part of it. Few people consider an oil spill from a tanker beautiful. It may glisten with an iridescent sheen, offsetting the colour of the ocean or the cliffs, its very viscosity an allure. But our knowledge of how harmful it is, and of our involvement (however remote) in how it came to be there, ruins any aesthetic pleasure we might take in it. In the same way, the familiar countryside is an ecological disaster, if only we could see it that way. Only when we understand what is going on in the landscape will our aesthetic, and our acquiescence, change. Beauty is in the eye of the beholder.

Rewilding hotspots

The artistic impression on the next page shows the first rewilding interventions; the first steps towards rebuilding nature in this landscape. There are two large-scale rewilding projects. One is in the top right-hand corner, in the distance beneath the low-lying hills, where a group of birds of prey are soaring in rising currents of warm air. Of the other, on the left of the picture, just one rewilded field is visible, the tip of a much bigger area that is outside the picture itself.

These hypothetical rewilding projects may be around 400 hectares or more, on a similar scale to Knepp. They are big enough to support herds of free-roaming animals all year round without supplementary feeding. They may have been set up by individual landowners, or by a group of smaller landowners – a 'farm cluster' – who have pulled down the boundaries between them to create a larger project. Various herbivores live there, including old breeds of cattle, horses, red and fallow deer, as well as wild boar (or an old breed of pig as a proxy for boar). Perhaps there are even bison or elk. The rewilded areas are ring-fenced so that the animals don't wander on to roads, railways, farmland and private property, or into commercial forestry and built-up areas.

A first step towards restoring nature in this landscape is to create new 'biodiversity hotspots' through large-scale rewilding. The storks and birds of prey riding thermals in the top right of the picture are flying over a Knepp-style rewilding project of 400 hectares. On the left of the picture is the tail end of another rewilding project with free-roaming animals, which extends out of shot for another 400 hectares.

These rewilding areas are now biodiversity hotspots. They have been left to themselves for at least five years and the resurgent complex habitats within them – including thorny scrub – now support burgeoning populations of both common and rare wild animals that have spontaneously colonised them because of the available food, and nesting and breeding opportunities. There would be a significant 'spillover effect' from these two areas, whereby birds, small mammals and insects expand into the surrounding landscape – if habitat were available. But at the moment these hotspots are islands in an eco-logical desert. The same is true of almost every site designated for nature in Britain, the majority of which are also incredibly small. Of England's 4,100 Sites of Special Scientific Interest (SSSIs), most are less than 100 hectares; the average size of the 2,000 sites run by the Wildlife Trusts is just 30 hectares.

Island ecology tells us a lot about isolated habitats.[14] Generally, the smaller and more remote the island, the fewer the species and the more vulnerable its ecosystem. Isolated populations, particularly those of species that can't fly or find it difficult to cross an inhospitable landscape with no food or cover, will be prone to inbreeding.

A small genetic pool is likely to decrease in variation over time, making a species ever less able to respond to both sudden and incremental change, such as disease or long-term drought. Some life forms are particularly vulnerable to isolation. Saproxylic beetles living on ancient oaks, for example, cannot travel far. If there are no other old trees within a kilometre or so of their host tree for them to colonise, they will simply die out when their host expires.

Even for those species blessed with mobility, wildlife hotspots can become species 'sinks'. With no competition from apex predators such as wolves and lynx to keep them in check, the UK has relatively high populations of versatile, medium-sized predators, animals with numerous prey species and that travel readily through the human landscape. The UK has about 240,000 foxes, 400,000 badgers and – often overlooked – 12.2 million domestic cats.[15] Cats alone catch up to 275 million prey items a year, mostly birds and small mammals.[16] The smaller and more isolated the habitat, the more conspicuous it is as a hotspot for life and the fewer the possibilities of escape from predation. Isolated habitats may be attracting endangered species, only to hasten their demise.

Another factor is the 'edge effect' – the impact of a surrounding hostile environment on an isolated habitat. A wood stranded in a cereal field often has a very different microclimate at its centre than at the edges, where it is exposed to wind, extreme heat and frost. The hard, linear boundaries of the modern landscape – with no messy, broad margins to soften the transition – mean that chemical sprays easily drift into a habitat, reducing its effective size. The larger the area of habitat, the smaller the relative area of its edges, and therefore the more protected the site.

Most importantly, though, isolation renders habitats and the species within them vulnerable to climate change. The UN's Intergovernmental Panel on Climate Change reports on climate zones and the 'action of climate velocity', a term used to describe the rate at which climate zones move through the landscape as the planet heats up.[17] A 3.2°C rise in temperature is the likely scenario if we continue to fail to make the radical changes required to keep to the targeted 1.5°C increase. At present, the environmental charity Rewilding Britain estimates that climate zones across Britain are moving north at a rate of 5 kilometres a year, hundreds of times faster than after the last ice age, 11,000 years ago.[18] This means that by 2050, Northumberland will be as warm as West Sussex is now.

Unless we knit our areas of nature together, they are doomed to fail. For animals and plants to be able to respond to the challenges of climate change, they must be able to flee in search of habitats that are still suited to them. They also need to connect with other populations if they are to retain the broadest possible genetic pool to help them evolve and so, possibly, survive.

Green corridors and land bridges

It is of the utmost importance to open up pathways between Britain's small patches of nature. By forming a web linking nature nationwide, green corridors can act as arteries, allowing animals and plants – including large herbivores – to migrate safely through the landscape, even through agricultural land and housing and industrial developments, from the coast to the mountaintops and into the heart of cities. This is where smaller-scale rewilding comes into its own. A small patch of land can have an exponentially positive effect if it operates as a link in the chain.

Creating green corridors between the rewilding project top right and off picture to the left is crucial. It allows wildlife to move through the landscape in response to climate change and enables breeding populations to meet each other, preserving genetic diversity. Another wildlife corridor connects the lefthand rewilding project with a National Nature Reserve in the far distance. Land bridges are also key, or the roads remain barriers to movement and, for many species, deathtraps.

This image shows a green corridor connecting the two rewilding projects. Another runs into the distance from the rewilding project on the left to another nature area that we can imagine on the far horizon. Free-roaming horses, bison and cattle are using the green corridors, replicating the effects of seasonal migration. Wildlife corridors would also facilitate access to nature for people, connecting walking and cycling routes.

Clearly, green corridors function only if they incorporate land bridges, also known as ecoducts, wildlife overpasses or green bridges, over roads.[19] Roads are deathtraps for any mammals, reptiles and even insects that try to cross them. Green bridges were pioneered in 1988 in the Netherlands, where there are now over 600 wildlife crossings.[20] They are usually planted with local trees, shrubs and other vegetation, and are often for wildlife only. Some even have chains of small pools across them, and little access ramps for frogs and newts. In Sweden one green overpass, combined with the construction of roadside fences, led to a 70 per cent reduction in accidents caused by roe deer and elk. In Canada, two green bridges over the Trans-Canada Highway in Banff have successfully reconnected two populations of bears, preventing genetic isolation. In Los Angeles, the world's biggest wildlife bridge, expected to be completed in 2025, will provide mountain lions with safe crossing over a ten-lane highway into the Santa Monica mountains.[21] Even birds like green bridges. In Brisbane, Australia, twice as many species of bird fly over the Compton Road overpass, which is characterised by native eucalyptus woods, than fly over the road itself.

Overpasses (which often incorporate paving for pedestrians and cyclists) and green bridges are used by a wide range of mammals, including bats, and can provide a vital connection for apex predators with vast territories, such as wolves and lynx.[22] The greener bridges are, the more animals use them. Wide overpasses near woods and scrub or water are the most successful.[23] Underpasses, which can be created by adapting existing structures, can also enable mammals and amphibians to cross roads, and it has been found that wider and shorter tunnels with a surface of natural substrate are used more frequently.[24] All types reduce the frequency of wildlife collisions with cars.

Land bridges, however, are still in their infancy, partly because of their expense but also because they remain a very low priority for planners. In the whole of the UK only two of significant size have so far been established: the green bridge spanning the A21 at Scotney

Castle in Kent; and the Mile End green bridge in London, which spans five lanes of the A11. Neither was designed with a specifically ecological objective.

Railways

Not included in this pictorial sequence is a railway. However, railway embankments and sidings can be effective wildlife corridors if they are managed well, providing important brownfield and scrub habitat. The tracks themselves – if they are not electrified – can be far friendlier to large mammals than busy roads.

The UK's rail infrastructure agency, Network Rail, manages nearly 32,000 kilometres of lineside across England and Wales. It owns 53,000 hectares of land, on which there are nearly 6.3 million trees – most of them less than fifty years old – and 200 SSSIs. The potential for wildlife is huge. Of obvious concern, however, is the danger of large trees falling on to the line, but, equally, some vegetation cover is needed to stabilise the soil on embankments and reduce the risk of flooding and landslip. Although the UK's is one of the safest railway networks in the world, the number of vegetation-related incidents rose from 11,500 in 2009–10 to nearly 19,000 in 2017–18, resulting in more than 1,750 train cancellations.[25]

Conventionally, Network Rail has approached vegetation problems reactively rather than pre-emptively, and nature has been a low priority. But, following an independent review in 2018, it is now committed to ending the net loss of biodiversity on its land by 2024 and achieving a net biodiversity gain by 2035.[26] A rewilding approach could amplify the ambition. Railway embankments, like riverbanks, could be managed by mimicking the rotational coppicing carried out by beavers. Coppicing embankment trees and scrub so that trees never grow too high or too big, while maintaining strong root systems of woody vegetation cover to hold back the soil, would increase trackside safety, reduce the cost of emergency management and repairs, and provide wonderful habitat for wildlife. Coppicing machinery could be designed to run alongside the embankments, cutting vegetation back, and the biomass from the chippings could even be used to fuel the machines themselves.

Hedges and crop margins

Hedges are a vital home for flora and fauna, being often the only remnants of thorny scrub habitat in the modern landscape. They are also, in effect, wildlife corridors for smaller species. Restoring them with a mix of native woody shrubs and doubling, or even tripling, their width are both relatively easy to do, and create a safe travel network for small mammals, reptiles, insects and birds, with food, nesting sites and protective cover. At Knepp we found that allowing our hedges to billow out to 7–14 metres deep attracted high densities of breeding nightingales – one of the country's most endangered birds, whose numbers across the UK fell 91 per cent between 1967 and 2007.

Leaving margins for wildlife around the edges of fields provides important habitat for birds and small mammals.[27] It also improves crop yields, even in conventional chemical-based farming, largely by providing pollinating insects, and the insects and birds that hunt crop pests.[28] This is true of urban agriculture, too.[29] When crop margins are aligned with hedgerows they create even more important foraging borders, rich in insects, for small mammals, birds and reptiles. This

Good hedgerows also act as wildlife corridors. Restoring multi-species hedges around field boundaries and allowing them to grow out wide with rough wildflower margins on either side, provides birds, insects and small mammals with food and protection.

is a crucial step and one that all farmers, conventional or otherwise, can implement to their own economic benefit, as well as to improve conditions for wildlife.[30]

Regenerative agriculture

Across the board, we must shift from chemical farming to regenerative agriculture, as shown by the changed field use in the centre of the picture on page 72. Regenerative agriculture (or 'restorative agriculture' as it is sometimes known in the US) involves rotating crops, using cover crops either between crops or on bare fields, and rotating livestock in order to sustain and improve soil fertility, avoiding the need for artificial fertilisers and pesticides.[31] It also avoids ploughing, which is a problem at the root of global soil depletion today.

The plough is one of humankind's most destructive inventions. Every time the earth is turned over, life forms in the soil – everything from earthworms to mycorrhizal fungi – are destroyed. Opened to the elements, the soil dries out, and begins to blow away in the wind and wash away in the rain. Eventually, there is no topsoil left in which to grow anything. Long-term ploughing caused the downfall of empires from ancient Greece and Rome to the Indus Valley and of civilisations in Central and South America.[32] Once renowned for its rich soils, much of the so-called Fertile Crescent, an area covering present-day southern Iraq, Syria, Lebanon, Jordan, Palestine, Israel, Egypt and parts of Turkey and Iran, is now desert. Time and again, deforestation, ploughing, agricultural irrigation and over-grazing has amounted to ecological suicide for civilisations. As Franklin D. Roosevelt famously said in 1937 in response to the agricultural depression in the US, 'a nation that destroys its soils destroys itself'. Yet we still haven't learned. Across the planet, fertile soil is being lost at the rate of 24 billion tonnes a year.[33] It is reckoned that half the world's topsoil has been lost in the last 150 years and that 1.3 billion people are trapped on degrading agricultural land.[34]

Across the world, from the USA to Australia, individual farmers – often those directly affected by soil degradation – are leading the agricultural revolution. The Groundswell farming festival, founded in 2015 by John and Paul Cherry, showcases the movement in the UK.[35] Regenerative agriculture involves a complete change in the farming mindset. The fundamental premise is working with nature, rather than battling against it. The remodelling of farming subsidies in the UK might encourage more openness among farmers to this way of thinking, but there is likely to be considerable opposition from big agriculture, manufacturers of conventional big farm machinery and chemicals, and the big food industry. The writing, however, is on the wall. As Gabe Brown in the USA has shown, unprofitable farms can become profitable if they switch to regenerative techniques.[36] They can also produce as much food as they did before – and that food is healthier because the plants take more nutrients from the soil, and contain few or no chemical residues.

By increasing rather than depleting topsoil, regenerative farm-land also becomes a powerful carbon sink, perhaps the most effective solution to climate change.[37] According to the Rodale Institute, an American not-for-profit organisation that supports research into organic farming, 'we could sequester more than 100% of current annual CO_2 emissions with a switch to common and inexpensive organic management practices, which we term "regenerative organic agriculture".'[38] In his book *Drawdown: The Most Comprehensive Plan Ever Proposed to Reverse Global Warming* (2017), the environmentalist and activist Paul Hawken identifies twenty effective methods of pulling carbon out of the atmosphere. Ten of them are techniques associated with regenerative agriculture.

One regenerative farming technique shown in the field beneath the rewilded area on the left of the illustration on page 72 is agrofor-estry, in which trees and/or shrubs are grown among crops or pasture for livestock. In this two-tier farming system – which goes back thou-sands of years and was once practised all over the world, including in Britain – the trees produce a crop of fruit or nuts. At the same time they increase the yield of the crops below by stabilising and increas-ing the fertility of the soil and encouraging pollinating insects and pest predators. In a wood pasture or 'silvopasture' system, such as the ancient *dehesas* of Spain and Portugal, the trees provide acorns or beechmast for pigs, or tree fodder for cattle. Agroforestry can provide an excellent buffer zone around rewilding projects, enabling many species to spill out into a wider area.

Another technique, depicted in the field beneath the agrofor-estry area in the illustration, is 'mob grazing'. This short-duration, high-density grazing system, pioneered by scientist and farmer Allan Savory, allows longer than usual for the grass to recover after a period of grazing.[39] The animals are corralled on to an area of pasture, then moved on, on average once a day. The pasture they leave behind is allowed to recover for between 40 and 100 days, depending on climate and soil type. Mob grazing restores the soil and produces healthier animals at a lower cost.[40]

In this way, regenerative farming works hand in hand with rewilding, eradicating pollution in the soil and watercourses, restoring water tables and soil biota, creating micro-habitats and increasing the permeability of the landscape for many species, including pollinating insects and natural predators of crop pests.

Shifting from industrialised, chemical-based farming to regenerative, nature-friendly agriculture is obviously a boon for wildlife. It also provides sustainable and more nutritious crops. In the field to the left of the picture, a herd of cows are being 'mob-grazed': moved on at frequent intervals, allowing the land to recover behind them. The field above has been allowed to 'scrub up', providing a nursery for naturally regenerating trees. Wild trees have far greater genetic diversity than commercially propagated trees and more resilience to climate change and disease. This natural wood pasture system also provides grazing for livestock.

Rewilding river systems and restoring wetlands

Rewilding river systems and restoring wetlands and floodplains bring tremendous benefits for biodiversity, water quality and carbon sequestration, as well as holding back water in times of big rains, thereby protecting land, roads and property downstream from flooding.[41]

Decanalising rivers and blocking or removing drainage structures by mechanical means can, however, be labour-intensive and expensive. At Knepp we partnered with the Environment Agency to rewild nearly 2.5 kilometres of the river Adur, taking out four weirs in the process. The Environment Agency couldn't even remember why the weirs were there, although they had cost a fortune to maintain. It took two summers with a digger and considerable expense to fill in the canal, return the river to its floodplain and restore shallow 'scrapes' and ephemeral ponds for water plants and invertebrates. At the time, this meagre stretch of river restoration was the largest on private land in the UK. At the end of the exercise, ecologists advised us to introduce 'woody debris blockages' into the river – dead trees and branches – to increase the braiding, dynamism and complexity of

Decanalising rivers and returning them to their floodplains, as in the bottom righthand corner, helps prevent destructive flooding of crops, roads and houses downstream, improves water quality and allows for the passage of migratory fish. Wetlands are fantastic habitat for wildlife, and sequester carbon. They also provide space for recreation. Beavers can create wetlands in months, at no cost (see page 112).

the water flows, and provide habitat for water invertebrates and fish fry.[42] It occurred to us that we were simply being a poor imitation of beavers, the experts in the field. Indeed, a pair of beavers could have rewilded that stretch of the river much more intricately in a quarter of the time, with far greater benefit to wildlife – and at no expense.

We hope beavers will soon be at large again in our landscape, as they are in this picture, undertaking their extraordinary hydrological engineering work for free. Here, the canalised river with its steep banks has been transformed into a complex system of pools, willow scrub, trees and flowing water. A bison is creating a sandy wallow, and water buffalo and elk – two more keystone species that could be introduced to enhance the wetland environment – wade among the reeds. This sort of habitat will be ideal for wetland birds including, quite possibly, black storks – a bird that very occasionally visits Britain but rarely stays because of the lack of suitable habitat.

Increasing wetland areas will also open up more opportunities for human enjoyment of nature. Kayaking in regenerated wetlands, seeing all these animals and birds up close from water level, would be an incredible experience.

Restoring water, whether a flowing river in open countryside or a small standing pond in a back garden, is a one-off endeavour. Generally, it requires little or no further management, but it can provide rich habitat for wildlife in perpetuity.

Woodlands and commercial forestry

The plantation monocrop in the bottom left of the picture opposite has morphed into a diverse, multi-generational, multi-species wood. Allowing trees to regenerate naturally wherever possible, as Rewilding Britain advocates, should be the default method for increasing tree cover in the landscape.[43] It costs nothing, and avoids the polluting and carbon-intensive process of propagation and planting. It allows the transitional phases of vegetation that are wonderful for wildlife, it ensures the genetic diversity of tree species, and it preserves soil biota and mycorrhizal fungi connections, which provide the healthiest start for young trees. We should plant trees only when local seed sources are absent or scarce, or where there is a specific aim, such as creating an orchard, windbreak, avenue or visual barrier, or replacing specific landmark trees.

We have to think more creatively about commercial forestry, too, not just for biodiversity but also to preserve the genetic diversity and adaptability of trees – their main long-term defence in the face of disease and climate change.[44] This means using local seed sources for native species. Also, rather than clear-felling vast areas, as is conventional in the UK, a common European practice known as 'continuous cover forestry' should be followed.[45] This means selectively felling individual trees to create a diverse, multi-generational commercial wood with understorey vegetation.[46] The trees then seed themselves, from mother trees.

The picture also shows dead trees, as is often mandated in Europe. Leaving a few dead trees standing provides nesting cavities for wild bees, and for bats, owls, kestrels and other birds whose droppings return nutrients back to the woodland soil.

Converting plantations of single-generation trees of just one or two species into a multi-generational, multi-species wood, as in the bottom lefthand corner, is hugely beneficial for wildlife. It also gives greater protection to the trees themselves from windblow, pests and disease. This wood is being allowed to expand through natural regeneration, and timber is extracted through the continuous cover cropping method.

Continue connecting

There will always be further improvements and connections to be made. For example, more areas can be linked using land bridges and green corridors (which could be towpaths, railway embankments, road verges, footpaths and cycle routes enhanced for wildlife, or land bridges). In the picture below we see how another wildlife corridor along the right-hand side of the central road has connected the wetland with the green corridor and the land bridges running between the two big rewilding areas.

There is no reason, either, why ribbons of rewilding couldn't run through the entire countryside, right into towns and cities. This would provide the framework to support nature everywhere – including in arable belts and deep into the heart of urban areas.

Adding more wildlife corridors to the existing network – as in the stretch along the road on the left – creates even greater resilience for wildlife and greater access to nature for people.

How rewilding can work in upland Britain

Changing the aesthetic

People have come to think of Britain's barren hillsides in the uplands and highlands – as seen in this image – as picturesque, the fell-walker's idyll. The short, springy grass feels like a carpet underfoot and the scale of the denuded hills rolling away into the distance can be awe-inspiring. There's a certain heroic grandeur to the shapes and forms; sunlight and clouds play across them and the very emptiness has a meditative quality and a sense of freedom.

Barren hillsides are the norm in the uplands and highlands of Britain, but they are the result of historic deforestation and over-grazing. The exposed peat 'haggs' in the foreground have been created largely by sheep. Drainage has also destabilised the soil, creating erosion fissures on the hillside on the right. Block planting of non-native conifers provides little natural habitat and adds to soil loss and water acidification. Moor-burning to stimulate the growth of heather for grouse shooting can destroy the ecology of peatland, one of the world's most important carbon storage systems. Without natural vegetation such as sponge-like sphagnum mosses, these hills can shed water in terrifying amounts in high rains, overwhelming the fields and canalised river in the valley below, and flooding the road and villages downstream.

But a closer look shows how degraded and unstable this scene is. The road through the centre of the valley has necessitated the canalisation of the river, destroying its natural meanders and increasing the risk of flooding downstream. Those emerald fields in the valley (or 'glen', in Scotland) are 'improved' leys for sheep, silage and hay – commercial and nitrogen-fertilised grass with no floral diversity. Once the river's floodplains, now drained, they no longer function as a 'sponge' to hold back flood water. On the hills above, deep erosion gullies caused by historic deforestation, state-sponsored drainage using shallow drains or 'grips' installed during agricultural intensification after the Second World War, and overgrazing and compaction by sheep and/or deer, shed vast amounts of topsoil, peat and water into the valley.[47] With such high grazing impact, there's no chance for trees, woody shrubs or other plants to regenerate on the hills and stabilise the topsoil, and the large number of sheep and deer, with their selective preference for palatable herbs and flowers, has reduced plant diversity to comparatively few species of coarse moor grass, sedge and heath rush.

Sheep

Feelings run high when it comes to sheep. There is no doubt that sheep farming has become a cherished part of upland tradition, something with an intrinsic cultural value that many people would like to be sustained at current levels. But it's important to consider the environmental impact, hidden cost and long-term future of this system, especially given that it is entirely financially underpinned by subsidies. In Wales, for example, the average subsidy for a hill sheep farm is £53,000, and the net income after subsidy is £33,000. In other words, without agricultural payments, the farmer would be making a loss of £20,000 a year.[48]

We've grown used to seeing large numbers of sheep on the hills, and it is easy to assume that this has been a harmless feature of the land for centuries. But, again, shifting baselines are at play. In medieval times, there would have been considerably fewer sheep in most places. There would have been more scrub cover for browsing and many more cattle on the hill, too, and shepherds would have controlled flocks much more closely, constantly moving them on to allow the land and vegetation time to recover, and bringing animals down from the hills for the winter.

Sheep farming on uplands increased radically from the late eighteenth century onwards, thanks largely to a boom in the wool trade. This shift in land use is epitomised by the infamous Highland Clearances, whereby thousands of people were evicted from their smallholdings in the Highlands of Scotland. And, although wool prices collapsed in the early nineteenth century, sheep numbers expanded even further in the twentieth century, driven by farming subsidies and 'headage' payments. These payments, which stopped in 2003, incentivised farmers to stock as many sheep as possible. Supplementary feeding also supported bigger flocks. In Wales, sheep numbers have more than trebled, from around 3 million in the late nineteenth century to 9.5 million in 2019 (almost three sheep for every person now living there).[49]

From a conservation point of view, sheep are problematic. They are not indigenous to temperate-zone Europe, and the vegetation here has not evolved to defend itself against their physical impact and their specific techniques of browsing and grazing. Sheep dung supports fewer dung beetles than cattle dung. Sheep were domesticated from mouflon, a wild, big-horned sheep native to Mesopotamia that lives in mountainous terrain on stony, dry soils. The mouflon's domesticated descendants were probably introduced into Britain by Neolithic settlers in about 4,000 BC. In their native lands, their tiny, sharp hooves allow them to balance balletically on rocks, and to evade predators. On the soft, wet soils and peatland of Britain, they drill into and compact the soil, thereby having a very different impact from the soft, spread-out hooves of the larger native ungulates. Imagine everyone attending a muddy music festival wearing stilettos. Now transfer this image to sheep, stilettoing their way across steep upland hills. In the foreground of the picture, human-built drainage and sheep traffic have exposed and compacted the bare soil, creating peat 'haggs' – blocks of peat that have become isolated from the peat mass (around 7 per cent of deep peats in England are hagged and gullied in this way).[50]

Like a goat, a sheep has a narrow face with mobile lips, a small mouth and a lower set of close-cropping teeth. This face is designed to reach into rocky nooks and pull plants from crevices, leaving few hiding places for flowering plants and herbs. Cattle are less dextrous and selective. They use their tongues to pull around longer plants, and in the growing season they focus on the most nutritious grasses, opening up opportunities for flowering plants.

As always, it's a question of how many animals there are and how long they remain in one place, as well as the overall vegetation and soil conditions. Even a relatively small number of sheep grazing loosely across a large area all or most of the year round can reduce biodiversity and cause erosion that releases soil carbon into the atmosphere. But a small number grazing a meadow or chalk grassland for a few months in the summer can be an important tool for conservation, preventing the development of grassy thatch that, without grazing, would smother floral diversity. It may well be that sheep have a useful role to play in upland conservation in low numbers, as one of several herbivore species. There are economic, historic and cultural imperatives to consider, too. In Chapters 4 and 5 we discuss the benefits of removing or significantly reducing sheep numbers for a period of time, allowing the land and vegetation time to recover, then sensitively allowing flock numbers to build again, together with complementary herbivores such as cattle and ponies, to levels no greater than the habitat can sustain.

Plantations

Few people would consider Britain's many solid blocks of non-native Sitka spruce plantations on hillsides scenic. They are the result of a post-war government policy of reafforestation to build up a strategic reserve of timber. This doubled forest cover in the UK – mainly in the uplands, and mainly of exotic conifers – between 1948 and 1995.[51] Virtually no flowers or, indeed, any plants at all are able to survive within these dark, acidic plantations, making them extremely poor habitat for wildlife. When entire mid-slopes of valleys are covered in plantations, high mountainous habitat becomes isolated, restricting the movement of birds, larger mammals and insects such as butterflies.

Conifer plantations also affect the natural hydrology of the landscape. Deep ploughing carried out before planting to improve drainage and tree growth has made water run-off worse, and increased the risk of flooding in valleys. Such brutal, large-scale ground preparation often destroys smaller water features, such as flushes and streams, altering natural drainage systems. As they mature, the intense blocks of trees can dry up natural springs and wetlands around them. Meanwhile, rainfall passing through the

canopy of Sitka spruce, particularly in the early stages of the trees'
growth, produces acid run-off, ultimately affecting fish and invertebrates in the rivers.[52] When the trees are clear-felled after thirty
years or so, the disturbance of extraction by large machinery leaves
unsightly claw marks, like a bomb site on the hill, and results in
further soil and carbon loss and run-off.

Peatland drainage and carbon loss

The plantations in the picture on page 77 will contribute to carbon
sequestration. But the carbon held in the timber is likely to be
exceeded by the carbon released from the deep, peaty soils by invasive ground preparation and drainage. Most timber products derived
from softwood plantations – biomass, pallets, cardboard, paper and so
on – have a short lifespan, too. They won't hold sequestered carbon
for long. Such dark, dense land cover will also absorb more sunlight,
leading to rising temperatures.

Peatland covers 23 per cent of Scotland's land area, 6 per cent
of Wales's and 11 per cent of England's. In the UK, there are
three main types of peatland: fens, raised bogs and blanket bog.
The UK has between 9 and 15 per cent of Europe's peatland area
(46,000–77,000 square kilometres) and about 13 per cent of the
world's blanket bog – one of the rarest habitats.[53] On the hillsides
and moorland above the plantations in our picture there is blanket
bog (classified as having a peat depth of over 400 millimetres), typical
of the uplands of Britain. In their undamaged state peatlands represent a vast store of carbon laid down in undecayed plant material
over at least 10,000 years. In Scotland, peatlands store around 1.6
billion tonnes of carbon – that is one-third of the carbon held in
the Amazon rainforest – despite being 250 times smaller than the
Amazon rainforest in area.[54] In England, peatlands store 580 million
tonnes of carbon. But only 4 per cent of them are in good condition.[55] When peatlands are drained, eroded and/or burned – as can
be seen on the hillside on the left – carbon is released into the atmosphere. The UK's damaged peatlands today emit around 3.7 million
tonnes of carbon dioxide every year, equivalent to the emissions of
660,000 UK households.[56]

Moorland burning

Rotational moorland burning is known as 'swaling' in the southwest of England (where it is used to provide grazing for livestock) and 'muirburn' in Scotland. It is now a common practice on sporting estates in the uplands, and principally aims to encourage populations of red grouse. Today, around 700,000 red grouse are shot in the UK every year, and roughly 1.3 million hectares of moorland in the UK are managed for grouse.[57] Burning grouse moors in rotational patches promotes the growth of young shoots amid differently aged stands of heather, to feed and shelter the birds. A study in 2019 of eighteen estates in northern England and southeastern Scotland suggested, however, that burning may not increase grouse numbers significantly.[58] Traditionally, landowners thought regular burning was necessary for the heather to remain dominant. But this has been disproved in areas where heather has remained dominant, rejuvenating itself naturally for at least forty years without burning.[59]

Many upland dwellers also believe controlled burning creates firebreaks and reduces the fuel load of dry vegetation, reducing the risk of catastrophic wildfires (which have been increasing in frequency and intensity on moorland as a result of past drainage and more recent drier summers). It is a practice with a strong cultural tradition among crofters, too, who use it to provide grazing for sheep, and is often seen as constituting responsible stewardship of the land.

However, the International Union for Conservation of Nature (IUCN) identifies burning as 'highly damaging to ecological, hydrological and soil processes'.[60] It also restricts natural regeneration by native trees and woody shrubs, which are a natural feature of moorland habitat.[61] Most importantly, it destroys sphagnum mosses, which are the main component of blanket bog, and the main plant material laid down over millennia as peat. Considered a keystone species, sphagnum mosses can hold up to twenty times their weight in water, providing an effective brake on rainwater and flooding. Burning dries out sphagnum, reducing its ability to hold water.[62] It also tips the balance in favour of fire-tolerant species, such as heather, and as the peat dries it becomes even more combustible. Indeed, half the wildfires in Scotland have been caused by muirburn running out of control.[63] Burning has become a vicious cycle. And fires, of course, release carbon.

Fire and herbivory are closely related. But the single-minded sporting focus on providing vegetation for a single species generally involves excluding large herbivores that may compete for that vegetation. In doing so, the habitat is deprived of the mouths that would consume more dry vegetation, reducing the ignitable fuel load. In the past, aurochs, elk and, perhaps, bison would have browsed the upland moors, performing this role.[64] It is important to break this cycle of fire dependency and propensity. Reducing fire and herbivory, and restoring natural vegetation and hydrology, returns moorland to a state where the right species of herbivore, in the right numbers, can be allowed in. Those herbivores then help to create diverse plant communities, reduce the dry fuel load and sustain a dynamic, functioning ecosystem – all of which substantially reduces the likelihood of fire.

Predator control

To protect grouse numbers, gamekeepers control predators.[65] For centuries across Britain, birds of prey (raptors) have been killed by farmers, the managers of sporting estates and egg collectors. But the rise of driven shoots (involving huge numbers of birds) around 1900 intensified the focus on raptors. By the end of the First World War, goshawks, marsh harriers, honey buzzards, white-tailed eagles and ospreys had been driven to extinction in Britain. Between the 1870s and 1970s golden eagles, hobbies, hen harriers, red kites and Montagu's harriers all declined to fewer than 100 pairs.[66] Peregrine falcons and sparrowhawks were also poisoned by the organochlorine pesticides widely used in agriculture throughout the 1950s and 1960s, with peregrines reduced to just 360 pairs in Britain by 1963.

By 1961 all birds of prey had been given full legal protection, and further protection was granted in 1981 under the Wildlife and Countryside Act. People also began reintroducing extinct birds. The numbers of reintroduced red kites, ospreys and white-tailed eagles are beginning to rise, while other birds such as common buzzards have been recovering on their own. But the now illegal culture of killing raptors remains deeply ingrained in some sporting estates and has so far been hard to shift, largely because the British authorities have not consistently enforced their protection laws for birds of prey. According to the Royal Society for the Protection of Birds (RSPB), there were eighty-five confirmed incidents of bird-of-prey persecution in the UK in 2019. Many more rare birds are suspected to have been killed by gamekeepers but have not been found, or their deaths were not reported. In 2020, when the countryside closed under COVID-19 restrictions and so there was less chance of detection, raptor persecution rose. The RSPB logged fifty-six potential offences in the two months following the first lockdown in March that year. Most were on or close to intensively managed sporting estates.[67] It is impossible, now, to imagine the quantities of raptors the countryside could hold, given that they have been persecuted for so long. But raptors were once numerous and an integral part of the ecosystem. After such a long absence, seeing birds of prey in the skies again can feel deeply 'unnatural' to some.

Hill farmers and the managers of grouse moors also sometimes target foxes and corvids, including ravens (which is illegal). They kill common and hooded crows, too, because these birds predate on chicks and newborn lambs. Those managing driven grouse moors also target jays, stoats, weasels, pine martens and polecats. Indeed, there is evidence that the control of foxes and corvids, in particular, helps wading birds such as golden plover, lapwing, redshank, snipe, oystercatcher and curlew, as well as skylark and meadow pipit.[68] The RSPB itself manages corvids and foxes on some of its reserves in order to protect lapwings and capercaillie.[69]

From a rewilding perspective, however, it is easy to see how the natural processes are out of kilter in such a system. Without apex predators and larger birds of prey, mid-ranking predators, such as corvids and foxes, thrive. Artificially high populations of grouse and lambs provide easy pickings. Ravens, if not killed by landowners, could help them by reducing the numbers of smaller, more populous corvids, such as jackdaws and crows. Killing foxes may actually cause the number of foxes to rise, because foxes are deeply territorial, and so – like wolves, lynx and leopards – keep out contenders. When a dog (male) fox is killed, its territory becomes a magnet for other foxes. In an effort to colonise the new space, an insecure vixen in contested land is likely to have five or six cubs, while a vixen in a secure territory will have just two or three. Foxes in secure territories with plenty of food also tend to live in family groups, with a dominant breeding pair; the other vixens in the group do not breed, but assist with feeding and looking after the cubs.[70]

Tick control

Another battleground for moorland managers, particularly on sporting estates, is the presence of ticks. Grouse can be affected by louping ill, a virus that is carried by ticks found on sheep and deer. High numbers of sheep and deer are likely to be inflating the presence of ticks (which can also carry Lyme disease, a bacterial infection that can cause long-term problems in humans). This is another vicious circle in an unbalanced, unhealthy landscape. To reduce the number of ticks on moorland, sheep are treated with insecticides, which can contaminate soil and watercourses.[71]

Mountain hares, a native species on the Scottish Biodiversity List (and therefore considered a species of principal importance for conservation), also carry ticks. Since the nineteenth century they have been widely shot for sport and are often culled to prevent damage to young broadleaf and conifer plantations. Between 1954 and 1999 mountain hares on moorland declined by more than 80 per cent.[72] Since the 1990s they have been killed on sporting estates, too, in order to lower the tick burden on the moor and so, supposedly, protect grouse from disease. The Game & Wildlife Conservation Trust, which represents shooting Britain, still recommends this practice, despite there being no evidence of any benefit to grouse.[73]

A different mindset

British uplands, such as the Lake District, the North York Moors, the Pennines, the Peak District and the Yorkshire Dales, evoke a strong sense of romance and cultural belonging among their inhabitants and visitors. Many consider these areas of marginal land as epitomising values and traditions that are worth fiercely protecting. But it is useful to look again at this 'cultural landscape'. Many of the land uses are, as we've seen, not that old. High numbers of sheep, driven grouse shooting, land drainage and canalisation of rivers are all products of the Industrial Revolution. Agricultural chemicals, 'improved' pasture, farming subsidies and headage payments for sheep, further hill drainage and forestry plantations are post-war interventions. Our great-great-grandparents would regard the uplands as new landscapes, not old. Indeed, they would barely recognise these hills. If we appreciate the heavy environmental cost and long-term unsustainability of these forms of intensification – in terms of soil degradation, carbon emissions, flood risk and biodiversity loss – it is worth digging deeper to seek the cultural landscape that was sustainable, and to identify land uses with a softer imprint.

Reimagine the scene in the picture on page 77 three or four hundred years ago, when traditions and attachments were just as strong, but there were fewer sheep and more cattle on the hill, as well as trees and shrubs, functioning river systems, healthy peatlands teeming with life, and more skilled and challenging field sports, such as roe-deer hunting. The Victorian pastime of walked-up grouse shooting involves a small number of 'guns' – people carrying guns – flushing up grouse in front of them as they walk. 'Driven grouse shooting', the modern equivalent, on the other hand, is far more intensive. It involves a long line of 'beaters' with flags, whistles and dogs flushing large numbers of grouse towards a row of stationary guns (often with loaders assisting them to increase the speed of shots) concealed in grouse 'butts' (roofless hides made of wood, stone and turf). We cannot turn the clock back to landscapes exactly as they were 400 years ago. But we can find a baseline that serves as inspiration for the future, and that recognises the cultural heritage and value of human participation in the landscape at the same time as preserving a functioning ecosystem.

Uplands in transition

Rewilding the river

In the scene opposite, the road has been moved to the right-hand side of the valley, with a bridge over the floodplain, freeing up the river for renaturalisation. Some of the drained, intensively farmed fields in the valley have been returned to floodplain, providing 'room for the river' and its meanders and braidings. The river is now able to respond to heavy rainfall running off the hills, soaking up water and slowing it down, significantly reducing the risk of flooding to villages and towns downstream. The road itself is now less likely to flood, too.

Reduced sheep numbers

The number of sheep has been significantly reduced. The remaining sheep paddocks in the valley are now permanent pasture for mixed grazing with cattle. They no longer receive artificial fertiliser, and so

levels of nitrate pollution in the river have declined. Instead, dung and bedding from animals overwintering inside are spread over these fields in spring. The enthusiastic earthworm response, retrieving the dung and straw from the surface of the fields, replenishes the soil with nutrients and other soil biota, and the presence of more worms attracts birds, such as oystercatchers, curlew, snipe and woodcock.

Hefted flocks

The sheep that remain are mainly the traditional 'hefted' flocks on the hills. Generations of learned behaviour have instilled in these old-breed flocks a natural instinct for their home range, and the ability to find the best grazing and shelter all year round, as well as resistance to the parasites, plant toxins and mineral deficiencies in their range. Low-maintenance, free-roaming and with no need for fences, hefted sheep – in low numbers – could be a useful tool for rewilding.[74] Although they are non-native, as one of a suite of herbivores their browsing, grazing and natural disturbance can, at the right levels, encourage the succession of plants and woody species on the hill, and ultimately improve biodiversity. In this image we see the first signs of

This picture shows the landscape in transition, on its way to becoming more integrated and functional. The first major interventions have been made with the removal of artificial drainage, the removal of plantations and the restoration of peat haggs. Some dynamic natural processes are back in play, and wildlife is returning.

vegetation responding to the overall drop in the number of sheep, with trees and shrubs regenerating first in the erosion fissures and gills, and gradually colonising open, disturbed patches on the hill. Their root systems and those of the recovering plants around them will help to stabilise the erosion zones that previously released landslides.

Hardy cattle

Hardy cattle have been introduced into the grazing mix. At high densities, livestock grazing, including cattle, is clearly damaging in the wrong habitats, such as peatland, shrubland or woodland.[75] Numerous studies have shown that reducing this pressure is beneficial. But the impact of cattle depends on the context. More than eighty studies have shown places where cattle grazing is beneficial, and, indeed, often crucial for the restoration of habitat.

In the past, old breeds of free-roaming cattle grazing and browsing on the hillsides, mimicking the actions of their ancestor the aurochs, would have contributed enormously to nutrient cycling and biodiversity in the uplands. Since the early nineteenth century, however, greater concentration on sheep in the uplands – and later the shift to larger, less hardy European breeds of beef cattle – has resulted in increased on-farm production and the importation of feed. The loss to the hill of traditional cattle grazing has been ecologically ruinous. Being ruminants, cattle can break down plant cellulose, even from poor-quality forage, very effectively. Their dung therefore reaches a higher stage of digestion than that of sheep or even horses, and supports more invertebrates, particularly dung beetles – a keystone species that pulls the dung down into little chambers in the soil. The 4 tonnes of dung a year produced by a single cow can sustain an insect population that weighs a quarter of her body weight, a vast food resource for birds and small mammals.[76] Cattle grazing helps to reduce thuggish *Nardus* and *Molinia* grasses and, as cows trample through the thatch with their soft, broad hooves (so different from the small, pointed hooves of sheep), they open up the sward for natural regeneration and push seeds into the soil.[77]

Peatland restoration

After several years, the peat haggs in the foreground of the picture are beginning to rewet and reintegrate into blanket bog. The cover of characteristic peatland plant species – including sphagnum moss and sedges – is increasing.[78]

Simply blocking drains and reducing or eradicating deer and sheep traffic over the peatland has a dramatic positive effect. But in some places, additional actions – such as covering the bare peat surface with vegetable fibre matting, liming, and actively introducing peatland vegetation – may be needed to speed up the revegetation process. Exactly which actions, or combination of actions, might be useful in a particular place will depend on the context.

This healing of peatland will contribute hugely to the mitigation of climate change. In England alone, the restoration of degraded peatlands through blocking moorland 'grips' or drains and reducing grazing pressure could reduce carbon dioxide emissions by 2.4 million tonnes each year. The UK Office for National Statistics' natural capital accounts estimate the cost of restoring the entirety of the UK's peatlands to £8–22 billion. But this cost is somewhere between one-fifth and one-tenth of the carbon-emissions benefits that would result.[79]

Plantation removal or adaptation

In our scene, the dark, regularly shaped plantations of Sitka spruce and Douglas fir, with their strongly defined edges, have been removed, and the whole area is more widely planted with thinner cover of mixed native broadleaf trees and Scots pine. In England and Wales, trees cannot be cut down without a licence or an agreed woodland management plan, while in Scotland up to 5 cubic metres of timber can be cut down per quarter year before a licence is required.[80] There is a strong presumption that once woodland is established it must remain as such in perpetuity.[81] But what 'woodland' actually entails is poorly defined.

One option to improve commercial upland plantations is to shift the harvesting regime from clear-felling to continuous cover forestry (CCF), whereby trees of different ages grow together and are felled individually as they reach the appropriate size.[82] This reduces soil loss and provides more even carbon storage, better-quality timber, and greater protection against disease and wind damage; it also improves the visual impact of felling, and creates habitat for wildlife.

Even better, from a rewilding perspective, is the approach shown in the image. A commercial forestry company will want to wait until the plantation is ready for harvest before felling and transitioning to mixed/native woodland. Indeed, a felling licence is unlikely to be secured until the plantation is mature, especially if it has been planted using funds from a government grant. It is unclear whether current regulations in England, at least, would allow natural regeneration to take place within the felled area to preserve the designation of woodland. But there may be an argument for establishing native trees by planting, anyway, if seed sources in the area are very limited. Regrowth from the highly invasive Sitka spruce, a North American tree, can be so vigorous as to require hands-on management for several years (a job that is crying out for volunteers).

Most commercial plantations have an existing fence, to protect the trees from herbivore damage. This can be used to keep sheep, hares and rabbits out of the new native plantings, at least until the saplings are established. In areas where there are too many deer, a deer fence will be needed unless the deer are culled – although fencing inevitably has a carbon cost, is unsightly and may harm other wildlife, such as capercaillie and black grouse (see page 177 for more on this).

The ultimate aim, however, is to make the fence redundant, so that it can be removed once the saplings are established and the density of livestock and deer on the hill has reduced such that they will not damage the new native woodland. Indeed, low-level cattle disturbance of the soil and trampling of grasses will hasten the process of natural regeneration as it extends from the new woods.

Uplands rewilded

The scene opposite is a hypothetical one, perhaps not too far in the future, in which the landscape has become fully integrated again, and natural processes, including apex predator trophic cascades, have been re-established.

The introduction of beavers, a keystone species, has increased the complexity of the river system, creating a series of interlinking ponds that slow the water even more, holding back heavy rains and releasing water slowly throughout the dry months. The water – no longer affected by soil run-off from the hills and pollution from agricultural chemicals, and enhanced by the filtration and oxygenation of the beaver system – now registers as the highest quality. Aquatic and terrestrial diversity has rocketed, with the woody beaver dams and lodges providing habitat and shelter for water invertebrates and fish fry. Atlantic salmon, long absent from this river, have been reintroduced as fry in pools upriver to reignite the salmon migration.

The landscape here is now stable, productive and biodiverse.[83] The restored peatland and naturally regenerating trees and vegetation on the hillsides are preventing soil loss, restoring carbon and protecting the valley from stormwater run-off. The braided, re-naturalised river has reduced the threat of flood destruction downstream, and migrating salmon have returned. The road has been moved to give space for the river, and a green bridge over it – just beyond the existing road bridge – allows passage for wildlife between both sides of the valley. The wildcat in the foreground is intrigued by a marbled white butterfly now colonising the uplands with climate change, and is one of a rebounding population of wildcats boosted by reintroductions. The lynx (the first missing predator likely to be returned to the UK) has not only reduced numbers of roe deer – its main prey species – but suppressed its smaller competitor, the fox. Having fewer foxes has released the pressure on small mammals, resulting in more prey species for owls and raptors. This restored landscape has revitalised the local economy and rural community, providing jobs in eco-tourism, fishing, stalking, foraging, hospitality and horticulture.

Downriver, weirs have been removed, facilitating the salmon's passage upstream as they return from the ocean to spawn. The salmon negotiate the beaver dams in heavy rains via rivulets cascading down on either side, as they have done for hundreds of thousands of years. Otters, feeding on salmon and other fish, have multiplied, along with waterbirds. On the hillsides, areas of erosion have completely healed and the former Sitka spruce plantations have successfully morphed into native woodland. The perimeter fences have been taken out, and

Scots pine, willow, birch, rowan and juniper are naturally regenerating in the gills and gullies and dotted about the hillsides in a scene resembling the upland wood pastures of medieval times. Small numbers of hefted sheep and hardy cattle roam free throughout the year. One day, perhaps, there will be wild Tauros cattle, bison and elk in this landscape. For the moment, the checking of the domesticated livestock is made easier by GPS collars on the leaders of the herds. A green bridge connecting the floodplain with the hill on the far side of the road enables livestock, wild herbivores and other animals to move between the two sides of the valley.

The increased habitat for small mammals and birds has provided food for raptors, such as hen harriers and Montagu's harriers, and, excitingly, wildcats, which are rebounding after successful reintroductions. Stoats, weasels and polecats are back, and the return of the pine marten has significantly reduced the population of non-native grey squirrels, resulting in the resurgence of native red squirrels.

In this imagined future, wild boar can also now be found in the area. Their rootling stimulates natural regeneration and colonisation by flowering plants. Their numbers are kept within manageable limits by hunting, the meat providing sought-after wild-boar sausages and charcuterie for local shops, with the odd piglet a hefty prize for a golden eagle. In this particular area of the uplands, which had been lacking in red deer, these animals have now returned, either reintroduced or attracted into the area by the increased availability of browsing and grazing. Along with roe deer, they provide another vital herbivore influence in the landscape. But in the absence of the wolf, their numbers must also be managed. This has created an opportunity for deer stalking, and for venison sales – another wild meat in demand from health-conscious consumers. Sturdy old-breed ponies carry the quarry off the hill in the traditional way, avoiding the destructive impact of quad bikes.

People still live and, indeed, thrive in this future landscape. The productivity and diversity of enterprises are higher than before. Farms in the valley have diversified into horticulture, with greenhouses providing vegetables and salad to the local market, and fruit from orchards. On the hill, foragers pick wild blueberries, fungi and herbs, some of which are made into products for the local farmers' market. Ecotourism supports a number of local businesses, including campsites, glamping, hotels, a farm shop, cafés, pubs and restaurants, wildlife guiding, birding, fishing and wildlife photography.

A particular draw for tourists in this vision of the future are wildcat and lynx, which have been reintroduced now that there's sufficient woodland and scrubland cover for these solitary stealth predators and the reconnected habitats are large enough to support their territories. Although they are extremely shy of humans and only rarely seen, the presence of lynx intensifies the feeling of wildness and life in the hills. Lynx prey primarily on roe deer, taking, on average, a roe deer a week. Since a lynx's territory is so large, however, and the roe quickly learn of its presence, the predator is constantly on the move. Human management is still needed to manage roe deer numbers, which are at an all-time high in Britain, providing another source of wild meat. The lynx has also had a marked impact on the fox population, meaning that fewer endangered species, such as capercaillie, are targeted by foxes. While lynx rarely target sheep in open pasture, especially in flocks, ambush cover has increased with the resurgence of woody shrubs, which makes it more possible. If a sheep is taken, the shepherd receives compensation.[84]

Reclaiming the countryside

This vision of a future landscape in both lowland and upland Britain is more ecologically complex, richer in natural resources and full of life. It has also revitalised local economies and rural communities. The younger generation, appreciating the quality of life and business opportunities, now opt more often to stay in these areas and bring up their families. Crucially, payments are now available to the managers of the land for providing ecosystem services such as soil restoration, carbon sequestration, flood mitigation, air and water purification, and providing wild spaces that improve human mental and physical health. Resilience and productivity are built into the land once more. But cultural heritage and social identity have not been lost; rather, people have reconnected with the living landscapes of the past and gained a safer, more sustainable future – one to look forward to with confidence and hope.

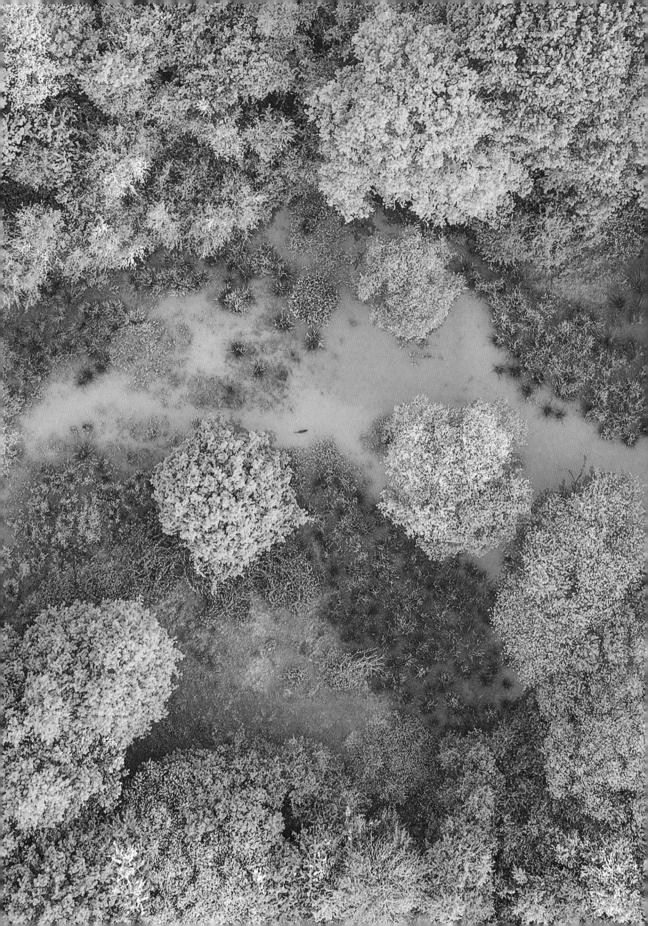

3

Rewilding Water
Recovering our natural water systems

Wetlands are incredible habitat, and vital for storing water. They also sequester carbon. But over the centuries they have been catastrophically lost to land drainage, mostly for agriculture. Restoring or creating natural water bodies is something most rewilders can do, at any scale, and the rewards for wildlife are enormous.

Beavers are the ultimate hydrological engineers, and do wetland restoration for free. This six-acre wetland at Knepp was created by a pair of beavers in a matter of months from a tiny stream. You can see one of the beavers, looking like a floating log, in the main body of water towards the left.

One of the quickest and most effective ways to re-establish natural processes in a landscape is by restoring or creating natural systems of water. This is something that both large- and small-scale rewilders are likely to be able to do to some degree, and it will bring huge benefits for wildlife. Although lakes, ponds, rivers, marshes and deltas cover just 1 per cent of the Earth's surface, they provide habitat for 10 per cent of all animal species and 30 per cent of all vertebrates. As we saw in the floodplain and peatland restorations in the image sequences in the previous chapter, wetlands are also vitally important for sequestering carbon.[1]

Like the introduction of keystone species, the restoration of water systems can be a one-off human intervention that kick-starts natural processes, enabling them to function on their own again – a premise that is at the very heart of rewilding. Free-roaming herbivore introductions, in restricted areas in a densely populated landscape, are likely to need some degree of ongoing human management, but a rewilded body of water can largely be left to its own devices, at least for some time. The larger the body of water, the more dynamic and autonomous it can be. A flowing river and its floodplain, once fully functioning and allowed the space it needs, may never need intervention again. A small standing pond, on the other hand – if it's in an area deprived of other natural influences, such as flooding, free-roaming herbivores, beavers and/or natural interaction with water-fowl – may, in time, need human intervention to act as a proxy for these influences, to keep it functioning and dynamic.

For ease of reference, we've divided this chapter into flowing and standing water bodies, although, of course, in nature there's no such clear delineation. In a connected, complex ecosystem, any water is part of a boundless and often invisible hydrological cycle. Water and wetlands, rivers and streams are connected to each other across catchments (the whole area of land that collects water) as well as through underground water systems, and through the atmosphere via the rain cycle. So, it's important to think about how water functions within an entire water catchment: how it materialises from water vapour in the air, falling to earth as raindrops, snow, hail, frost or dew; how it permeates through vegetation and the earth, or runs off the top of compacted soil, or evaporates and returns to the sky; how it accumulates in watercourses, gathering capacity and decreasing in speed as it flows from the mountains to the lowlands, filling underground aquifers or flowing out to sea.[2] And think about how that flow can vary dramatically from season to season. This is crucial for understanding how water can behave on land, and the natural processes that will help a water system remain biologically rich and restlessly dynamic.

Restoration models often focus on larger water bodies, on the assumption that larger means more important. But smaller water bodies are far more abundant and, in many cases, more important for biodiversity.[3] Lakes and ponds of less than 10 hectares represent 30 per cent of global freshwaters and are vital for carbon and nutrient cycling, and the restoration of small streams, ponds and wetlands positively affects waters and biodiversity downstream.[4] Rewilding water systems at all scales is key to the recovery of nature.

Flowing water: rivers, streams and floodplains

The state of our waterways

Rivers and streams run like arteries throughout the landscape, with small running waters – the burns, bournes, rills, runnels and brooks – as capillaries. The dense network of blue lines on maps often define county, parish and landowner boundaries, and they provide ecological highways for wildlife. Compared with continental Europe, most of the UK's rivers are little more than large streams, although together with their associated watercourses they run for 200,000 kilometres in total.

In England, 44 (around 2,500 kilometres in total) of the 1,500 or so rivers are legally protected on paper (although mostly not in reality) as Sites of Special Scientific Interest (SSSI), as important river types and for their associated habitats and species.[5] Even so, the condition of the country's rivers has been dire for decades and has recently been getting even worse. None of the rivers are designated 'high status' (in near natural condition) – something that is rarely mentioned by water managers. Figures released by the Environment Agency in September 2020 indicate that only 14 per cent of English waterways (rivers, lakes and streams) are in good ecological health (a status designation that already reflects degradation, including the loss of rare or sensitive species).[6] This percentage is considerably lower than the 27 per cent reported in 2010. In 2020 no English river achieved 'good chemical status' – meaning that in *all* English rivers concentrations of one or more listed chemicals exceeds the Environmental Quality Standards Directive. Despite pressure from the EU Water Framework Directive to improve water quality, no improvements have been made in England since the last publication of data in 2016. England is still – as the EU described it when it joined the Common Market in 1973 – the 'dirty man of Europe'. Waterways in the devolved nations, where populations are lower and intensive agriculture less dominant, fare much better. Some 65.7 per cent are classified as in good health in Scotland, and 64 per cent as good in Wales.

But England's rivers are dying, primarily thanks to agricultural run-off and sewage discharges. In 2020 alone there were 400,000 raw sewage discharges by water companies into English rivers, double

The water cycle

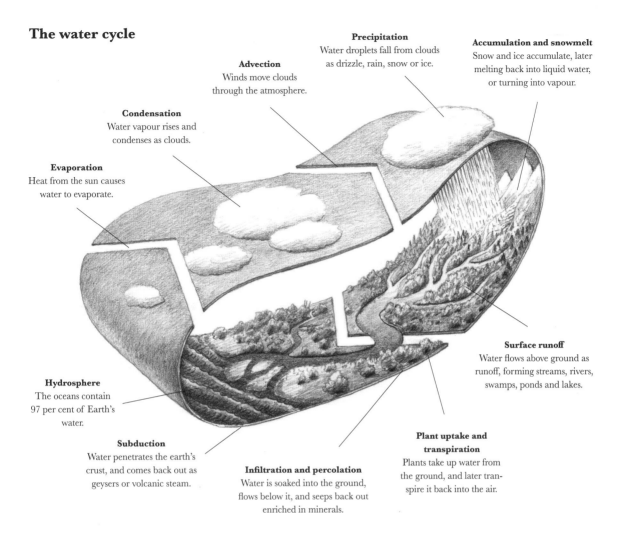

Advection
Winds move clouds through the atmosphere.

Precipitation
Water droplets fall from clouds as drizzle, rain, snow or ice.

Accumulation and snowmelt
Snow and ice accumulate, later melting back into liquid water, or turning into vapour.

Condensation
Water vapour rises and condenses as clouds.

Evaporation
Heat from the sun causes water to evaporate.

Surface runoff
Water flows above ground as runoff, forming streams, rivers, swamps, ponds and lakes.

Hydrosphere
The oceans contain 97 per cent of Earth's water.

Subduction
Water penetrates the earth's crust, and comes back out as geysers or volcanic steam.

Infiltration and percolation
Water is soaked into the ground, flows below it, and seeps back out enriched in minerals.

Plant uptake and transpiration
Plants take up water from the ground, and later transpire it back into the air.

that of the previous year.[7] Blueprint for Water, an initiative run by the Wildlife and Countryside Link (a coalition of fifty-seven conservation non-governmental organisations in England), questions whether the government's target in its 25-Year Environment Plan, published in 2018, for 75 per cent of water bodies in England to be in good (note, not even 'high') condition 'as soon as possible', is achievable.[8]

In 1996 the UK government designated its first Nitrate Vulnerable Zones (NVZs) in an effort to prevent water pollution from agriculture, and these now include about 55 per cent of land in England. Farmers within these zones must comply with strict regulations on the application of fertiliser and organic manure and the storage of slurry, silage and fuel. Of course, the rules should apply to farmers outside

Water moves around our planet by the processes shown here. The water cycle shapes landscapes, transports minerals, and is essential to most life and ecosystems on the planet.

these zones, too. But within NVZs, the Environment Agency and Catchment Sensitive Farming officers have made concerted efforts to educate farmers who have been slow to wake up to the overuse of fertilisers and inadequate storage of slurry and manures. With increased education and the threat of enforcement action, many more farmyards in these areas are now fit for purpose. However, even with grants available, this has not been enough to persuade some marginally profitable livestock – and particularly small dairy – farmers to get their houses in order. Compare this approach to the Dutch government who, in 2021, allocated 25 billion euros to radically reduce livestock pollution and incentivise their farmers to transition to low-nitrogen farming.[9]

NGOs continue to take the UK Environment Agency to task (and court) over the lack of prosecutions for continued pollution of watercourses. The problem is a lack of political will at the highest levels of government. The UK government does not ensure compliance or bring prosecutions, whether the criminals are farmers, industrialists or water companies. Recent budgetary and staffing cuts to the Environment Agency have severely affected its ability to monitor water quality. A Greenpeace study revealed that, in 2018, the agency's water-quality sampling and sampling points had fallen by nearly 50 per cent since 2013.[10] The country's rivers are, in effect, being killed on purpose, although why the government has determined to do so is mysterious.

Another profound English problem is the historic modification of watercourses, which has been undertaken in earnest from the Industrial Revolution onwards.[11] The natural functioning of rivers has been severely restricted by historic dredging, and the straightening, deepening and widening of watercourses for transport, drainage and the management of agricultural land, the removal of gravel beds and bankside trees, and the creation of dams and weirs, which block migrating fish and other creatures. Not a single river in the UK now flows fully freely. We tend to ignore this because – in another case of shifting baselines – we've grown used to how the streams and rivers look. Also, nature itself can put a reassuring veneer over the adaptations.

So what can rewilders do at this larger flowing-water scale? Given that a watercourse is likely to cross several land-ownership boundaries and be affected by numerous negative influences both upstream and downstream, from land possibly some kilometres away, is there any point trying to restore a stretch of stream or river? And what if that stream or river borders land owned by two different people? If the landowners have incompatible views, is it worth rewilding one side only? Would it help wildlife and natural processes, or would it simply antagonise the neighbour?

Rewilding a stretch of river or waterway

Rewilding any part of a watercourse involves a considerable degree of collaboration. In the UK, any modification to a stretch of river or a stream requires permissions and licences.[12] One cannot change the banks of a watercourse, remove material from the bed, alter flood defences, remove a fence, dam, weir, bridge or culvert of any size, or do anything that affects the flow, capacity or habitat of the watercourse, without permission. In Scotland, the Scottish Environment Protection Agency has authority over rivers and bodies of water; in Wales, it's Natural Resources Wales. In England, the Environment Agency has authority over main rivers, and on any other watercourse it is the lead local flood authority or internal drainage board. If tides affect a watercourse, a marine licence from the Marine Management Organisation may be required before any changes are made. A local Piscatorial Society may have fishing rights. Many techniques of watercourse improvement, especially weir removal, channel narrowing, re-meandering or floodplain reconnection, will also affect surrounding reaches. So, any work must be done in collaboration with the authorities, neighbouring landowners and, possibly, angling groups.

Still, there is now more support in terms of funding and advice for individual landowners who want to improve the quality and function of watercourses on their land.[13] While water quality fundamentally involves the whole catchment, there should be considerable benefits to wildlife and possibly even water quality, even if neighbours can't be persuaded to join in.

The land management on the right-hand bank of the river in the illustration overleaf, for example, provides huge benefits to wildlife. To some degree, that mitigates the effects of poor land management on the left. The lower stocking density of animals on the farmland on the right prevents soil erosion and run-off from the fields, and the well-maintained farm buildings, silage clamp and slurry lagoon

Effects of land management on a river system

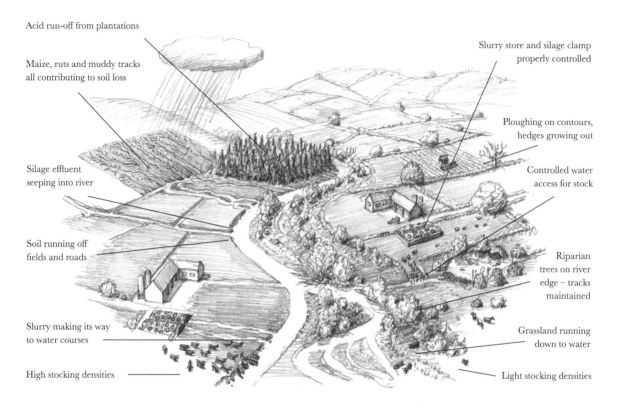

Acid run-off from plantations

Maize, ruts and muddy tracks all contributing to soil loss

Silage effluent seeping into river

Soil running off fields and roads

Slurry making its way to water courses

High stocking densities

Slurry store and silage clamp properly controlled

Ploughing on contours, hedges growing out

Controlled water access for stock

Riparian trees on river edge – tracks maintained

Grassland running down to water

Light stocking densities

Irresponsible farming on the left bank contributes to pollution and run-off, choking life in the river. The land management on the right bank mitigates these effects to some degree. The floodplain and wetland scrub neutralises some of the pollution and helps aerate the water. It also provides wildlife habitat and a green corridor along the riverbank. But clearly, if both sides of the river were restored, the benefit to this stretch of river would be dramatic.

ensure that pollutants do not leak into the river. The fields closest to the river are now flower-rich permanent pasture and never ploughed. Encouraging broadleaf native trees along the river has stabilised the bank, reduced erosion and shaded the water, cooling it for fish and invertebrates. The trees' shedding of leaves and woody material into the river feeds aquatic invertebrates and detritus-shredding crustaceans such as *Gammarus*, which are important food for trout and wading birds. A rewilded margin of wetland scrub along the river provides a rich and dynamic habitat for wildlife, receiving sediment and slowing down floodwater in heavy rains. All these improvements increase the oxygenation of the water, neutralise acidification caused by run-off from the pine plantation, and filter the chemical pollution from the maize field, farmyard slurry, leaking silage clamp and degraded, artificially fertilised pasture on the left. The improvements have also created an effective green corridor, facilitating the movement of riverside and other species.

Obviously, if the farmer on the left were to do the same, the impact on this stretch of river and the waters downstream would be exponential. Those who own both sides of a river are in a position to make really positive changes, especially if they can go one step further and rewild the land on either side by removing drains, restoring a dynamic floodplain, and linking streams and ditches back to the floodplain to 'rewet' the land. However, extreme care must always be taken to ensure that, in any rewetting of a floodplain, it is not by water that is continuing to be polluted by sources upstream or around it.

The report *River Restoration and Biodiversity* (2016) by the International Union for Conservation of Nature (IUCN) promotes methods of restoration that encourage natural processes to create the self-sustaining, dynamic, physical habitats of a river system, and contains a wealth of information and recommendations.[14] It sets out six key principles that should underpin the efforts of all river restoration initiatives. First, from the very outset, it's important to think holistically and focus on the restoration of the whole river system, rather than a single species or aspect. Engage all the people with different interests in the river as early as possible, while still at planning stage. Get a good grasp of the relationship between the river and its floodplain, looking at the whole river catchment. Target the root causes of the degradation of the river, not just the symptoms (this will almost invariably mean looking at the big picture). Use minimal intervention wherever possible, so that rivers can recover their natural processes by themselves. And, finally, having taken a baseline survey before doing any work, use robust monitoring techniques to measure the results over at least the next five years. This will help to inform other river restoration projects, as well as the future of the one in question. See Further Resources for more excellent publications on river restoration.

Rewilding a floodplain

Floodplains in the UK have been drained for agriculture since Roman times, but their comprehensive drainage happened during the Industrial Revolution. In 1860 Knepp received a loan from the government of £2,500 (equivalent in purchasing power to more than £300,000 today) to drain water 'laggs' on the estate. They never grew a successful crop. There are many similar former floodplains in the

UK, drained for agriculture, that are still too wet to produce crops reliably. Such areas have great potential for rewilding.

By now, one would have thought the folly of building on flood-plains would be understood, but despite stark warnings from the Environment Agency, tens of thousands of homes continue to be built every year in areas that are at high risk of flooding.[15] Between 2013 and 2018 more than 84,000 houses – one in ten of all new homes – were built in zones at high risk of flooding.[16] Concreting over the natural flood basins of a river not only puts lives and property at risk. It also increases the force of floods downstream by preventing the land from absorbing water and speeding the run-off down tarmac roads, driveways, patios and pavements, increasing the likelihood of rivers bursting their banks downstream. On average, flooding costs the UK economy £1.1 billion a year. This cost is only likely to increase as extreme weather events become more common, danger-ously high tides occur once a decade rather than once a century, and the sea level rises 15–50 centimetres this century.

Never has it been more important to reclaim floodplains for nature. This is where it pays to think big, bold and ambitious.[17] The very nature of a floodplain is a challenge to the conventional manage-ment mindset. In many ways, it epitomises the concept of rewilding. It is a place of constant flux and often dramatic metamorphosis. One minute, there is a lush meadow thrumming with insects; the next, a sheet of open water. Often the changes are seasonal. Sometimes they appear out of the blue, affected by sudden heavy rains kilometres upstream. This very dynamism creates endless opportunities for life, sustaining water invertebrates, wetland plants, birds and amphibians with complex habitat and life-cycle requirements. When looking at a floodplain that is entirely underwater, it's easy to assume that life within it has drowned. But wetland species, especially invertebrates, have evolved to survive and thrive in even prolonged periods of inundation. Most of the floodplains of the Pripyat River – the Pinsk marshes – covering about 270,000 square kilometres of Polesie in southern Belarus and northwestern Ukraine, are underwater for between three and six months of the year. Known as the 'Amazonia of Europe' and home to 827 species of higher plant and nearly 250 species of bird, they remain one of the world's most vital feeding grounds for resident and migratory birds. Before their floodplains were drained and claimed, every river in Europe and much of lowland Britain would have looked like this.

Historically, natural floodplains have been productive for agriculture as hay meadows and seasonal grazing. The Domesday Book records hay meadows on floodplains throughout England and Wales.[18] Only about 3,000 hectares of these remain today, most of them designated SSSIs for their ancient, diverse communities of wildflowers and grassland.[19] Unlike modern agricultural rye grasses – which drown when flooded, leaving a soggy, rotten mess – native grasses and wildflowers recover from flooding immediately. Hay meadows are a valuable way to manage floodplains, enabling them to continue functioning as water and carbon storage systems while producing eye-catching, flower-rich meadows that are wonderful habitat.

Although prone to seasonal flooding, most hay meadows, nevertheless, retain drain-runs within the fields themselves, and ditches around them. They are still a significant step away from rewilding, but this might be the land-use of choice for those who have control of only one side of a river and who must preserve the status quo for the sake of the landowner on the opposite bank. The Floodplain Meadows Partnership offers advice on the management and restoration of floodplain meadows.[20] The production of hay from such a meadow can be a useful asset for a rewilding project, on those unpredictable occasions when it becomes necessary to provide supplementary fodder for free-roaming animals (during prolonged snowfall,

The Pinsk, or Polesie, Marshes, at the confluence of the Horyn and Pripyat Rivers in Belarus. Once, every river in Europe would have looked like this.

for example). In normal years, hay from flower-rich meadows is much in demand as premium fodder for horses and livestock.

Rewilding a floodplain, encouraging it to express its full potential and allowing the river to engage with it dynamically, is really possible only when all land managers likely to be affected by the changes are in agreement, or if there is a single owner for all or most of the land basin in question. Essentially, it involves getting rid of artificial drainage, filling in ditches and smashing up the drains running underground towards the river so the surrounding lands become wetter. It links streams and land.

Many river restoration projects focus on 're-meandering' a straightened or canalised stream or river: filling in the straightened route and returning it to the original meanders. These are often still visible on the floodplain, especially on LIDAR (Light Detection and Ranging) images. This is something we did at Knepp in partnership with the Environment Agency in 2008, on 1.5 kilometres of the upper reaches of the Adur, which had been canalised during the Industrial Revolution. We also accentuated other low-lying areas of the floodplain that had been filled in and levelled off for agriculture by creating 'scrapes', or temporary ponds, with a digger. These natural indentations retain standing water long after floods, providing important temporary shallow water bodies for aquatic plants, animals and wetland birds.

Looking back, we were, perhaps, too intent on the idea of a single-thread meandering channel. The excavations by Reg, our enthusiastic digger-driver, were over-engineered, too deep and steep-sided, creating little more than another canal but, this time, a wiggly one. It has taken several years of further interventions with woody debris blockages and animal disturbance to soften these hostile banks.

A different approach has recently been offered by the 'Stage Zero' philosophy pioneered in the American Northwest, so called because it aims to restore rivers to the stage before human disturbance.[21] This creates a more resilient mosaic of habitats than does simple in-channel restoration.[22] Its process-based system aims to restore the river to its wider floodplains by raising the channel bed and returning the valley floors to their original form, allowing the flow of numerous braided channels, and therefore greater complexity. The river Nairn at Aberarder in Inverness-shire, the Allt Lorgy (a tributary of the river Dulnain on the Spey, also in Scotland), and the National Trust's Holnicote Estate and the Riverlands Porlock Vale Streams Project in England are highly successful examples of this new Stage Zero approach to floodplain restoration.[23] Small streams and ditches have been filled in, allowing the water table to rise and complex streams to thread their way across the valley floor.[24]

Depending on location, allowing some or all of a floodplain to develop into wet woodland – often known as 'carr', and one of the UK's rarest habitats – will encourage a plethora of insects associated with water-saturated, decaying wood, as well as otters, bats, amphibians and reptiles. Deer-fencing may be needed, at least to begin with, to enable natural regeneration, and the process may be kick-started by planting damp-loving woody species such as alder, birch and willow. Willow supports more species of moth and insect than any native tree other than oak, and provides cover for marsh tit and willow tit. Wet birch and alder favour birds such as siskin, redpoll and crossbill, and sedges, ferns and mosses flourish beneath. If a combination of different, interconnecting habitats – a wetland mosaic – is enabled to arise, including transitional habitats such as emerging scrub, and possibly involving irregular disturbance by free-roaming animals, the benefits for wildlife will be exponential.

One word about fencing on floodplains, however. Open floodplains can be wonderful habitat for wading birds. It is wise to be cautious about providing trees or putting up fences, which can become perching posts for raptors and corvids hunting wetland birds and their chicks.

Rewilding a stream

In many ways, a stream has the same components and influences as a river – just on a smaller scale.

Restoring even tiny streams in an upper catchment can have a positive impact on the main body of a river several kilometres away. Shaded by trees, they can become a source of cold clean water, lowering water temperatures and helping to refresh and aerate water downstream. They can also provide spawning grounds for fish and breeding habitat for other wildlife that may colonise the river elsewhere.

A critical rule is always to work with the existing hydrology and geology. Trying to create a chalk stream in a clay catchment is simply not going to work.

Dense bankside trees provide food and shade for river-dwelling organisms and habitat for terrestrial wildlife.

Pools are a refuge for invertebrates and fish during low flows.

A wide river corridor is an important ribbon of unmanaged land that supports many different organisms, and connects habitats for shelter, feeding and migration

In many ways, a stream performs like a river on a smaller scale. The more natural and dynamic it is, and the better the habitat on its banks, the more niches it provides for life.

Gravel bars are important for specialised invertebrates.

Riffles are home to many different invertebrates, providing food for fish.

Underwater tree roots provide shelter for fish and stabilise banks and sediment.

Removing weirs

There's something enormously satisfying about removing a weir to enable the passage of fish, although, inevitably, it takes time to secure permission to do so. At Knepp, the Environment Agency needed little persuading to remove the largest weir and decommission three others. They had been costing a fortune to maintain and no one could even remember what they were for. Although the largest weir cost £250,000 to remove, the Environment Agency estimated this was negligible compared to the costs of ongoing maintenance – meaning, ultimately, considerable savings to the taxpayer. Only a year later, sea trout were migrating up the river in numbers, for the first time since the weirs were built in the nineteenth century.

Angling clubs have commonly installed low weirs in streams and rivers to create 'pools' for fish, especially in the summer months

The influence of humans on river catchments

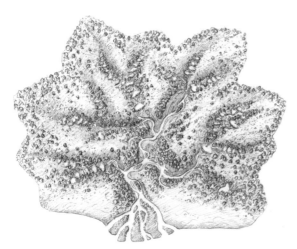

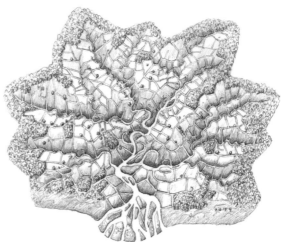

Before agriculture: This natural river catchment, shaped like a maple leaf, has braided rivers, thick, riparian vegetation and undulating terrain. The upper reaches are wood pasture populated by large herbivores. Water is held in the landscape by beaver ponds, intact soil, scrub, grassland and functioning floodplains.

Medieval agriculture: Under land ownership, the river catchment now comprises small fields with livestock. Without beavers, woodland has become static and closed. Ditches help get water off the land and the rivers run quickly. Rotational farming allows fields to lie fallow for three years, restoring soil fertility, but ploughing is already depositing silt in the delta.

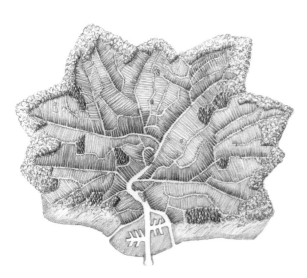

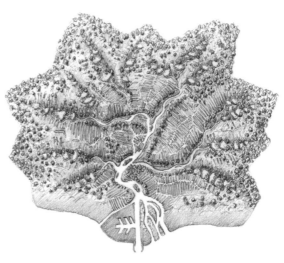

Modern agriculture: The river and streams have been canalised and the large fields drained by underground pipes. The land surface has been levelled by ploughing. No fields are left fallow, and much topsoil has been lost. Plantations of softwoods add to soil run-off. Flash floods are common. The delta wetlands, drained for farmland, now incorporate harbours that need regular dredging. Few species survive in the polluted river.

Regenerative farming and rewilding: The fields are no longer ploughed and the topsoil is deepening. Field margins have been given to nature. The upper reaches are wood pasture, producing meat and timber. The drains still work, but with the soil more stable, and beavers restored, flooding is almost unheard of. The estuary prevents the build-up of silt. With no agricultural pollution, the river is now heaving with life.

when water levels are low. Initially, the idea of creating deeper water – the kind of deep, clear pool that one might find at the base of a waterfall – seems a good one. However, after a few years, that pool above the weir begins to resemble what it truly is: an impoundment. It is a blockage to flow and movement, an architect of deeper, slow-flowing water where fine sediment settles out, builds up and covers the bed. Over time this changes the habitat, reduces the diversity of species able to occupy it and often leads to the erosion of banks around the structure. The misconception is that a pool is just a depth of water. In truth, it is an active feature formed and maintained by the energy of a water flow, and the scouring, abrasive effects of turbulence. When a pool is formed by a beaver dam, the impact of the blockage is mitigated by the leaky nature of the dam itself and its temporary nature; the dam will eventually be abandoned by the beaver and broken down by the river.

In the case of a waterfall, the pool is below the drop rather than above it. It is true that weirs can also create pools in the turbulent water beneath them, but in a lowland landscape that small benefit comes at the price of hundreds of metres of silty, slow-flowing, poor-value habitat above the weir. In a lowland valley with a slope of 1 metre per kilometre, a small (50-centimetre) weir produces an impoundment effect for 500 metres upstream.

Keystone species, nutrient cycling and river dynamics

The presence of a particular species or group of species can affect an entire river system and its surrounding environment. This can provide another way of looking holistically at a water catchment and the role of a piece of land within it. There'll be myriad species that have positive knock-on effects for the habitat and diversity of a river system. For now, let's consider just three key influencers – trees, beavers and salmon – and look at how they affect river dynamics and the distribution of nutrients, as well as interact positively with each other. This will help us appreciate not only what we've lost over the centuries, but also how powerful the restoration of species can be for the renewal of natural water systems.

Trees

Trees are vital to how rivers function. It is now considered best practice to establish trees or allow them to return to riverbanks, with the aim of shading about half the water. The cooling shade is crucial for fish, which have very little control over the temperature of their bodies. Water temperature affects the distribution, migration, survival, physiology, feeding, growth, reproduction, ecology and behaviour of all fish species.[25] Rising temperatures in rivers whose banks have been deforested, principally for agriculture, are now thought to be one of the major causes of declines in salmon populations.[26] In the Cairngorms in northern Scotland, a £5.5 million project has planted 250,000 native trees on the upper stretches of the River Dee, one of the UK's last remaining salmon rivers. It aims to mitigate the catastrophic impact that high water temperatures have had on spawning salmon and salmon fry in recent years.[27] With snow on the hills diminishing every year, trees will be critically important for cooling rivers for salmon and, indeed, all river fish.

Fallen trees and woody debris in rivers create more dynamic water flow. They also generate algae, which provides food for aquatic invertebrates and crustacea that feed fish. The apex fish predator, the pike, is king of this food chain. Tangled roots and branches create a 'mangrove effect', sheltering small fish from predators such as cormorants, kingfishers and otters.

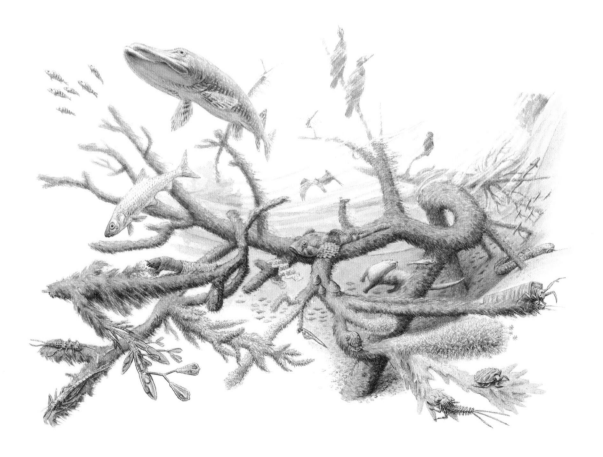

Woody debris and rotting leaves in the river, and the algae that grow on them, are food for larger aquatic invertebrates and crustacea, which, in turn, feed fish and wetland birds. Branches and roots that reach over and into the water – known as the mangrove effect – provide perching and basking sites for birds and reptiles above the water, as well as shelter for aquatic invertebrates, fish and fry underwater. A stream or river also acts as a transport system for seeds, which, falling from overhanging trees and vegetation, can be carried downstream to germinate far from the parent plant.

Dead trees also make a vital contribution. As obstacles, fallen trunks and branches help to connect the river with its floodplain, forcing water in times of high flow back into riverside areas, along with any sediment and nutrients it may be carrying. Large woody blockages slow the flow of the river, and as the speed of the water is reduced, so is its ability to erode riverbanks. As water flows over and around the blockage, localised scouring of the riverbed creates pools that provide additional shelter and resting areas for fish, as well as exposed areas of finer material, such as gravel, which are the favoured 'redds' or spawning grounds for salmon.

At Knepp, it was clear that the estate's 2.4-kilometre stretch of the Adur was still distinctly lacking dynamism, even after two years of restoration work with a digger shifting the river from its Victorian canal back to its original meanders on the floodplain. Adding blockages of large branches and tree trunks instantly began to excite the movement of the water, generating turbulent riffles over the shallows, shifting and softening the steep banks, scouring out the riverbed to create deeper pools, and depositing sediment that built mini-sandbanks where green sandpipers – never before seen at Knepp – began to feed.

Beavers

Of course, at Knepp we were simply imitating – and poorly at that – what a pair of beavers would do in half the time and at no expense. Beavers create the ideal woody debris blockages that all river restoration projects now try to incorporate.

Before they were hunted to extinction for their pelts, for meat, for castoreum (a medicinal oil they produce from scent glands) and because they interfered with land drainage, beavers in the UK would have numbered many millions. Their ghosts live on in place names from Beverley and Bewerley in Yorkshire to Beverston in Gloucestershire and Beverley Brook running through Richmond

Park down to the Thames. The British landscape under the influence of this extraordinary ecological engineer would have looked entirely different.[28] Valley bottoms would have been transformed by beavers and their dams into myriad pools of open, clean, standing water, with boggy areas connected by braided streams and rivulets. Complex, impenetrable wetland corridors would have reached every part of the landscape, shaped by these remarkable animals.[29]

Scrub and trees on the banks of rivers and streams would also have been felled and/or regularly coppiced by beavers. It is important to have a good percentage of trees along rivers and streams, but too many is as detrimental as too few. For this reason – counter-intuitively, perhaps – the Forestry Commission is one of the beaver's most ardent advocates. It spends significant amounts of money every year clearing trees from riverbanks. Plantations of spruce or fir are particularly damaging to rivers because their falling needles acidify the water. But dense, overhanging shade in general prevents light from reaching the water, depriving the aquatic ecosystem of energy. It is light that enables plants and algae to grow, feeding the invertebrates that form the basis of the food chain. Too little shade, however, and the water can become too warm, reducing the oxygen available for fish and insects. The ideal stream ecosystem is a mosaic of dappled shade and open areas, and this is what the beavers provide for free, their rotational coppicing of trees and shrubs providing that ideal combination of light and shade over the water.

In the modern context, beavers mitigate both climate change and pollution. Studies in the USA and in continental Europe, where beaver populations are rebounding, have identified beaver ponds, which store sediment and soil run-off, as powerful carbon sinks.[30] Bacteria in the organic matter in beaver ponds also filter out nitrogen and other pollutants resulting from agricultural run-off.[31] The beavers' leaky dams both store water, holding it back during flood peaks and preventing destruction downstream, and allow for the slow, continued release of water during droughts.[32]

A fundamental consequence of beaver activity, however, is the creation of soil itself. Pools created by beaver dams silt up over time. Eventually the dam is abandoned, leaving the rich sediment to dry out, forming 'beaver meadows': nutrient-rich floodplains with a rich diversity of flowers.[33]

By increasing the volume of decaying wood in the water, beavers increase the natural nutrients (as opposed to the artificial fertilisers used in farming, which often overwhelm water systems) that feed aquatic invertebrates and fish fry. They also provide intricate woody nurseries to protect small fish from agile predators such as kingfishers, cormorants, mergansers, herons and otters, and larger fish.[34] Beaver ponds can dramatically increase the number of fish, compared to the same stretch of water before beavers.[35]

Ironically, anglers often oppose beavers. Some worry that beaver dams may obstruct the migration upstream of salmon.[36] However, beavers and salmon thrive together in rivers in North America and Russia.[37] In Europe, published data is lacking, perhaps unsurprisingly. After all, the beaver became extinct here, and the salmon has declined cataclysmically in all European rivers. But in Norway, Eurasian beaver and Atlantic salmon still live alongside each other, as they have for thousands of years. Norway supports a large proportion of the world's wild Atlantic salmon, and every year 65–70,000 recreational anglers fish for them.[38] Beavers, which number about 80,000 in Norway, are common on six of the top ten salmon rivers and are not perceived as any problem by any of Norway's fisheries organisations.[39]

Once, beavers and salmon coexisted in now unimaginable numbers in British rivers. The complex micro-habitats created by beavers formed an integral part of the salmon's life cycle and the water cycle and movement of the river. Where beaver dams are today perceived as problematic for the passage of fish, there are generally other, human-caused influences at play. Low flows of water caused by draining the surrounding landscape for agriculture, and the disconnection of rivers and streams from their floodplains, driving the flows deep into incised, heavily engineered channels, all prevent beaver dams from functioning as they should.

Where farmland extends to the river's edge, the flow of water cannot respond to beaver dams by extending sideways. However, if a stream or river is allowed the space to function, when it is reconnected to its floodplain, the work of the beaver restores the natural features on which the whole aquatic ecosystem, including numerous fish species, depend. Deep pools are formed behind the dams, and riffles and spawning gravels are left behind as they erode. As riverbed levels rise, numerous side channels in the surrounding floodplain emerge or are created, allowing fish to swim up past obstacles, with the river then depositing silts back on to the floodplain and away from the spawning areas.

The return of beavers is one of the most positive ecological stories of recent times. Having been almost hunted to extinction in continental Europe (by 1900 there were just 1,200 individuals in twelve relict populations) they now number more than 1.2 million, thanks to 161 reintroduction programmes in 24 European countries.[40] In the United States, where they were hunted mercilessly by fur trappers to around 100,000 individuals by 1900, beaver protection in the late nineteenth and twentieth centuries has seen the population rise to around 10–15 million today (still a fraction of the estimated 60–400 million before the arrival of Europeans).

Beavers were probably functionally extinct in Britain by the sixteenth century. Their reintroduction – the first native mammal to be restored to Britain – has been slow in coming, and is still half-cocked. A few hundred beavers are now back in Scotland, thanks to an official reintroduction programme begun in 2009 in Argyll, and earlier illegal releases and escapes in the River Tay in Perthshire. Despite being protected in Scotland in 2019, beavers are still being shot by farmers concerned about flood damage to crops.[41]

In England, a small population of wild-living beavers on the river Otter in Devon, originally from an unlicensed release, became the subject of a trial reintroduction in 2015. The five-year monitoring and public consultation programme managed by the Devon Wildlife Trust ended with the DEFRA announcement in August 2020 that the beavers on the Otter, now reckoned to consist of up to fifteen family groups, have leave to remain.[42] Since 2017 Natural England has granted licences for beaver releases into tiny enclosures in North Yorkshire, Cumbria, Cornwall, Devon, Dorset, Somerset, Gloucestershire, Essex and West Sussex – including a six-acre pen at Knepp. Meanwhile, in Wales, a small-scale, unfenced reintroduction is being developed by the Welsh Wildlife Trusts. There are also an undisclosed number of beavers, thought to be the result of illegal releases, living in the wild in some rivers in England and Wales.

The return of beavers as ecosystem engineers in the UK is fundamental to rewilding. They perform almost all the tasks required for river and stream restoration at virtually no cost, and the environmental benefits are far-reaching and unequivocal. The sooner the UK government allows – even encourages – beavers to live freely again throughout the country, the better. But there needs to be a clear, proactive management system in place to reassure farmers and other land managers that their interests will be protected. The Beaver

Trust, set up in 2019, is working hard to achieve this. The UK could take lessons from Bavaria, which has more than 25,000 beavers in an area smaller than Scotland, much of it intensively farmed, and where an effective management system has reduced opposition by farmers and anglers to virtually nil.

Hearteningly, the public response to beavers in the UK has been overwhelming. In January 2020 a YouGov poll found that 76 per cent of Britons supported beaver reintroductions.

The rotational coppicing of riverbank vegetation by beavers provides an ideal combination of sunlight to stimulate the growth of aquatic plants, and shade to keep the water cool for fish. Innumerable species benefit from these watery kingdoms, including the black stork standing on the dam – currently an occasional visitor to England – which might well start nesting in the UK with the beaver's return.

Salmon

Just as trees feed salmon, most notably through the agency of beavers, scientists have recently discovered that salmon feed trees.[43] In Alaska and in Kamchatka, Russia, where salmon runs are still prolific, hundreds of thousands of salmon decompose every year on the riverbanks – a phenomenon not seen in Europe for many centuries. Both the male and the female of most Pacific salmon species die after spawning. These mass die-offs every year transfer vast amounts of nitrogen, phosphorus, potassium and calcium to the river and – via scavengers such as bears, foxes, jackals and birds of prey – the land. In Alaska, where more than fifty different mammals, raptors and other birds, amphibians and insects feed on salmon, trees can grow up to three times faster than trees on streams where there are no salmon. Up to half of the nitrogen taken up by trees within 150 metres of a salmon-rich riverbank originates in the sea.[44]

Female Atlantic salmon (*Salmo salar*), the species that lives in Europe, generally survive spawning and manage to return to the sea (where, much weakened, they are generally eaten by larger creatures). But males rarely make the return journey. In Scotland, a rotting kelt – a post-spawning cock salmon, bright red, with its distinctive hooked lower jaw – can sometimes be seen wrapped around a rock or cast up on a shingle spit close to where it spawned. It's the merest whisper of the scene that would once have been part of the rivers' annual cycle. Europe once had the greatest density of salmon rivers in the world. As they still are in Alaska, the salmon runs would have been an upstream tsunami of ocean-rich protein and minerals. Daniel Defoe, writing in the Scottish Highlands in 1724, described 'salmon in such plenty as is scarce credible'. Wild salmon is now extinct from most European rivers. The Thames surrendered its last in 1833. Overfishing also devastates the populations at sea. For every hundred salmon that leave the UK's rivers, fewer than five return – nearly 70 per cent fewer than just twenty-five years ago.[45]

The loss to rivers of this natural system of nutrient cycling as a result of declining salmon and other migratory fish has been catastrophic. In the Cairngorms, the River Dee Trust is attempting to address the problem of nutrient loss by tossing deer legs (which are removed as waste during butchery) into the river. Where nature falters, we make work for ourselves.

The loss of salmon from Britain's river catchments and the surrounding landscapes has also dramatically affected the food chain.

Fatty salmon eggs are an important resource for invertebrates, other fish and birds. Mammals, including otters, foxes and pine martens (and in the past bears and wolves), and birds of prey, such as ospreys and sea eagles, as well as gulls and crows, feed on salmon, dead and alive.[46] All these animals transfer the oceanic nutrients to the land via their waste, by dropping bits of flesh and bones, and by eventually dying themselves. Dung beetles and other waste-eating insects, fungi and bacteria then break down the nutrients into the soil, where mycorrhizal fungi eventually transport them into the root systems of plants and trees, sometimes many kilometres from the river. The nutrients can end up being highly concentrated around the favourite perching sites and nests of such birds of prey as ospreys, for example, creating micro-habitats that add to the complexity of the environment.

Once, living salmon would have supported the life cycle of another important and now threatened species in Britain, the freshwater pearl mussel. The larvae of the mussels attach themselves to the gills of juvenile salmon. This enables the mussels to stay in the freshwater pools without being flushed downriver into salty waters where they wouldn't survive. There may be a symbiotic relationship here. The mussels don't harm the fish. Indeed, the antibodies the salmon develop when the larvae attach themselves may well help protect the fish from sea lice and other parasites. The loss of salmon in rivers, made worse by the suffocating effects of agricultural run-off, has caused a catastrophic decline in this mollusc, many millions of which once filtered the rivers.[47]

Both anglers and commercial fisheries have spent time and money trying to address the problems of salmon conservation. Much has been written on salmon 'management' and finding the illusory magic bullet to restore their fortunes. However, it is not the species that needs managing, but human attitudes and our relationship with the environment. Almost all the solutions required to restore extinct or dwindling populations of salmon are not about managing fish but restoring ecosystems at catchment scale to allow river systems to function again. The salmon-centric attitudes of the past are at last being gradually supplanted by a much more intelligent 'whole catchment' approach. Attempts have been made to 'reintroduce' or augment salmon populations, particularly in urban or industrialised river catchments which they deserted during the ravages of the industrial revolution. However, it is more than likely that once water quality and access issues have been resolved, salmon stock recovery will be largely the work of the species itself.

Standing water: small lakes, ponds and puddles

Whether in a suburban back garden or across thousands of hectares of countryside, standing water is an intrinsic part of the 'freshwater landscape'. If unpolluted, it is a biodiversity hotspot for plants and animals, both terrestrial and aquatic. The beautiful reflections of lakes and ponds also provide a focal point for people. Identifying the type of standing water habitat, and the quality of the water itself, will help in the decision to protect, restore, create more or, perhaps, do nothing at all. This section focuses on ponds and small lakes, which can be more easily created or restored than larger lakes. But many of the principles discussed here can also be applied to larger bodies of water.

Pond basics

There is no globally agreed definition of a pond. When does it become a lake or a puddle, for example? Nature – especially water – is in constant flux, defying the definitions that people like to impose on it. However, conservation policy relies on categories. In the UK, a pond is generally taken to mean a body of water (normally fresh water but occasionally brackish) varying in size between 1 square metre and 2 hectares (roughly equivalent to 2.5 football pitches), and which holds water for at least four months of the year.

This definition covers everything from the tiny puddles in tank tracks on Salisbury Plain – a critical habitat for the rare fairy shrimp (*Chirocephalus diaphanus*) – to large, ornamental 'Monet-style' lily ponds and small lakes. It also includes temporary or 'ephemeral' ponds, which dry up annually; these are extremely rare in the modern landscape, but provide vital habitat for some of the most endangered creatures (see page 135).

Ponds are everywhere in the UK, from glacial depressions on the Cumbrian fells and oxbows on lowland floodplains to dew ponds in the drier landscapes of chalk downs. Globally, ponds and small lakes represent about 30 per cent of all standing waters.[48] The small size of ponds compared to large lakes, river systems and oceans means they are often ignored, despite their collective importance for climate and ecosystem processes.[49] Continental Europe, for example, has between

5 and 10 million small lakes and ponds, but they are largely excluded from national monitoring programmes and poorly represented in conservation policies. This saliency error ('one big is more important than many small') is slowly being recognised and addressed, supported by relatively recent evidence of the importance of small water bodies for wildlife and people both locally and, even more critically, in the wider landscape.[50]

In the UK, half the ponds have been lost since 1900, mainly because farmers filled them in but also because they have been built over. Ponds are a nuisance for large machinery in arable fields, and agricultural drainage has done away with numerous temporary ponds and puddles. Any remaining ponds are degraded by pollution and overwhelmed by vegetation, trees, shrubs and reeds because of a lack of grazing.[51] Indeed, many of Britain's countryside ponds are now full of silt and overshaded as a result of the conversion of pasture to arable. The good news is that ponds are mostly relatively easy and cheap to create. Often, however, new ponds lack aspects that could make them even better habitat. A rewilding approach can take pond restoration one step further by helping to protect, restore and create small patches of clean water in ways that provide the variety and complexity of vegetation that are vital to freshwater wildlife.

The importance of ponds for wildlife

In the UK about a third of aquatic plants and animals live in both standing and running water.[52] From the perspective of plants or animals, the slow edges of a river or stream can be very similar to the edges of a pond or small lake. Many very mobile aquatic species, such as water beetles and dragonflies, travel readily between them.

Overall, however, at a landscape scale, ponds support more species of plant and animal than any other type of water body.[53] Current thinking in conservation attributes this to the diversity of pond types in a landscape (again, think of the complexity of ponds, pools and channels created by beavers over time). By contrast, rivers and streams provide a more homogeneous environment, and often suffer from pollution.

Ponds are also stepping-stone habitats, essential for amphibians, such as the great crested newt and the common toad, which need plentiful clean water in which to breed, and good wild terrestrial habitat linking ponds (since as adults they spend most of their time on land). In semi-natural landscapes in Britain – such as conservation areas and extensively grazed heathlands and commons – one in five ponds supports at least one endangered 'red list' species of plant or animal. Particularly good examples of pond-rich, semi-natural landscapes in lowland Britain are the Lizard Peninsula in Cornwall, the heathland commons of Pembrokeshire and the New Forest in Hampshire. There, depending on the season, there can be as many as one hundred pools per square kilometre.[54] In contrast, only one in one hundred ponds in the wider farmed countryside supports red-list species.[55]

Pond myths

Despite being common and much loved (just look at the many garden pond groups on social media), ponds are relatively poorly studied and understood. There are several common misconceptions about them, such as the idea that for a pond to dry up is terrible for the wildlife in it, and that water levels must be constant. Similarly, it is often (wrongly) thought that fencing from livestock is needed to protect freshwater wildlife, that a bigger pond is better than a smaller one, that 'choking' vegetation must be removed, and that a pond should not be shaded by trees.[56]

One entrenched myth is that natural succession has an end point: that without human intervention, a new pond will inevitably become dry land. While this is correct in some cases – ponds, being generally dynamic, can change very rapidly – the state of play is much more complex.

Depending on the local hydrology, geology and natural processes, a pond may change into wet woodland, fen or bog, perhaps reverting now and again to a temporary pond. It can then remain relatively stable for a very long time. For example, sediments accumulate very slowly in ponds that dry up annually. With suitable levels of grazing to prevent encroachment by trees and scrub, temporary ponds can live on for millennia. Pingos, for example – ponds that were created at the end of the last ice age – are now largely restricted to parts of East Anglia and Herefordshire because they have been destroyed else-where, mostly by agricultural intensification (see page 141).

All types of pond – large, small, shaded, grazed and ungrazed – can contribute to landscape-scale freshwater biodiversity, as long as they are not polluted. Ensuring that ponds contain clean water is by far the most important consideration in rewilding projects at any scale. This depends, to a great degree, on the quality of the surrounding land and vegetation.

The evolution of ponds is more complex than commonly thought, and new ponds may take one of a number of successional routes.

Pond evolution

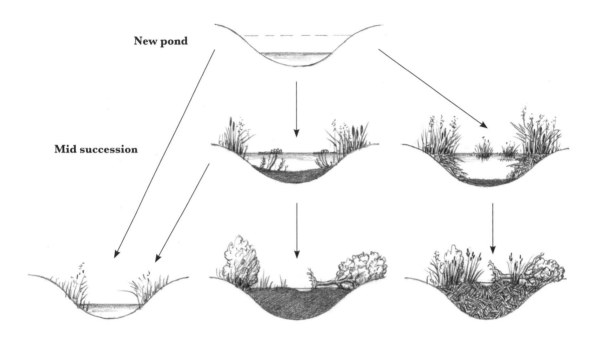

New pond

Mid succession

1. Temporary or ephemeral ponds develop where water is shallow and there is little chance for sediment to build up, as in grassland, moorland and sand dunes. The pond sites themselves tend to be relatively stable and so can exist for long periods of time.

2. Generally, new ponds tend to fill with sediment over time. Eventually, fallen trees and, perhaps, livestock disturbance, can create smaller ponds in the water-logged soils.

3. In small ponds where the water is particularly still (where there is no water turbulence generated by wind or dynamic, incoming streams), rafts of vegetation can develop and may even cover much of the pond's surface. If the nutrient levels are low, these rafts can develop rich communities of fen or bog plants.

Natural and artificial ponds

Most British ponds today are artificial. They generally came about from abandoned open-cast mines, quarries and mills. Others were created more recently for conservation purposes, or for public enjoyment in gardens.

However, the geological record shows that ponds have always been common, forming where water collects in surface irregularities. Studies of relatively undisturbed ancient woodland in the UK show that, in the past, the surface of the land was probably full of seasonal and permanent ponds. Human activity, not least grading and ploughing lands, has effectively smoothed over the land, so that only a fraction of these natural ponds remains.[57]

Natural ponds can be created by very long-term processes, such as glaciation, or random, short-term processes. When a tree falls, for example, water collects in the hole left by the uplifted root ball. The excavations of wild boar, and the compactions made by large herbivores gathering in one spot – beneath a favoured branch for scratching, say – can create shallow ponds. On the floodplains of natural river or stream systems, ponds are created as part of the scouring effect of flooding or from meanders and channels that become cut off from the main 'stem' or stream of water.

The master pond-builder is the beaver. Water is the beavers' protection from foxes, domestic dogs and, in the past, wolves and bears. Their woody dams create deep ponds where they can build safe lodges for their kits. Extending from this, they make a matrix of further pools connected by channels to give them easy access to fodder and browse. In very cold climates, beavers maintain deep water to prevent a pond from freezing over – a huge benefit to wildlife in winter. Beavers can exist in an area for many years, but eventually, if ponds begin to silt up and local food resources dwindle, they will move on. This dynamic of beavers creating and abandoning ponds is part of a long-term cycle that encompasses an extraordinary range of aquatic and terrestrial habitats and a vast spectrum of wildlife.

Artificial ponds, if created with these processes in mind, can be just as biologically rich and interesting as those created by nature. 'Thinking like a beaver' and trying, as much as possible, to replicate its activities through pond creation is one of the most important and rewarding things a rewilder can do – at any scale.

Putting it into practice

Streams and rivers

The tree-beaver-salmon entanglement described on pages 110–18 is just one example of how life forms depend on each other. It is important to think holistically when thinking about restoring a flowing natural water system. Only if we 'see' and understand what *isn't* there can we envisage the possibilities for the future and raise our game.

At the very outset of a project, though, the immediate task is to familiarise yourself with the here and now: the watercourse itself and the constraints currently imposed upon it. Whether you're considering an extensive rewilding project involving a mosaic of different habitats around the main stem of a river and its tributaries or the restoration of just a few metres of one bank, an overview of the catchment, reach and local scale will be important in informing your approach.

There will be many complex interacting processes at play in the river, and many of those will be outside your control. Consider:

- Where you have control (ownership and legal responsibilities)
- Where you might have a detrimental, as well as beneficial, impact (consider extremes, such as flood risk to infrastructure)
- What is outside your scope (consider extending your potential influence by working with neighbours and existing catchment partnerships)

Existing data and information

It's possible to find out a fair amount of information about your water system if you ask or search the following sources:

- River Basin Management Plans, part of the EU Water Framework Directive, which set out the current state of water environments in each catchment and aspirations for their future management.[58]
- Previous monitoring data may be available through statutory organisations and will give a good idea of what has been done before.
- Historical maps can be useful to see if your river has been modified in the past.
- Local expertise from landowners, local catchment-based organisations, angling clubs and canoeing/kayaking groups as well as agencies can help you to understand the current issues affecting your river.

- Contact the local Rivers Trust, Wildlife Trust, Fisheries Trust or local community group to find out what work they are doing and talk about how your rewilding project could benefit the wider river system.

Walk and survey your river or stream

You will doubtless have spent a lot of time near your section of river or stream, but have you cast a critical eye over what seems so familiar? What is that yellow flowering plant at the edge? Why does it grow there? What lives in and around it? Why does it generate a small eddy? Why is the stream silty at the edge and gravelly in the middle? Why are most of the small fish in that patch? Are those low-hanging branches causing too much shade? Is that eroding bank a problem? Is floodwater in the adjacent field bad for the meadow flowers, and how long will it stay wet?

It is crucial to look at the broader hydrology of your water system, and the dynamism – or lack of it. Are natural processes occurring, and if not, why not? Are there any obvious modifications to your stream or river? Are there any obvious features or structures and what influence are they having?

The best way to learn is to engage with people in the know, to question and start discussions. It's easy to commission a report from the River Restoration Centre, for example, but this can distance you from the process of understanding how the report arrives at its conclusions.

There are many survey methods available to help you analyse your river. Some are very detailed, but others are designed for 'citizen scientists' and provide a good starting point for your journey.[59] The list of topics you'll want to know more about once you start looking may well grow lengthy, but the more you can understand how your river ecosystem functions, the better you'll be able to help it, even if – as is so often the case – the answer is simply to give the river its space and leave it alone.

If you are going ahead with a restoration or rewilding project on a watercourse, it's vital to carry out a baseline survey before you begin (see page 368). Monitoring is commonly an afterthought, but it's essential for documenting and understanding change, adapting your long-term approach and providing evidence for requests for funding future work. Without a baseline, there is no comparison for subsequent monitoring. You'll need to factor in time and funds for this.

Planning and consents

Because rivers and streams have a significant effect a long way upstream and downstream, consents relating to flood risk, planning permission, local-authority byways and even waste disposal (site-dug spoil) may be required. It's advisable to allow plenty of time for gaining consents. Consider what activities are easy to get approval for, and which may be subject to lengthier approval processes. To avoid any surprise costs or delays, get consenting organisations on board early in the planning phase. They will be able to advise what is required, and may suggest design ideas that fit well with their priorities and will be more easily approved. It is common for work on restoration projects to take three years or more to get started because there are so many aspects to consider.

What makes a good river or stream?

Flowing freshwater systems and floodplains require certain hydrological conditions, which, in natural systems, are governed by natural processes. Generally, dynamic natural processes can be re-established by reversing historical human interventions and, where appropriate, introducing missing keystone species. The degree to which this is possible will depend on the size, unique circumstances and logistical constraints of your watercourse and its environment. For some habitats and in some situations natural processes can be generated artificially. The degree to which natural processes can be restored depends on the scale of the project and the opportunities available, but aim to tick off the following key points:

- Understand what you have now and what you could expect
- Work with natural processes to assist natural recovery
- Maximise the available habitats by retaining a mosaic of physical features
- Remove unnecessary historic modifications, to let the river flow freely
- Consider the potential for connecting or improving connections with other stand-alone (clean water) features, such as ponds and headwater streams (see 'Standing water' section, above) – provided, of course, that these water bodies are not polluted or loaded with sediment
- Give space to the river/stream
- Create or enable a broad corridor of typical trees and marginal plants
- Remember that erosion and deposition are natural processes

- Improve the interaction between valley, channel, bank and flood-plain, or simply establish stronger links between the water's edge and your land – think like a beaver!
- Discourage or fence out livestock, for a period, to allow vegetation to recover; once that has happened and the banks are stabilised, you may consider allowing some large herbivore disturbance to generate complexity of vegetation and habitat
- Accept seasonally wet areas as valuable habitat
- Encourage carbon cycling by leaving trees, branches and blockages as habitat and food
- Avoid 'tinkering' and 'gardening'; think 'untidiness = biodiversity'; less intervention is often much more productive

Ponds and standing water

Without any natural disturbance at all, a pond may become over-grown with reed mace, bulrush, alder and willow carr (wet scrub). Eventually it may disappear. Even in the absence of beavers, large free-roaming herbivores (particularly elk, water buffalo and red deer, but also cattle, ponies and wild boar) can prolong or even prevent this process. They hinder the vegetation growth in standing bodies of water through browsing, puddling and coppicing plants at the margins. The trampling of water margins by cattle can be beneficial to an array of insects.[60] Natural populations of geese, ducks and other waterfowl also help to keep areas of free-standing water open by eating the reeds and other aquatic plants in the shallows, while swans, with their long necks, can uproot plants from deeper water. But where populations of waterfowl are over-inflated by feeding, they can damage new habitat. The bread tossed to ducks, for example, pollutes the water, either directly by decomposing or in the form of the guano generated by so many birds.

At Knepp, the new ponds we created and old ponds we restored fifteen years ago by mechanical digger need no ongoing management. Our large herbivores, freely browsing and poaching (trampling) some of the pond margins, control the scrub and maintain early successional habitat. Wetland birds do the rest.

As always, it's a question of how many herbivores are causing how much disturbance, and over what period. Too many, and their impact may destabilise banks and decimate vegetation; too few and, depending on the nutrients in the pond, vegetation can quickly smother it. The stocking density of large herbivores and their impact on rewilding are discussed in later chapters. But bear in mind that even dramatic disturbance, if not sustained for long periods, can be beneficial, opening up new and surprising habitats for wildlife. Conversely, the slow encroachment of vegetation on a pond provides opportunities for many species. Scale is also significant. If you have a number of ponds, it doesn't matter if some are excessively grazed and others barely grazed at all. Indeed, some wetland birds prefer bare ponds, and the mosaic effect of different grazing impacts will create a really interesting mix of habitats that will benefit biodiversity across the board.

At smaller scale, where natural processes are unavailable or not functioning, or if the level of disturbance on a pond is too low or too high, or if all the ponds in one area have become similar to each other, human interventions – such as dredging, tree-felling or coppicing – may be needed to keep ponds dynamic and diverse.

But there is also merit in not being over-enthusiastic with pond clearance. The old adage still holds: Don't just do something. Stand there. Allow different successional stages to play out over time, with different levels of shade and vegetation structure that will benefit different communities of species. Of course, the longer you leave it, the higher the cost and effort of extracting trees and silt to return the pond to open water. One option is to allow the pond to slowly silt up altogether, while creating or restoring another pond in a different place. This can be particularly productive if you have space for several ponds. Many species, particularly insects, have complex life cycles during which they require different habitats. When ponds in close proximity to one another are at different vegetative stages, the opportunities for wildlife rise exponentially.

Similarly, if you have a single pond, there is great merit in managing areas of it in rotation, so as to encourage a mosaic of different habitats within it. Different stages, including the early ones, provide opportunities for different communities of aquatic invertebrate. As ponds mature, they provide habitat for priority amphibians, such as great crested newts and common toads, and different plants. A mosaic of pond habitats is also highly beneficial for birds and bats.[61]

Even in larger-scale rewilding projects, the active creation and restoration of ponds may be needed. Wild boar can create wallows, tree-fall pools may eventually form and, of course, beavers will create wildlife-rich ponds on floodplains without any help from humans. Away from floodplains, however, or on land that was previously intensively managed for agriculture and where natural depressions have been filled in and levelled, most ponds – if they still exist – are in a sorry state and need mechanical excavation to kick-start functioning systems again.

As with the restoration of any water system, it is important to look at the wider context. What wet areas and ponds (including puddles) already exist? Are they polluted? What was the landscape like before it was intensively managed? Was there much water? This will help you decide whether you need just to restore some ponds, or create more. Generally, consider both creating and restoring ponds, unless you have many already.

Keep in mind that you may need planning consents to excavate a pond. If your pond links to a stream, for example, you need an Ordinary Watercourse consent, and if you're digging on a floodplain, you will need a Flood Risk Activity Permit.[62]

When restoring a pond, make sure you understand its current value for wildlife, rather than working on looks alone. Depending on where they're located, some old muddy-looking ponds can support exceptional wildlife. The Freshwater Habitats pond-management risk assessment can help you decide whether to carry out a survey before you start.[63] At the very least, this will provide you with a baseline for future surveys. Measuring how the life in your pond improves is one of the most enjoyable rewards. Certainly, if you're near a site that is already important for nature, for instance the New Forest, Anglesey, the Pevensey Levels or even a small wetland nature reserve, you should seek professional advice before embarking on extensive restoration, such as dredging or removing shading or vegetation.[64]

If you're considering creating a pond from scratch, you'll be 'nesting' it in another habitat, such as woodland, heathland, grassland or floodplain, so make sure it won't be replacing habitat that is already of high quality. Look at existing high-quality ponds in the landscape for inspiration, and make sure your design is appropriate to the local hydrology and geology. For example, it is easy to create ponds on clay, where even relatively small ponds can be permanent as long as you have a large surface-water catchment. On gravel and sandy soils, in contrast, you may need to create trial holes and monitor them for a year or more to understand the minimum and maximum levels of the groundwater. Only then can you design your pond.[65]

These principles are well illustrated in a successful rewilding project on what was once an arable farm in Suffolk, eastern England.[66] Initial pond surveys revealed little of interest, except a very small population of tassel stonewort (*Tolypella intricata*, a critically endangered plant that requires bare, muddy ground for its spores to germinate) and a small population of great crested newt. Six ponds were restored and nine new ponds created as part of a phased restoration programme over a number of years.

Seven years later, and with no other intervention, follow-up monitoring indicated that the ponds supported a rich aquatic invertebrate community, including nine nationally notable invertebrates and sixteen dragonfly species. Clustered stonewort (*T. glomerata*), a rare species not previously recorded in any of the ponds there, has also appeared.

Clean water

Clean, unpolluted water, with no added nutrients or pesticides or road run-off, is key for healthy freshwater wildlife. Clean water is now rare in lowland landscapes, except in extensively managed, semi-natural areas. Nutrient overload is the most common cause of water pollution. This is when water has been contaminated with agricultural fertilisers (nitrogen and phosphorus), livestock waste or slurry, or discharges of human sewage. Nutrient overload gives rise to a process known as 'eutrophication'. This can cause blooms of algae, which feed on nitrogen and phosphorus. Algae block the light that aquatic plants need for growth, and when algae die and decay, they use up the oxygen in the water, which can kill fish, crabs, oysters and other aquatic animals. In the worst cases, algae blooms poison drinking water and cause dead zones in rivers and even vast areas of the sea. Water quality can be analysed using field kits or through a certified laboratory.[67] In the UK there are cheap kits available from the Freshwater Habitats Trust that anyone, including budding citizen scientists, can use. Laboratory analysis is more expensive, but will be wider-ranging. One-off analysis of the water chemistry of your existing water bodies may not be easy to interpret, so you may need professional advice or ongoing monitoring.

You can make some general assumptions about water quality based on local geology. On areas of heavy clay, water bodies are supplied by surface water. If land use in these surface-water catchments includes intensive farming, industry, roads and urban areas, it is very likely that the pond water will be polluted. On the other hand, water bodies that are fed by groundwater can have good water quality even in polluted landscapes, because the water has filtered through sand and gravel. There is evidence for this in areas such as the Lower Windrush Valley in Oxfordshire and some intensively farmed parts of Norfolk.[68] Yet, springs that look gin clear may contain high levels of nutrients, particularly nitrates (there are many examples of these in the Chilterns and the Cotswolds). As soon as the water becomes still in a pond, thick algae colonies begin to grow.

When restoring ponds, it is important to target those that have clean water, or where the source of pollution can be fixed. Similarly, only create ponds where the source of water is clean. Avoid linking ponds to streams or ditches, unless their entire catchment is under semi-natural land use and the water is unpolluted. Ponds connected to running water silt up faster, too, because of higher levels of sediment, so generally avoid these water sources unless you have no other. For ponds in gardens or allotments, or near buildings, harvest rainwater from roofs. That is the best source of water there is. Another source of nutrient overload is topsoil. Keep loose topsoil away from new and restored ponds at all costs; it can always be used in raised beds or on another part of your land where you need fertility.

The most important thing when restoring a pond or creating a new one is to ensure it is protected from pollution.

How pollutants get into water systems

Ditches and streams can carry polluted water directly into a pond, as well as silt which settles and can quickly fill a pond.

Run-off from roads, urban areas and ploughed fields can seep and flood into ponds.

How to protect water systems from pollution

Woodlands and grasslands with low-level grazing are good places to locate ponds to ensure the water remains clean.

Ponds can be created near streams and other watercourses, but don't link them directly to polluted streams and rivers.

Proximity to other wetland habitats

Water bodies near extensive wetlands tend to be particularly diverse and become more quickly colonised after creation or restoration. For example, the complex of some forty ponds at Pinkhill Meadow near Oxford, created from 1990 onwards, is now one of the richest pond sites in Britain for aquatic plants and animals.[69] The wetland mosaic is in a meander of the river Thames but not connected to this desperately polluted river, so water quality is good. Its location on the floodplain means that the site was colonised quickly, including by such rare species as clustered stonewort.[70]

Various studies have shown that creating multi-pond complexes or networks is more beneficial for wildlife than single ponds. However, in certain landscapes, single ponds can be significant. For example, single, remote dew ponds are really important for isolated populations of amphibians and fairy shrimp in the South Downs of southern England.

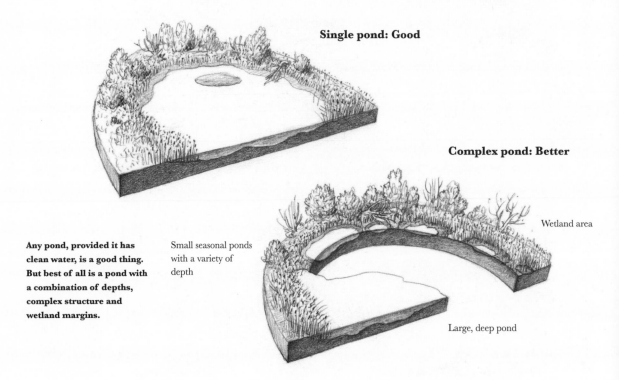

Single pond: Good

Complex pond: Better

Wetland area

Small seasonal ponds with a variety of depth

Any pond, provided it has clean water, is a good thing. But best of all is a pond with a combination of depths, complex structure and wetland margins.

Large, deep pond

Design for wildlife

There are no set designs for new ponds. Design is secondary in importance to clean water, and every new pond must be adapted to its local geology, type of water source and landscape and site characteristics. But some thought can help to minimise long-term management and ensure the scheme benefits plants and animals already living on the site. Grazing will help to keep a pond open, with a varied vegetation structure.

For ponds in woodland, or where there is little or no grazing, consider creating some steeper edges. And in larger ponds you can create underwater islands away from the edges. This helps maintain bare substrate for aquatic plants, and slows down colonisation by trees and in-filling with leaves.

Shallow water is all-important. If you have the space, create some very long margins with low bank angles. A gradient of 1:10 is ideal.

If the water in the pond fluctuates, there will always be some shallow water, whatever the overall water level. This is essential, because most plant and animal species in ponds live in less than 10 centimetres of water. The area that wets and dries according to the water level is known as the 'drawdown zone'. It is a habitat for semi-aquatic beetles and many other insects during their larval and adult stages, as well as mud plant species. If you love ponds and wildlife, you must welcome these muddy margins. They provide habitat and building material for many plants and animals.

Designing the drawdown zone

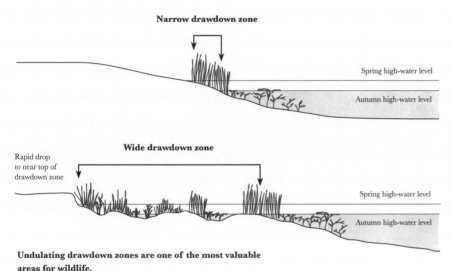

Narrow drawdown zone

Spring high-water level

Autumn high-water level

Rapid drop to near top of drawdown zone

Wide drawdown zone

Spring high-water level

Autumn high-water level

Undulating drawdown zones are one of the most valuable areas for wildlife.

In traditional pond design (top), the margins of the pond between the winter (high) and summer (low) water levels (known as the drawdown zone) are rarely considered. Few plants can survive on steep banks. But a wider, more undulating drawdown zone around a pond creates wonderful marsh and mud habitats in summer (bottom).

Do not plant 'wild' ponds. This speeds up vegetation succession, denying the early habitat stages that are important for so many species, including stoneworts, which need bare substrate, and specialist beetles and dragonflies that have adapted to early conditions when competition and predation are low. Planting can also import invasive, non-native species such as New Zealand pygmyweed, which are then almost impossible to eradicate. There is a rewilding point to be made here, too. Wherever possible, do not disrupt natural colonisation. Humans generally presume to know what plants should go where, but nature often shows us a very different picture. Plants will colonise a new pond remarkably quickly if there are other good-quality waterbodies in the vicinity. Certain seeds, such as those of willow, can travel huge distances on the wind. The majority of aquatic plants hitch a ride on the feet of wetland birds or the hooves and coats of herbivores. At Knepp, at the start of the rewilding project, we had one small pond containing the rare and very pretty water violet. As water quality improved across the site, the water violet – aided by the herbivores and rising numbers of wetland birds – colonised numerous other ponds.

We must take care, too, to respect localised adaptation. Wild trees and plants in a particular area are likely to have evolved traits to make the best of local conditions. Allowing flora to colonise ponds naturally is the best way of ensuring the continuity of this evolutionary process. The same species imported from elsewhere or – worse – propagated in a commercial nursery will not have these advantages.

That said, as in all things to do with nature, there are exceptions. There may be wetland plants that have become extinct in your area and/ or are in dire need of conservation. It may be good to restore those. Always take advice from conservation professionals, and if you do plan a reintroduction, try to find a source as local to you as possible.

Ponds in gardens and urban areas may not be colonised naturally very fast, since they are likely to be far from other ponds and wetland habitats. Aesthetics is perhaps more pressing in towns than in the countryside, and if urban ponds are to be cherished, they may need to feature some attractive, preferably native, wetland plants from the beginning.

Lining ponds is common practice in urban areas or gardens. Liners come in many guises, from butyl and EPDM (ethylene propylene diene monomer) rubber to plastics such as PVC and HDPE (high-density polyethylene). Clay, the traditional way of lining artificial ponds – such as dew ponds on chalk downland – is much kinder to nature. Unless you have a very strong reason to do so, avoid lining wild ponds with non-natural liners.

The most important thing to remember when restoring or making ponds is that different plants and animals have different requirements. Sometimes it is better to have several ponds with varying characteristics than to try to make the right pond for everything.

Creating ponds by removing drainage

Land other than free-draining soil is likely to have drainage infrastructure. Creating ponds by destroying drainage is hugely satisfying, once you overcome any nervousness about demolishing infrastructure that may have been in existence for centuries. Naturally, the type of soil dictates how easily it will hold water. Clay is a dream for this. At Knepp, on heavy Weald clay, it's been great fun, smashing up Victorian clay land drains and watching ponds and wetlands emerge as if by magic, and wildlife pouring in. But most soils will wet up if there is enough water coming in from the surrounding catchment. If getting rid of drains does not create open, standing water in your conditions, the resulting boggy wetland will still be wonderful for nature.

Breaking land drains can be trickier than you might think, however. Just finding them can be difficult. We're still trying to hunt down some of the main drains leading into our stretch of river at Knepp. Drains should be recorded on plans belonging to the property – on drawings, if Victorian, and on detailed surveys from the 1960s onwards. But often what appears on paper is very different from what has happened on the ground. The easiest way to locate a land drain is to find the outlet at the ditch, river or water channel – often a concrete block holding the pipe in place – and follow the angle of the pipe backwards.

Breaking a drain in one place will usually result in water circumventing the break and finding the drain on the other side, so use a digger to smash up a considerable section of drain(s) in order to get the water to sit on the land. If the drains are Victorian they will be made of clay and the shattered remnants can simply be left in the field. If they are more recent, they are likely to be plastic. Pull out the plastic fragments as the digger unearths them.

Remember that water works in mysterious ways. As is the nature of rewilding, your pond or wetland may not materialise quite where you expect it to.

Ephemeral ponds

Removing drainage may also create the phenomenon of temporary or 'ephemeral' ponds, which we've already mentioned in the floodplains section. A similar habitat reveals itself in the drawdown zones of pond margins, of course, but the ephemeral pond is such an important component of wetland habitat – and so neglected – that it merits a section in itself.[71]

Ephemeral ponds are a universal component of global biomes from equatorial forests to arctic tundra. But they are increasingly under threat because of drainage for intensive agriculture and development, the altering of watercourses and over-extraction of water for domestic and industrial use. The sharpest decline is in western Europe. In a few areas of southern and eastern Europe and North Africa decent numbers of temporary ponds remain, but they are unprotected by law and have a bad reputation as spawning grounds for mosquitoes.

Just like floodplains, ephemeral ponds illustrate what is unique and important about the concept of rewilding. Ephemeral ponds expand the way we think. Dramatically different from season to season, they may, at some points in the year, not resemble a pond at all. Indeed, they are likely to dry up completely. Like so many systems in nature they contain the potential for surprisingly complex food webs. Living with, and even restoring, this type of dynamic, unpredictable, shifting habitat is a challenge to the conventional conservation mindset, which is generally reassured by habitats that can be kept in some kind of stasis. Conservationists can be nervous of change; rewilders welcome it.

Keystone species

Over millions of years, species have evolved to suit these ephemeral ponds, including pioneer species and a rich invertebrate fauna that cannot tolerate fish predation in permanent ponds. Many of these species are now extremely rare, thanks to the disappearance and degradation of ephemeral ponds. Their loss has profound implications for the wider environment.

Recent studies, for example, have identified fairy shrimp and tadpole shrimp as keystone species. These specialists of ephemeral ponds survive dry periods using dormant resting eggs. Fairy shrimps are large filter-feeders that can become very numerous, generating a biomass rich in protein and fatty acids and thus constituting an important food for amphibians and wetland birds, including long-distance migrants.[72] But they are also – like the beaver – considered to be 'primary consumers'. This means that how and what they consume has an exponential effect on their environment.

Fairy shrimp can control algal biomass, helping to establish the kind of clear water habitat that benefits submerged aquatic plants and invertebrates.[73] Tadpole shrimp, on the other hand, are keystone predators that bring to bear strong effects on the ecosystem through trophic cascades, much like the wolf on land. By consuming their prey species, mostly small aquatic invertebrates which predate on creatures further down the food chain, they release pressure on the very smallest invertebrates, algae and aquatic plants. The idea of temporary ponds being breeding grounds for mosquitoes may be because the ponds in question are missing these keystone species and no longer optimally functioning. Tadpole shrimp may even ward off mosquitoes simply by their smell.

Putting it into practice

Species of fairy and tadpole shrimp are likely to have real potential as reintroductions. At present, Britain has only one species of tadpole shrimp (*Triops cancriformis*). The oldest known animal species in the world, dating back at least 220 million years, it is found in only two locations in the UK – the New Forest in Hampshire, and a site near the Solway Firth in Scotland. There may well be other tiny creatures worth considering for (re)introduction, as we begin to learn more about the neglected phenomenon of ephemeral ponds.

Creating ephemeral ponds

It is relatively easy to create shallow, temporary ponds that are prone to annual drying out and freezing, particularly on heavy clay. Floodplains also lend themselves beautifully to the creation of shallow ponds. When artificially created by a digger, temporary ponds are known as 'scrapes'. They are a typical feature of some of Britain's most wildlife-rich, extensively managed habitat, such as the New Forest in Hampshire and the Lizard Peninsula in Cornwall, where they support specialist and rare plants such as the beautiful water fern pillwort and the three-lobed water crowfoot, as well as tadpole shrimp and a plethora of wetland birds.

Puddles in muddy, rutted tracks are also surprisingly important ephemeral pond habitat. In the UK, they are key sites for fairy shrimp – as in the tank tracks on Salisbury Plain, mentioned earlier. It's easy to overlook habitat like this, and even easier to 'improve' it to the detriment of species living there. Studies in Belgium exploring the potential for the reintroduction of fairy and tadpole shrimp in ephemeral ponds, for example, have noted that adding gravel to puddles on tracks has a detrimental impact on fairy shrimp populations. As with ponds in general, mud and clean water are the most important components.

Restoring waterways: what to remember

The basic principles behind creating, restoring and rewilding rivers, streams, ponds and small lakes are simple. You must first understand what you have, and only then decide what to do, if anything. If in doubt, take advice from professionals and consult the wealth of information online (see Further Resources for suggestions). Consider the different geographical scales, too, from the landscape to the single body of water, as well as your time scales. Above all, think clean water. This is not easy in hyper-polluted Britain. But it is essential.

4

Rewilding with Plants

How to create dynamic habitats with vegetation

Plants are often studied in isolation, without thinking about their evolutionary relationship with animals. Large herbivores, in particular, are important drivers of habitat creation, shaping woody shrubs and vegetation through browsing and physical disturbance.

The way the plants react, rebound and flourish in response to this disturbance shows how plants need animals just as much as animals need plants.

Restoring a piece of land is, in the first instance, about encouraging the return of a wide range of plant life, including shrubs and trees. No matter where the land is, a complex community of plants is an indication of biodiversity and a functioning ecosystem. How this vegetation looks, and how many species it contains, largely depends on the soil type, topography, geology, climate and altitude. In mountainous areas, at high altitudes, in the shearing winds above the treeline and near the snow, vegetation will be slow to grow, subtle and low, clinging to rocks, scree and thin soils. It will tend to be a mixture of small sedges, rushes, arctic-alpine flowers and dwarf herbs, and bryophytes (liverworts and mosses). Obviously, vegetation there is completely different from that in the lowlands, where, with warmer weather and deeper soil, trees, shrubs and plants grow thick and fast, and flowering seasons are much longer.

Scientists have done a lot of work, categorising plants into communities according to where they live. The UK's National Vegetation Classification comprises 286 community types subdivided among 12 major types of vegetation and further sub-communities.[1]

This has been useful in preserving rare habitats and species. But rewilding demonstrates that where things currently are may not reflect where they once were, or could be in the future. The prevailing wisdom that, in Scotland, trees will not colonise terrain above an altitude of 650 metres, for example, is refuted by experience in southwestern Norway, a landscape that was once as barren as the Scottish Highlands and with almost exactly the same geological and climatic conditions.[2] After a century or more of land abandonment and – crucially – no grazing pressure, Scots pine, rowan and aspen and, higher up, dwarf willows, birch and juniper have naturally regenerated up to an altitude of 1,200 metres.[3]

Within a century, the sheep-ravaged landscape of southwest Norway – with a climate, topography and ecology similar to the Scottish Highlands – has naturally regenerated with native woodland. On the left, an old photograph from 1927 shows a tightly grazed valley in the region of Stavanger, before it was abandoned during the agricultural depression of the 1930s. The photograph of the same scene (on the right) was taken in 2014.

The fact that we're not used to seeing trees higher than 650 metres above sea level in Scotland does not mean they can't grow there. After centuries of logging, burning and intensive grazing, we have simply grown used to the barren hillsides and assume the barrenness is natural or irreversible. What is holding back spontaneous Norway-style reafforestation in the Highlands is, principally, too many deer. On estates, deer – which are fed over the winter to keep numbers high for stalking – browse off any emerging scrub and saplings before they have a chance to establish.

Meanwhile, arctic-alpine plants and subarctic bryophytes – plants typical of the last glacial period, such as mosses – cling on to commons in lowland Norfolk, where one would never expect to see such plant communities. They exist on grassland around 'pingos' (also called kettle lakes): ponds created when Ice Age glaciers retreated, leaving hard lenses of ice pressed into the ground. As the climate warmed, these ice lenses melted and the soil above them collapsed, forming a depression that filled with water. The grazing of horses has been crucial to the survival of these pockets of 'arctic' habitat, which may well have been common in much of the UK not so long ago. But traditional grazing in East Anglia has been largely abandoned, and the remnant grassland communities of small fen plants, orchids and subarctic-type bryophytes are rapidly disappearing as coarse and competitive grasses, trees, bracken and bramble move in and shade them out.

Both too much grazing, as in the Scottish Highlands, and lack of grazing, as in the commons of East Anglia, clearly define plant communities. Yet the role of animals as drivers of vegetation has been largely overlooked over the last century. This is partly because of the powerful influence of 'forestry'. Trees and shrubs are now artificially propagated and planted instead of naturally regenerated. Also, in the absence of herds of free-roaming wild herbivores interacting naturally in wild areas, we simply don't see their influence any more. We may notice the agency of wind, birds and insects, and perhaps even of small mammals and flowing water, when we think of soil disturbance and pollen and seed dispersal. But we've become blind to the complex range of interaction between flora and megafauna (many species of which are now, of course, extinct). This has fundamentally distorted our understanding of the dynamics of vegetation and how best to recover and conserve it.

Looking at vegetation more holistically, as expressing many influences, including those of large herbivores, apex predators and other wildlife, is a fundamental aspect of rewilding. It is vital that those influences are naturally dynamic. Plant communities need to change and shift, incorporating boom-and-bust scenarios. Responding to such events as outbreaks of pests, disease, extreme weather and long-term climate change is also vital. As ever, it's a question of moving the focus from what we think *should* be there and doing everything we can to sustain that (the target-led approach of conventional conservation) to allowing nature to show us what *can* be there and accepting, or encouraging, the natural ebb and flow (the process-led approach of rewilding).

A new way of thinking

From vegetation succession to animal disturbance

As any European farmer knows, if you leave a patch of land aban-doned, it reverts to scrub and, eventually, trees. Until recently, this 'vegetation succession' was considered the primary driving force of the development of nature in temperate regions of the world. The twentieth-century botanists Frederic Clements and Arthur Tansley talked of 'climax vegetation' – a destination that nature was supposedly endlessly struggling to reach.[4]

This view dominated scientists' thinking until very recently, and it has profoundly influenced the way we imagine our landscapes looked thousands of years ago. Before human impact – the prevailing theory goes – any land with the climate, soil and hydrology for trees to grow would have been covered with closed-canopy forest. In temperate-zone Europe, it is often believed, only the tops of mountains, the very steepest slopes and some raised bogs would have been devoid of tree cover.

The idea of green, primal forest has come to represent the antithesis of our nature-depleted, built-up, polluted, chopped-up landscape. When we think of ways to recover nature we often imagine re-establishing deep forest – by planting trees. But closed-canopy woodland supports relatively few species. Almost all plants require the energy of full or at least dappled sunlight, especially if they are to produce fruit, nuts, seeds, berries and pollen – high-energy food sources for birds, invertebrates and herbivorous mammals. Dense stands of trees shade out most shrubs and smaller plants from beneath them. A rich variety of fungi may thrive in these dark conditions, especially those that specialise in decaying wood. But the greatest floral, and therefore wildlife, diversity in woodlands is to be found in direct or partial sunlight, such as open glades, coppiced areas, artificially maintained 'rides' and clearings, and around the edges. Animals and birds may use closed woodland for cover, but most of their active lives is spent in more open habitats. Possibly only three birds in the UK – marsh tit, pied flycatcher and wood warbler – breed and feed exclusively in closed-canopy woodland.[5]

If we take ubiquitous closed-canopy woodland as a baseline for nature, it is well-nigh impossible to explain biodiversity itself. After all, most of the UK's native flowers, lichens, insects, birds, mammals and

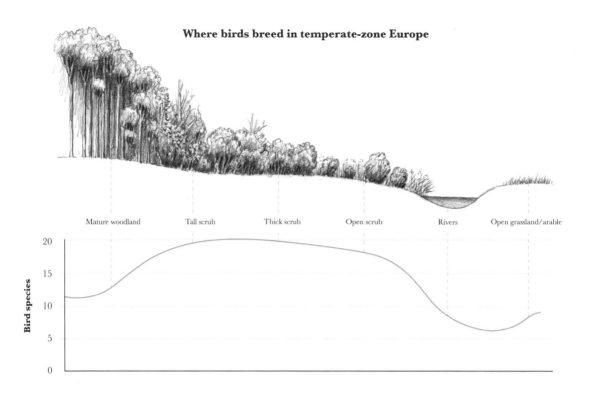

Where birds breed in temperate-zone Europe

Mature woodland   Tall scrub   Thick scrub   Open scrub   Rivers   Open grassland/arable

The number of birds that breed in closed-canopy woodland are very few. High scrubland favours the most species, with species numbers falling as scrub becomes lower and sparser.

reptiles need sunlight for at least part of their life cycle. The native trees themselves signal that they evolved in much more open conditions. Light-demanding trees such as oak, Scots pine, aspen, birch, crab apple, rowan, sallow, wild cherry and wild pear cannot regenerate in closed-canopy woodland. Common shrubs, such as juniper, sea buckthorn, dogwood, field rose, dog rose, hawthorn, blackthorn, wild privet, gorse and hazel, also require open, light conditions to thrive.[6] So where did they occur in the past if all was closed-canopy forest?

Certainly, natural processes like disease, insect infestation, storms, drought, floods, landslides and even the domino effect of big trees toppling when they die can open up lighter spaces, or glades, in woodland. But unless these openings are really extensive, they are swiftly dominated again by shade-tolerant species, such as beech and hornbeam, which out-compete any light-demanding saplings that manage to germinate.[7] These disturbances are also relatively rare. Before modern forestry, when we began artificially propagating trees in commercial nurseries and transporting them all around the world, tree diseases were probably less common than they are now. Major outbreaks of disease among trees would have occurred hundreds, if not thousands, of years apart.[8] Extreme weather events, such as floods, drought and tree-levelling hurricanes, were also less common, and often local. Fire is often quoted as an important forest-opener. But, unlike those of the southern Mediterranean, native British woods are not prone to catching fire. As the historian and ecologist Oliver Rackham observed, 'broadleaved woodland burns like wet asbestos'.[9] On their own, these influences are not enough to explain the evolution or the survival, let alone the dominance, of the oak and all the other light-demanding trees in the landscape.

What is missing from this list of natural disturbances to vegetation is the most common of all: that of large, free-roaming grazing animals. It has been overlooked by natural science for more than a century.

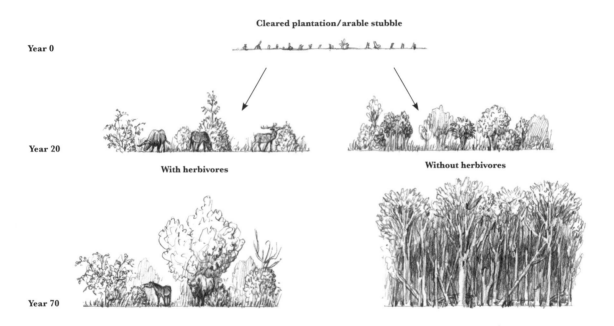

Cleared plantation/arable stubble

Year 0

Year 20

With herbivores **Without herbivores**

Year 70

As in the rest of the world, Britain's flora co-evolved with an extraordinary range of megafauna, the giant herbivores and apex predators that contributed to Earth's peak in biodiversity about 100,000 years ago.[10] Hippopotami wallowed where the river Thames flows today. Straight-tusked elephants weighing 13 tonnes, a close relation of the diminutive African forest elephant, thundered through the landscape, alongside narrow-nosed rhinos and herbivorous cave bears bigger than grizzlies.[11] In Britain, some megafauna, among them the aurochs (the original wild ox) and the tarpan (the Eurasian wild horse), may have survived until just a few thousand years ago. Woolly mammoths died out only 11,000 years ago. Woolly rhinos

Vegetation succession with and without herbivores. Starting from a baseline of cleared woodland or arable stubble, vegetation on its own will eventually develop into closed-canopy trees. But with animals present, a much more complex, open and dynamic system evolves.

145

– once hunted by scimitar-toothed cats, gigantic cave lions and cave hyenas twice the size of the laughing hyenas of Africa – survived until 8,000 years ago, and the last giant deer is known as the 'Irish elk' after a fossil record in Ireland that was dated to only 11,000 years ago.

These giants existed in numbers hard to conceive of today. Some, like the tarpan and aurochs, lived in 'super-herds'. They would have pushed over trees, snapped branches and moved seeds over great distances. Their dung, urine and rotting carcasses re-energised the soil and fed plants and other life. With vastly different grazing and browsing preferences and techniques, often working symbiotically with vegetation, their collective disturbance would have driven a dynamic, complex, shifting mosaic of open wood-pasture habitats, woody wetland and grassy savannah.

The ghosts of Pleistocene megafauna haunt the behaviour of Britain's native flora today. The extraordinary abilities of shrubs and trees to sprout new shoots from broken stems, grow protective scar tissue over stripped bark and deep incisions, and produce unpalatable tannins and aggressive thorns, Rackham deduces, must be 'adaptations to recovering from the assaults of elephants and other giant herbivores'.[12] This is why oak, ash, beech, lime, sycamore, field maple, sweet chestnut, hazel, alder and willow respond so readily to the forester's billhook when coppiced, pollarded or spliced and laid as a hedge.

The demise of these herbivorous giants, mostly at the hands of spear-throwing Stone Age hunters, heralded a 'catastrophic shift' to an ecosystem that became, inevitably, less complex, dynamic and biodiverse. 'The extermination of the great tree-breaking beasts in Palaeolithic times', Rackham writes, 'may have been mankind's first and farthest-reaching influence on the world's forests'.[13]

But not all megafauna became extinct then. Some survived in Britain into the Holocene, the post-Ice Age era beginning about 11,700 years ago. Indeed, some continued to exert considerable influence on vegetation into historical times. Until the Bronze Age in Scotland and northern England, elk – the largest surviving species of deer in the world – browsed on riverine and submerged aquatic plants in the summer and on leaves, twigs, conifer needles and bark in the winter. Brown bears, living on a predominantly herbivorous diet of grasses, roots and berries, were relatively numerous in Roman Britain.[14] Wild boar wallowed and truffled and rootled the soil until the thirteenth century. Hundreds of thousands of beavers, arguably the world's most transformative keystone species, would have been tree-felling, coppicing and building dams in wetlands until the Tudor Vermin Laws finally exterminated them.

All these animals were eventually extinguished from Britain, alongside their predators, the wolf and lynx. As the islands became ever more intensively managed and densely populated, there was less tolerance for free-roaming animals, and there were no places left for them to hide. The last record of a beaver killed was in Yorkshire in 1786. By then, of all the megafauna that had once roamed the British Isles, only red and roe deer and small herds of semi-feral ponies – descendants of the tarpan – remained in the wild.

More megafauna did, however, survive in continental Europe, despite being similarly hunted and persecuted, because areas of wilderness and mountain ranges provided them with refuge. Now, thanks to the large-scale abandonment of land, legal protection, reintroductions and a growing public tolerance for wildlife, populations of beaver, elk, bison, brown bears, wolves, lynx and wild boar are rebounding in European countries.[15]

100,000 years ago, Britain was a land of giants: hippopotamus, elephant, rhinoceros, aurochs, tarpan and giant elk preyed upon by bears, lion-sized scimitar cats and spotted hyenas.

In this chapter, the emphasis is on understanding how vegetation responds when animals eat it, and how flora and fauna co-evolved and are therefore designed to function together. Conjuring up the megafaunal ghosts of both the distant and not-so-distant past helps us to envisage the kind of vegetation and treescape of the future that will assist us in recovering biodiversity and functioning ecosystems. We must also consider how to manage existing woodland and encourage the expansion of trees, shrubs and other vegetation.

On larger areas of land, introducing free-roaming domesticated animals as proxies for missing megafauna is key to both establishing and sustaining plant diversity. What the large-scale rewilder needs to decide is which species to introduce and at what stage of vegetation growth. It may also be beneficial to vary the numbers and species of animal from time to time, in order to encourage a greater variety of plants, and greater complexity of plant structure. On areas of land that are not large enough to sustain free-roaming animals, it will be crucial to mimic their activity.

Of course, myriad smaller creatures play a role in creating habitats, and it's important to appreciate how declines in these species also affect vegetation dynamics. Hares and geese are grazers too. And the smallest mammals and birds – vital for their role in dispersing seed – have declined dramatically, as have pollinating insects such as night-flying moths, whose importance as pollinators is often forgotten.[16]

Thorny scrub and natural regeneration

Perhaps the most important thing of all is to understand the role of thorny scrub – which is largely missing in the modern landscape – and how it serves as midwife for new trees in a landscape with large herbivores. Thorny scrub is one of the UK's most biodiverse habitats. Brambles, hawthorn, blackthorn, dog rose and gorse provide flowers for pollinating insects, buds and fruits for birds and small mammals, and thorny cover from predation. Scrub is both fortress and larder. At Knepp, this is where we find nightingales and turtle doves, and other rare species such as redwings, fieldfares, lesser redpolls and yellowhammers, alongside all the more common songbirds, including record-breaking numbers of blackcaps, whitethroats, lesser whitethroats and dunnocks. We rarely acknowledge the full-scale loss of this habitat, but it is one of the main factors in Britain's catastrophic decline in biodiversity.

Once, thorny scrub would have been a common feature of the landscape, prized for the raw materials it produced, to make dyes, gunpowder, tool handles, brooms, walking sticks, hurdles, thatching spars, basketry, furniture, fodder, fuel, charcoal, berries and medicine. But today scrub products have been replaced by mass-produced, often plastic, alternatives. Supposedly now useless, scrub has been demonised, an affront to our sense of orderliness. Areas where it appears are considered 'wasteland', and, armed with motorised tools, we are swift to hack it down. The last bastions of scrub in the modern landscape tend to be industrial wastelands, slag heaps, spoil tips, railway sidings, gravel pits and abandoned quarries. Ironically, these brownfield sites are now notable for wildlife. They are far 'greener' than most greenfield sites and, indeed, many nature reserves.[17] In the UK, they've become sanctuaries for species on the verge of extinction, such as cirl bunting and natterjack toad.

Despite its conservation value, even conservationists have struggled to understand scrub as habitat. Part of the problem is its ephemeral nature. Scrub doesn't stand still. The more you cut it down, the more prolific it becomes – responding as energetically and defensively to the billhook and shears as it does to herbivory. Unlike other habitats, it doesn't fit our tidy-minded scientific definitions. Where does it begin? At the margins, as low-lying bramble in grassland, as shrubs in bare ground or as reeds in marshy areas? Where does it end? When the young trees are taller than the shrubs around them, or as the understorey in closed-canopy woodland? Scrub is endlessly morphing – a discomfiting notion for the modern mind. Conservationists, bent on keeping a landscape in stasis to preserve targeted species, have for decades regarded encroaching scrub as the enemy. Vast sums have been spent on its eradication. Indeed, scrub-bashing is a staple weekend activity of conservation volunteers.

In this mindset, we have also forgotten the vital importance of thorny scrub in the regeneration of open-grown trees.[18] It is nature's barbed wire. Birds, squirrels and mice stash acorns and other seeds in larders in the ground to feed themselves through winter. Where they're protected by thorny scrub, forgotten seeds (and those hidden by predated creatures) are able to grow up protected from deer, rabbits and other browsing animals. Wind, too, can blow the seeds of ash, lime and sycamore into the protective embrace of thorny scrub.

Ellenberg values for the light preferences of plants native to the UK[19]

Common name	Taxonomy name	Height (cm)	Light value	Common name	Taxonomy name	Height	Light value
Common juniper	*Juniperus communis*	500	8	Field elm	*Ulmus minor*	3100	5
Sea-buckthorn	*Hippophae rhamnoides*	300	8	Common ash	*Fraxinus excelsior*	2500	5
Purple osier willow	*Salix purpurea*	300	8	Small-leaved lime	*Tilia cordata*	2500	5
Scots pine	*Pinus sylvestris*	3000	7	Common lime	*Tilia cordata x platyphyllos*	2500	5
English oak	*Quercus robur*	3000	7	Black alder	*Alnus glutinosa*	2000	5
Silver birch	*Betula pendula*	2500	7	Field maple	*Acer campestre*	1500	5
Downy birch	*Betula pubescens*	2000	7	Common holly	*Ilex aquifolium*	1500	5
Apple	*Malus sylvestris sens. lat.*	1000	7	Bird cherry	*Prunus padus*	1500	5
Crab apple	*Malus sylvestris sens. str.*	1000	7	Midland hawthorn	*Crataegus laevigata*	1000	5
Goat willow	*Salix caprea*	1000	7	European spindle	*Euonymus europaeus*	600	5
Grey willow	*Salix cinerea*	800	7	European hornbeam	*Carpinus betulus*	3000	4
Bay willow	*Salix pentandra*	700	7	Common ivy	*Hedera helix*	3000	4
Common buckthorn	*Rhamnus cathartica*	600	7	Large-leaved lime	*Tilia platyphyllos*	3000	4
Wayfaring tree	*Viburnum lantana*	600	7	Wych elm	*Ulmus glabra*	3000	4
Common dogwood	*Cornus sanguinea*	400	7	Wild cherry	*Prunus avium*	2500	4
Tea-leaved willow	*Salix phylicifolia*	400	7	Common yew	*Taxus baccata*	2000	4
Short-styled field rose	*Rosa stylosa*	300	7	Wild service tree	*Sorbus torminalis*	1900	4
Harsh downy-rose	*Rosa tomentosa*	300	7	Common hazel	*Corylus avellana*	600	4
Black poplar	*Populus nigra sens. lat.*	3000	6	Common box	*Buxus sempervirens*	500	4
Sessile oak	*Quercus petraea*	3000	6	Common beech	*Fagus sylvatica*	3000	3
Aspen	*Populus tremula*	2000	6				
Whitebeam	*Sorbus aria*	1500	6				
Rowan	*Sorbus aucuparia*	1500	6				
Common hawthorn	*Crataegus monogyna*	1000	6				
Black elder	*Sambucus nigra*	1000	6				
Rook whitebeam	*Sorbus aria agg.*	1000	6				
Alder buckthorn	*Frangula alnus*	500	6				
Sorb tree	*Sorbus domestica*	500	6				
Blackthorn	*Prunus spinosa*	400	6				
Plymouth pear	*Pyrus cordata*	400	6				
Guelder rose	*Viburnum opulus*	400	6				
Wild privet	*Ligustrum vulgare*	300	6				
Dog rose	*Rosa canina agg.*	300	6				

Scale

Description	Value
Light-loving plants, rarely found where relative illumination in summer is below 40%	8
Plants generally found in well-lit places, but also in partial shade	7
	6
Semi-shade plants that rarely thrive in full light	5
	4
Shade-loving plants, often found where relative illumination in summer is below 30%	3

German botanist Heinz Ellenberg devised a scale for assessing the amount of light different plants need in order to thrive, known as Ellenberg values. Here, we can see that the native woody shrubs that are hugely beneficial to wildlife, and which should be playing a much greater role in nature restoration in the UK, are predominantly light-demanding. They thrive in the kind of open landscape created by rewilding with large free-roaming animals. The Ellenberg values for light in the right-hand column range from shrubs that like a combination of shade and light (4) to light-loving shrubs rarely found where light exposure in summer is less than 40 per cent (8). There are no examples of plants in Britain that like deep shade.

The scrubby encasement supplies the growing sapling with fungal networks underground, connecting it with mature trees of its own and other species in the vicinity. This provides the young tree with nutrients and, via ingenious chemical circuitry, protects it from pests and disease. It also provides a microclimate that shields the growing sapling from wind, excessive rain, snow and burning sun. Even bracken and outbreaks of creeping thistle can act as nurseries for saplings, protecting them from deer, hares and rabbits. Once the trees have grown, they shade out the light-loving thorny scrub, brambles, thistles and bracken beneath them.

The role of thorny scrub as a nursery for future generations of trees was well known in the past. In the days of commons grazing, medieval foresters recognised it as a vital resource. An old forest proverb proclaims 'the thorn bush is the mother of the oak'. In the seventeenth century the agricultural writer Arthur Standish described how forest officers would 'caste acornes and ashe keyes into the straglinge and dispersed bushes; which (as experience proveth) will growe up, sheltered by the bushes, unto suche perfection as shall yelde in times to come good supplie of timber'.[20]

In the eighteenth century, however, changing demand for timber led to the invention of artificial plantations: continuous stands of single-generation trees that were propagated in nurseries and planted out en masse. With the trees set close together so that competition for light would induce long, straight trunks with no lateral branches, plantations eventually eclipsed Britain's biodiverse wood pastures dominated by open-grown trees – the 'forests' of old, such as Epping Forest, the New Forest and the Forest of Dean. In human-created plantations thorny scrub became a hindrance, and without thorny scrub to protect the young saplings, grazing and browsing animals caused devastation. Livestock and wild ungulates such as deer had to be kept out of the plantations at all costs using surrounding ditches and fences. Soon the idea of 'wood pasture' became a contradiction in terms. The 'forest' had become a place of trees and nothing else, while 'pasture' was a place of grassland with livestock and no trees.

Separated in this way, the biodiversity of both wood and pasture was considerably diminished. Dynamic, shifting margins, where one habitat merged, evolved and collapsed into another, gave way to a world of human control, of hard edges. And, with no place for thorny scrub in the landscape, we soon forgot how trees had ever regenerated without us. 'Natural regeneration' was redefined as simply the germination of seeds falling directly from mature trees, typically in clear-felled areas within a plantation.

It is only now, with the advent of rewilding, that we are beginning to appreciate the immeasurable benefits of allowing trees to regenerate as nature intended, and the potential damage that planting can do to areas of nature and trees themselves. A review led by the Royal Botanic Gardens, Kew, in 2020, focusing on the 'ten golden rules for reforestation', urges us to 'protect existing forest first' and 'use natural regeneration wherever possible'.[21] The charity Rewilding Britain, in its report *Reforesting Britain* (2020), advocates natural regeneration as the default system for re-establishing trees in the wider landscape.[22] But this will happen only if we move away from the cultural mindset of planting that has prevailed in Britain for more than 200 years. Given the huge vested interests of commercial nurseries and big forestry business, this will be a challenge. Even conservation charities, such as the Woodland Trust, are geared towards tree planting, and often the funding on which they depend demands it.

The New Forest in Hampshire has changed little over the past 1,000 years. Its wood pasture landscape is characteristic of what was originally meant by a 'forest'.

How should we bring back more trees?

There is, undisputedly, a desperate need for more trees in Britain's landscape. Following centuries of deforestation, neglect and agricultural intensification, only 13 per cent of the UK's land area is 'wooded' today, making it the second least wooded country (after Ireland) in Europe, where the average tree cover is 44 per cent.[23]

But the terms 'wooded' and 'reforestation' can be very misleading. Anyone thinking of the UK's existing 13 per cent woodland cover might envisage something wild, old and biodiverse. But 83 per cent of this is in fact commercial forestry managed for production.[24] Only 2.5 per cent of the UK (just over 600,000 hectares) is ancient woodland. The destruction of ancient woods has been going on for centuries. Large areas were lost in the seventeenth century, when Charles II sold off the royal forests, and still more ancient woods were cleared as a result of the Enclosure Acts – the appropriation in the eighteenth and nineteenth century of commons land for agriculture. Vast areas succumbed to the axe during the agricultural boom of the nineteenth century, and again in the twentieth. The few remnants are still under threat. The new high-speed railway between London and the West Midlands (HS2) alone is poised to destroy or irreparably damage 108 of them.[25] In addition, the majority of commercial forestry – 51 per cent of the UK's total woodland (1.6 million hectares) – is made up of conifers, or softwoods, principally non-native Sitka spruce, which provide notoriously poor habitat for wildlife.

In the race to get trees back, we must be clear about what kind of treescape we want for the future. What kind of woodland should we establish, and why? We must question whether the prevailing planta-tion model is really the most effective, cost-efficient, sustainable way of re-establishing trees. What are the ecosystem services or public benefits that it actually brings? Above all, we must consider the best way to replace the ancient woodlands that have been lost.

Trees for carbon capture

In the desperate search for solutions to climate change, capturing carbon through planting trees has been hailed as the ultimate quick fix. In 2011 the Bonn Challenge was launched by the government of

Germany and the International Union for Conservation of Nature (IUCN), aiming to restore forest to 3.5 million hectares of land by 2030.[26] Even so, deforestation continues apace. The planet lost 3.8 million hectares of tropical primary rainforest in 2019 – an area almost the size of Switzerland.[27] But tree planting has also gathered momentum across the globe. Seventy countries pledged to restore a total of 210 million hectares to trees. Ethiopia alone claims to have planted 350 million trees in one day in 2019.[28]

But the extent to which tree planting can tackle global warming is highly contested by scientists. Almost half the area pledged in response to the Bonn Challenge is for commercial monoculture plantations. In hot countries, where trees grow faster, these will be harvested every ten years or so. Often, too, they are made into short-term products, such as paper, cardboard, pallets and biofuel, releasing much of their stored carbon back into the atmosphere. According to recent studies, plantations store just one fortieth of the carbon of naturally regenerated forests.[29] Only one-third of commitments under the Bonn Challenge aim to restore natural forests.[30]

Planting trees alone cannot be the answer to climate change. If we could plant all the land that used to have trees on it before modern agriculture, it would still capture only about 200 billion tonnes of carbon – about twenty years' worth of emissions at present rates.[31] There is simply not enough land in the world to plant trees to offset carbon emissions from fossil fuels.

Indeed, in our rush to plant in the name of carbon capture, we may be doing more harm than good. The wrong trees in the wrong place can actually exacerbate climate change. Reforesting seasonally snow-covered zones, for example, can have a warming effect. Dark trees in the Arctic absorb more winter sunlight than snow-covered ground, which tends to reflect the sun's heat back into the atmosphere – a phenomenon known as the 'albedo' effect. Likewise, the conversion of broadleaf forests to coniferous forests across Europe over the past few decades has resulted in changes in albedo, the roughness of the forest canopy, and water evaporation from plants and the surface of the land – all of which have added to warming rather than lessening it.[32]

In 2019 the UK government pledged to plant 30 million trees each year between 2020 and 2025 in an effort to meet its net-zero carbon-emissions target by 2050.[33] In the highly unlikely scenario that this promise is kept, it may not be as beneficial as predicted. In the year up to March 2019 approximately 22 million trees were put in the ground, but a great proportion of them were conifers in commercial plantations, and mostly in Scotland. Planting conifers on peatland is now widely recognised as having a detrimental impact on climate because it releases vast quantities of stored carbon from the peat.[34] The drainage of functioning bogs may also increase emissions of the long-lived greenhouse gas nitrous oxide.[35] The Natural Capital Committee (an independent body set up in 2012 to advise the government on how to manage England's 'natural wealth') itself warned in a report from 2020 against focusing tree-planting subsidies solely on the cheapest land available – such as wetlands and upland farming areas – because this can increase emissions of greenhouse gases both at source and elsewhere.[36] Planting on uplands currently used for grazing, for example, risks increasing the importing of meat from tropical countries, which would then have an incentive to fell more carbon-sequestering rainforest to make way for animal pasture. Planting semi-natural grasslands with trees may not result in a significant overall gain in carbon storage, anyway.[37] It also risks destroying other important habitats. Saplings recently planted in Cumbria are now having to be uprooted following the discovery of rare orchids on the site.[38]

Generally, carbon calculations focus solely on the storage capacity of trees, without considering either the end use (in the case of forestry) or the potential for carbon capture of the environment in which the trees are growing. The soil's potential to store carbon (likely to be considerably reduced in plantations) is rarely taken into account. Indeed, there are still significant data gaps in how we measure soil carbon. The Countryside Survey data used to assess soil carbon stocks, for example, includes topsoil samples only to a depth of 15 centimetres.[39] Significant quantities of carbon are held below this depth, particularly in peatlands.

Little, if any, research has been undertaken into the carbon storage of naturally regenerated woodland, scrubland and open wood pasture in the UK or, indeed, in temperate-zone Europe, although recent research in the tropics shows that naturally regenerated woodland is six times better than agroforestry and potentially forty times better than plantations at storing carbon.[40] In the rewilding scenario, the overall carbon capture in deeper soil, increased soil biota and mycorrhizal fungi, open-grown, long-lived trees, groves of trees, thorny scrub, understorey shrubs, deep-rooting herbs and restored wetlands is likely to be enormous. In the presence of herbivores, under the right grazing and browsing pressure, it could be even higher. The agricultural journalist Graham Harvey explains in *The Carbon Fields: How Our Countryside Can Save Britain* (2008) how, as grass roots die back and regrow under successive cycles of grazing and recovery, the carbon in the soil increases.[41] Woody shrubs, too, invest heavily in their carbon-storing root system when their leaves and stems are browsed, waiting for the chance to release that energy into swift above-ground regrowth whenever a pause in herbivory allows.

As we balance the positives and negatives of different systems of tree establishment, studying the carbon-storage potential of naturally regenerated open-grown trees, scrub and woodland on open land – the scenario of rewilding, as opposed to natural regeneration within forestry plantations – will be crucial. We must also think about eco-systems as a whole and what other benefits trees, woody shrubs and plants may provide, such as biodiversity, water storage, and pollution and flood control. Ellie Crane points out in her report *Woodlands for Climate and Nature* (2020) that 'There are few studies assessing the effect of afforestation on more than one ecosystem service at a time. This is a significant unmet need for evidence-based policy making.'[42] Fixating on trees for carbon storage alone could lead us down a blind alley that allows us to address few, if any, of the other environmental challenges we face.

The financial and carbon costs of planting

As we seek the speediest solutions to these pressing environmental concerns, we must also keep in mind the question of cost and effi-ciency. The upfront costs of propagation and transport are rarely factored into the carbon-sequestration calculations of tree plant-ing. Saplings are usually produced hundreds of kilometres from the plantation, so there is the carbon cost of transportation, as well as of equipment (spades, fence-post hammers, polypropylene or plastic mesh tubes, plastic ties, preservative-treated wooden stakes and tree guards). Often the site is sprayed with one or more applications of herbicide to kill off competition for the saplings as they grow – another carbon cost, and destructive to the soil. Sometimes the saplings need watering, too.

When the trees are finally established, there is the labour-intensive task of removing the tree guards, and the carbon cost of disposing of or recycling them. Most tree guards are supposed to degrade with exposure to sunlight, but in practice this doesn't often happen, as we know from planting woods at Knepp before rewilding. If the trees grow well, the tree guards are not exposed to enough sunlight to rot. If the young trees die, the cylinders topple over and are subsumed by thickets of grass. Even if the tubes do decay as intended, they still leave plastic residue in the soil.

Then there is the cost of failure. A high percentage of planted trees die, but the numbers are rarely accounted for. Most funding models focus on saplings put into the ground, rather than survival rates. As part of the upgrade of the A14 road in Cambridgeshire, Highways England felled 400,000 trees and shrubs. In March 2021 a council report announced that a 'large proportion' of the nearly one million saplings planted as replacements had died.[43] Even if all are replaced, there is no guarantee the success rate will be any better.

The upfront financial cost of these carbon- and labour-intensive tree-planting systems is clearly considerable, and indeed without government subsidy the plantation paradigm would be uneconomic.[44] By contrast, natural regeneration by its very nature avoids almost all the carbon and financial costs of tree planting. Where there is a large presence of deer and/or livestock, it may be necessary to put up fences or to use the new technology of invisible fences (see page 208) to keep animals out of an area for a time, to allow shrubs and trees to establish. As we've demonstrated at Knepp, however, on lowland arable fields thousands of oaks, ash, wild service, crab apple and thorny scrub can regenerate without the need for any fencing at all, even in the low-level presence of roe deer and rabbits.

Natural regeneration can happen almost anywhere without intervention and at a scale that human planting cannot hope to replicate.[45] An impressive example is the 58 million hectares of former croplands in Russia and Kazakhstan where the collapse of collective farms in the 1990s resulted in 'spontaneous reforestation'.[46]

Biosecurity, natural resilience and genetic diversity

When we consider how to increase tree cover, we need to think of natural resilience and biosecurity. How can we ensure the best chance of survival for trees in the face of climate change and the increased frequency of extreme weather, pollution and disease?

With global commerce and travel and a warming climate, new diseases spread faster than ever. Using saplings from commercial nurseries carries a risk of importing disease. Recent outbreaks in the UK of sudden oak death, sweet chestnut blight, oak processionary moth and ash dieback are all linked to the plant trade.[47]

Of course, stopping all plant imports would not prevent disease from spreading in other ways. Perhaps more effective is to safeguard trees' natural resilience. Artificial planting renders trees much more vulnerable to disease. A block of trees of the same species and age will be much more susceptible to an outbreak of disease than a multi-generational wood of many different species. The natural immunity of trees grown in commercial nurseries is also compromised from

(Top) The discovery that a tree can supply nutrients to another tree via underground networks of mycorrhizal fungi has revolutionised our understanding of tree ecology. We can no longer think of them as isolated individuals.[48]

(Bottom) Trees can even use this mycorrhizal network to send out chemical warnings alerting other trees to attack from pests or disease, stimulating their neighbours to protect themselves.

the very beginning. Bare-root whips are not as well connected to the soil as naturally established seedlings, and often lack the appropriate fungal associates to provide nutrients and protection from pests and pathogens. Nurseries have begun to inoculate the roots of saplings with a few species of mycorrhiza during propagation to try to establish this function artificially, but the invisible underground world of mycorrhizae in nature is, as yet, poorly understood. There is no guarantee that the mycorrhizae applied in nurseries are the right ones, let alone in the right combination for that particular tree in the location for which it is destined.

The death rate of planted trees is high. The whips are vulnerable and can easily dry out and die before or after they're replanted. The tree guards themselves are poor protection against wind and act as a barrier to flooding, which means they often get knocked over. Even if they're fenced off from deer and livestock (which would simply push them over), young trees are still prone to disturbance from rabbits, hares, voles and badgers, and colonisation by ants. Water that collects inside the plastic cylinder can induce rot and mildew, and harbour insect pests. If neglected, the tubes rub against the saplings' etiolated stems and inflict damage of their own. All this weakens the immunity of the young trees.

In contrast, if a self-willed tree germinates by itself in good, functioning, chemical-free soil, it establishes early mycorrhizal connections that supply it with nutrients as it grows. Studies suggest that some trees supply their offspring with nourishment through these underground networks, and even nurse ailing neighbours.[49] A tree under attack from pests or disease can also send out chemical warnings via the mycorrhizal network to alert other trees (not only those of its own species) in the area, stimulating them to protect themselves by, for example, producing repellent tannins and chemicals.[50]

Above all, genetic diversity underpins the survival of all native trees. It gives them their best chance of surviving climate change. In the wild, trees have an astonishing ability to respond to change, both sudden – such as extreme weather and disease – and incremental – such as pollution and a warming climate.[51] Because they produce a vast number of seeds over a long and productive life, and because each of these seeds is produced by pollen that has arrived from diverse sources (trees generally do not self-pollinate), the genetic variation inherent in trees is enormous. Natural selection promotes the survival of individuals best suited to their location. At the same time, because these localised individuals are exposed to pollen blowing in from trees both near and far, they retain great genetic variance, giving them the ability to adapt. Their lavish numbers of offspring, of which only a few reach maturity, provide ample opportunity for natural selection to operate over comparatively short timescales. While they have evolved to suit their location, trees continue to adapt to new threats and conditions as they arise.

The response of trees to the recent spread of introduced chalara or ash dieback (caused by an ash-specific fungus, *Hymenoscyphus fraxineus*, that is thought to have been imported into the UK in 2012 in a consignment of saplings from a commercial nursery in Holland) is a good example. Initially it was thought that no common ash would survive in the UK, and scientists were suggesting hybridisation with other exotic ash species in order to introduce resistance. However, we now know that almost all wild ash stands in Britain contain significant numbers of resistant trees. The response of trees to attack by pest or pathogen can happen in just one generation. As disease wipes out non-resistant individuals the survivors breed with each other, creating a new generation of disease-resistant trees. It may be a while – perhaps several hundred years – before ash regains its numbers (the Woodland Trust has estimated that 80 per cent of the estimated 150–180 million ash trees across the UK will be lost in the coming years). But in the long term it will survive – as long as space is given for the natural regeneration of wild trees.

Saplings propagated in commercial nurseries, on the other hand, are, inevitably, derived from a limited number of sources and therefore cannot approach the immeasurable genetic diversity of saplings grown in the wild. Nor can they embrace the selective adaptations that trees in the wild evolve over time to suit the local conditions.

It is worth underscoring this point, because the future of all trees in the landscape depends on it. No method of seed-collecting and propagation by human hand can come close to replicating the inherent diversity and resilience of wild trees born through natural regeneration. Genetic uniformity – the product of the entrenched system of artificial planting – severely compromises trees' ability to adapt to rising temperatures, pollution, extreme weather and disease. The absence of widespread populations of wild trees jeopardises the long-term future of all trees.

A different concept of 'woodland'

The current spades-in-the-ground approach, however well intentioned, almost invariably results in closed-canopy, single-generation woods that are plantations in all but name and notoriously poor habitat for wildlife. The 'woodlands' of mixed native species planted at Knepp in the days before our rewilding project started are lifeless compared with the naturally regenerated scrub and wood pastureland around them.

This 60-year-old beech plantation at Knepp, thinned for timber and fenced out from animals, is – like most closed-canopy woodland – relatively poor habitat for wildlife.

Planting trees is obviously appropriate in some circumstances. Commercial forestry will continue to be part of the mix of treescapes in the UK, and there is an economic and environmental argument for increasing commercial forest coverage (in the right place) and using more home-grown timber, rather than imports, for building, ensuring the long-term storage of carbon. According to many foresters, the UK's current building regulations are unnecessarily stringent, preventing the use of much locally grown softwood timber, which, being raised in the UK's relatively warm, wet conditions, is of a wider grain than softwoods grown in, say, Scandinavia.

Agroforestry – integrating trees and nature margins within agricultural cropping systems – can increase yields.[53] Trees provide a beneficial microclimate for crops as well as improving the soil and mitigating against pests.[54]

More enlightened practices, too, are gradually being adopted by commercial forestry.[52] Continuous-cover forestry (a practice that is much more common in Europe than in the UK) involves planting mixed species in multi-generational plantations, and encouraging self-seeded trees. Instead of clear-felling, it uses targeted extraction on only small areas. Better commercial forestry, together with natural regeneration wherever possible, can make a really positive contribution to biodiversity, carbon sequestration and other ecosystem services.

Agriculture can also play a useful role in bringing back trees, as it begins to adopt regenerative practices such as agroforestry, forest gardening, orchards and silviculture. The presence of more trees in towns and cities supports urban wildlife, makes the inhabitants happier, cools the warming streets and cleans polluted air.

When it comes to increasing tree cover in the wider landscape, however, natural regeneration must be the default approach. This will require a much broader vision of how treescapes could and should look. Future woodlands need to be much more open and complex, closer to the ancient 'forests' of the past: shifting mosaics of open-grown trees, groves of trees, thorny scrub and understorey, grazing lawns, heath and wetland vegetation, with large herbivores (or a proxy for them) keeping systems dynamic and resilient.

Knepp's wildland today resembles the 'forests' of old, the kind of wood pasture landscape immortalised in Jean Baptiste Camille Corot's painting *Forest of Fontainebleau*, 1846 (top).

Open-grown trees

We've already mentioned the need to rehabilitate scrubland in the cultural mindset. Another, often overlooked, aspect when considering reforestation is stand-alone, open-grown trees: light-demanding species such as Scots pine, and the giant oaks of wood pasture, parkland and royal hunting 'forests', as well as trees that have been allowed to grow out of ancient hedgerows.

An open-grown oak throws out gigantic lateral limbs and a vast canopy producing an abundance of acorns and 360 degrees of habitat for wildlife. The same species in closed-canopy woodland fights for the light, growing slender, straight trunks with no lateral branches and tiny canopies at the top where they reach the sunlight. The open-grown tree not only produces much more pollen than its woodland cousins, but is also far more exposed to the wind, allowing the pollen to travel.[55]

Definitions of woodland are so narrow that open-grown trees are almost never included in surveys even when they exist in old parkland or ancient wood pasture. Yet they are huge contributors to carbon sequestration and biodiversity. The English oak, which, like the Scots pine, cannot regenerate in closed-canopy conditions, supports more life than any other native tree species in the UK. Some 2,300 species – not including fungi and micro-organisms – live on oaks. A mature, open-grown oak produces 700,000 leaves a year. These break down after they fall in autumn to form a rich leaf mould that provides habitat for scores of fungi, including many kinds of bolete, brittlegill and truffle. Over its lifetime an oak will produce millions of acorns, giving food for deer, badgers, wild boar and scores of birds and small mammals in the run-up to winter. Even shade-tolerant trees, such as lime, produce far more pollen, nectar and foliage when growing out in the open. The open space allows them to extend their limbs laterally, providing a vast, spreading habitat with myriad niches for nesting birds, bats and small mammals.

Individual trees are particularly vulnerable to destruction by humans. Countless thousands of mature trees that once stood in the middle of fields have been grubbed out to make life easier for agricultural machinery, or have slowly died from the effects of ploughing and chemicals. Few people protest against the felling of individual trees for roads and developments. Often, we simply mistreat them until they die, tarmacking up to the trunks and parking cars on their roots. In our mania for tidiness, large trees are often removed from public spaces and private gardens simply because their leaf litter is considered a nuisance, or because their roots wrestle with pavements and patios.

Our future treescapes need to incorporate these open-grown giants, which means thinking of the landscape a century or two ahead and committing to open space for trees in perpetuity. That is challenging for a culture entrenched in short-term thinking.

Dead trees

'The man of science and of taste', the English landscape designer Humphry Repton wrote in 1803, 'will … discover the beauties in a tree which others would condemn for its decay.'[56] Taste has changed dramatically since the eighteenth century. Then, dead trees in a landscape (often shown in old master paintings) signified the positives of life in death and continuity. But to the mechanised mind dead trees signal neglect and uselessness, and of course we now have the motorised tools to make felling them temptingly easy. In fact, dead and dying trees provide habitat for thousands of species. Simply not cutting them down – since standing dead wood is better habitat than fallen trees – is a positive act for nature.

Partly, the prejudice against dead trees originates in the modern mania for tidying up. But we also misunderstand the natural processes of decay. Often, signs that we consider harbingers of death for a tree are, in fact, agents of longevity. A tree that we might consider to be nearing the end of its life may actually have barely reached middle age. Fallen branches (which we customarily remove) decompose around the tree's roots, feeding the veteran tree with nutrients. Beetles and saproxylic (dead wood-eating) invertebrates and fungi break down the burden of dead tissues, creating another reservoir of nutrients as the tree ages. The longest-lived trees, such as oak, hollow

Dead trees are vital habitat, and largely missing from the modern landscape. Rotting wood provides food for stag beetles and insect larvae, which provide food for birds such as, in Europe, the large black woodpecker. Lynx and wildcats make their dens in the protection of fallen trees. The hollowed trunks of standing dead trees provide habitat at the right height for wild bees, owls and bats, and their skeletal branches are the perfect vantage point for birds of prey.

out as they grow old, creating roosting and nesting sites for bats and birds. Bat and bird guano provide a further energy boost, and the hollowed, buttress-like structure becomes more stable than the tree's younger self, able to withstand the fiercest storms – something that will be increasingly important as climate change forces more frequent extreme weather events upon us.

One of the species most affected by the loss of hollow trees in our landscape is wild bees. These insects need cavities of 40–60 litres, far bigger than, say, a woodpecker's hole, which is on average 3.3 litres. It's because of this loss of natural habitat – 200 years of hollow trees eradicated from the landscape – that wild bees have started to colonise lofts and roof spaces, where, being a nuisance, they're often fumigated.

Obviously, it's important to remove unstable trees from areas of public access. But safety is often used by local councils as an excuse to get rid of mature trees, which require more maintenance, and replace them with saplings – as in the case of Sheffield City Council's controversial felling of 5,500 street trees in 2018. At Knepp we remove trees on public footpaths and overhanging roads and property if a trusted arborist considers them a risk (beware of tree surgeons touting for work!). We leave dead trees everywhere else. It's astonishing how many skeletons of all species there are after just fifteen years – something we're simply not used to seeing. Naturally, with ash dieback, the frequency of dead trees is rising. But, being a relatively soft wood, ash hollows quickly and wild bees have been swift to colonise them. Turtle doves also love to use them as territorial markers. Sad as we are to lose so many, we now see the silver lining in the life they provide after death.

A new model for funding

It is difficult to predict how many trees and which sorts will arise through natural regeneration. This is clearly a problem for current funding models, which target the number of trees put into the ground. But as funding systems for the public good and ecosystem services develop, bio-credits and carbon credits for natural regeneration will, it is hoped, be part of the mix. One of the biggest obstacles to promoting natural regeneration more broadly, however, will be the vested interests in tree planting – everything from big forestry and commercial nurseries to grant aid.

Putting it into practice

Natural regeneration on arable fields

One of the easiest ways for trees and shrubs to regenerate naturally is on former arable fields, as has happened in New England in the USA, and vast areas of Russia and Kazakhstan.[57] At Knepp, we realised this purely by accident. Because, at first, we failed to attract a conservation grant to erect a deer fence around the 450-hectare Southern Block of the estate to enable us to introduce free-roaming animals into the area, we simply abandoned the arable fields after their last harvest. We couldn't afford to seed them with native grasses and wildflowers as we had done in the Middle and Northern blocks, so they were left as stubble. By force of circumstance, it was an unstrategic withdrawal from farming, happening piecemeal over five years. The least productive fields were abandoned first, until, eventually, in 2006, all the fields had been let go.

This haphazard process of freeing the land in stages, combined with no reseeding of grass and leaving the fields 'open', proved to be rocket fuel for vegetation succession and, in particular, thorny scrub. Completely unintended, it has produced some of the richest areas for biodiversity in the whole project. The presence of hedgerows was hugely advantageous. After three to five years blackthorn suckers began to feed out into the stubble and, with no cover of thick grass sward to hold them back, hawthorn, blackthorn, dog rose and bramble, their seeds excreted by birds feeding in the hedges, began to establish independently in the middle of the fields (although not easily visible in the satellite photographs). We soon realised the importance of not driving around the field margins, as we had when there were crops in the fields, since this prevented the spread of shrubs from the hedgerows.

The speed of growth was astonishing. Within a year, the young scrub seedlings were 15–20 centimetres tall. The following year, they were almost double that. Within four or five years the thorny thickets in some fields had become nesting habitat for yellowhammers, bullfinches and dunnocks, and were providing berries and invertebrates for fieldfares, meadow pipits and redwings – winter visitors we had rarely seen at Knepp before.

Once they had billowed out to a depth of nearly 8 metres, the hedgerows themselves attracted large numbers of nightingales, a bird that had by 2001 disappeared entirely from the estate, in line with the national trend of this species' decline. Vast numbers of jay-planted oak seedlings, as well as sallow (naturally hybridising willow), wild service, birch,

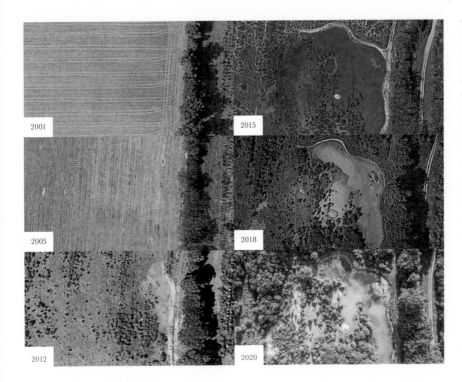

crab apples, hornbeam and hazel, germinated in the abandoned fields, although any seedlings that found themselves outside the protection of thorny scrub were soon browsed off by roe deer and rabbits, which were present in the area in relatively low numbers.

2001

2015

2005

2018

2012

2020

The rewilding of an arable field in the Southern Block at Knepp. It's difficult to see on a satellite image from Google Earth, but already in 2005 shrubs and oak saplings had self-seeded and were beginning to grow. We smashed up drains in the wettest corner of the field (bottom right) and used a digger to excavate a shallow scrape. The resulting pond has spontaneously increased in size and become a complex shallow wetland with 'mangroves' of sallow.

All images except bottom right © Google Earth

The very wettest fields, compacted after years of farming and now deprived of the rotavator and starved of oxygen, have been very slow to change. Fifteen years on, some have started to evolve with vegetation. Others have barely moved at all. We imagine that, if soil invertebrates are still unable to colonise and aerate them, these waterlogged pans will eventually form shallow ponds – a very different type of habitat. But in all the other arable areas, scrub of some description has taken off to a greater or lesser degree. Some fields are now dotted with enormous roundels of bramble. Some have become dense stands of sallow. Others feature tangles of hawthorn and blackthorn, some with an unusually high proportion of dog rose, within which vigorous, jay-planted oaks are now, after sixteen years, 4–5 metres tall. A number of fields in the Southern Block are characterised by self-seeded crab apples, protected by thorny scrub but now reacting to browsing by producing their own sharp, thorn-like twigs. In spring these fields are speckled with delicate, scented blossom.

We cannot know the exact recipe that has produced such different results in every field. Topography and varying soil types are clearly major factors, as is timing. Which species colonise an abandoned arable field will depend on the weather that season, especially whether it's wet or dry, or if there's a frost. The seed sources nearby, and the agents of their distribution (birds, small mammals, prevailing winds), play a part. Whether the field's withdrawal from agriculture coincides with a 'mast' year – when trees produce a bumper crop of seeds – is also hugely influential. All trees require specific conditions to seed. For example, the fluffy seeds of sallow, which, in a mast year, drift on the breeze like snow, are viable for only two weeks. They need to find wet, bare ground during a two-week window in March if they are to germinate. At Knepp this was provided by the bare soils when certain fields were left after harvest in a year preceding a mast year for sallow. Nowadays, the rootling of our free-roaming pigs provides sallow with patches of open soil in which to germinate.

The history of field use also has a bearing on what, if anything, survives in the seed bank. Ploughing and chemicals will have affected soil function. However, the last application of chemicals may be particularly significant. At Knepp, fields where maize was the last crop showed very little woody, scrubby growth for many years. We suspect that this was because of applications of the extremely toxic (and now banned) herbicide Atrazine. It is only now, twenty years on, that vegetation is beginning to emerge in these fields, except now it has less chance of establishing because of the presence of large numbers of browsing animals.

By the time we finally attracted Higher Level Stewardship funding for the Southern Block, in 2009, enabling us to build our deer-fenced perimeter and introduce free-roaming herbivores, the thorny scrub had grown mature enough to defend itself from herbivory. While tender and young, and putting all their energy into setting their roots, emergent shrubs are extremely vulnerable to browsing. But if they survive for four or five years (at which point they may be around knee-height) they are able to produce thorns and repellent tannins. Indeed, the more they are browsed as mature shrubs, the more tannins and thorns they produce. Perhaps, also, chemical signalling through underground networks of mycorrhizal fungi re-establishing in the soil helped to provide community protection by stimulating even those shrubs that had not yet been browsed to protect themselves against herbivory.

The thorny scrub environment we had inadvertently allowed to establish in the Southern Block at Knepp was, in effect, similar to a vegetation pulse that happens in nature after a collapse in herbivory through disease or starvation. Those eight years of growth without browsing were crucial for setting a baseline of vegetation complexity that could then do battle with the grazers and browsers.

Meanwhile, the mosaic of plant combinations, unique in almost every field, has been an added boon for biodiversity. Numerous species of bird, insect and small mammal have particular requirements in terms of nesting, display sites, and water and food sources, particularly if they have complicated life cycles. Mixing it up – something we had not consciously planned – showed us the value of aiming for a more complex and varied habitat by factoring in randomness over time. We would now always recommend rewilding in stages. Even if it's just a single field, it's worth withdrawing a portion from cultivation at a time – say, one-sixth to one-quarter of the field per year – in order to take advantage of the different weathers and seeds available across a number of years.

It also became clear to us that wildflowers and native grasses had no trouble colonising the Southern Block on their own from seed sources in the hedgerows. But it is only now, twenty years on, that we're beginning to see such plants as fluellen and fumitory, species that are not found in hedgerows. One way of increasing floral diversity in the early stages would be to scatter locally sourced wildflower seeds. Or spread out 'green hay' from a local flower-rich meadow in the fields as they are left fallow.[58]

The speed of vegetation succession in the open arable fields of the Southern Block was alarmingly fast for many of our neighbours. We're able, now, to demonstrate the astonishing biodiversity of thorny scrub and the breeding success there of some of the UK's most endangered birds, including nightingales and turtle doves. This is now by far the most popular area for walkers and has drawn considerable attention from national and international media. Most people around us seem to have grown accustomed to, or even appreciative of, the wild new look. But in the early stages of succession, when we had no rare birds or insects, the rapid transformation of the landscape was a challenge, especially for those whose houses overlooked it.

Times have moved on a little. Rewilding as a concept is becoming more acceptable, thorny scrub is on its way to rehabilitation as habitat, and the public is more aware of the need to care better for our land. But change – particularly dramatic change – will always attract controversy. Depending on where your land is, and how prepared you are to deal with any fallout, you may be inclined to adopt measures for a gentler transformation overall, or consider mixing approaches, perhaps planning a faster transition in some areas and a slower one in others.

Natural regeneration on lowland grassland

At Knepp, the pace of change in the Northern Block – an area of 235 hectares that was, at the time of rewilding, a mix of permanent pasture and short- to medium-term leys (land put down to grass) – has been markedly slower.

The colonisation by woody plants of the grassland of the 235-hectare Northern Block at Knepp has been far slower than in the open arable fields of the Southern Block. This is the state of play after fifteen years (with a relatively small herd of cattle).

There we simply took up the internal fences, erected a boundary fence and introduced a free-roaming herd of old English longhorn cattle (originally 23, now numbering about 110). Their trampling, we hoped, would open up the sward for vegetation succession. With much lower grazing pressure than in the 280-hectare Middle Block – where there are 90 longhorns, 20 Exmoor ponies, 50 red and 300 fallow deer, and occasionally some Tamworth pigs – we had expected thorny scrub to begin colonising the fields within about five years. In fact, the grassland has been a far more powerful suppressant than we anticipated. It's only now, twenty years on, that blackthorn suckers are beginning to advance into the grassland from the hedgerows. Out in the fields, small bursts of hawthorn, blackthorn, dog rose and gorse are beginning to take hold. Plenty of oaks are germinating throughout the fields – from acorns stashed by mice or jays – but these are browsed off continually by the cows, rabbits and roe deer. The tiny, stunted oak saplings will fail until the thorny scrub embraces them or a significant decline in browsing pressure allows them to shoot up.

In the open grassland of the Northern Block at Knepp emerging thorny scrub begins to embrace naturally regenerated oak saplings (circled), protecting them from browsing.

Although we had hoped for a more dramatic transition, the Northern Block has usefully demonstrated a pace of change that might be easier for neighbours and onlookers to come to terms with. This softly-softly approach might appeal to some land managers.

There are, however, several options for speeding up the process in grassland if quicker succession is desired or local seed sources are lacking. First, it's important to determine the quality of the permanent pasture. In the Northern Block we had 0.4 hectares of complex, wildflower-rich sward known as Crabtree Platt, containing plants such as dyer's green-weed and salad burnet. This little patch is so precious that we treat it very differently, taking care to keep it clear of trees. It is still unfenced and, as part of the rewilding area, open to the large herbivores. But volunteers pull out seedlings, brambles and bracken, and it is 'topped' (mown) in late summer or early autumn. We're now sowing wildflower seed from it in surrounding areas.

If you're lucky enough to have a complex, wildflower-rich sward that has never been ploughed, take a sympathetic approach, disturbing the soil as little as possible in order to protect the soil structure and biota. The aim here should probably be a wood-pasture landscape with a relatively small ratio of trees and scrub that enhance rather than overwhelm the existing species-rich grassland and increase rather than decrease biodiversity. It's important here to consider carefully which large herbivores to introduce (in low numbers) to allow some natural disturbance while preserving floral complexity, and then to wait for tree and shrub seeds to establish gradually. It may even be worth excluding stock altogether to start with, removing animals at certain times of the year, or moving them around using invisible fencing, giving the vegetation time to recover and in effect mimicking animal migrations in nature.

There is really no reason to be in a hurry. But if you wish to establish trees sooner, you could establish rabbit-proof fenced polygons in the pasture. You can plant tree saplings within them as an initial seed source, or simply let the area inside the polygon lie fallow to receive seeds naturally, in which case the result will be varied and unpredictable. Fifteen years ago, in areas of our rewilding project that had previously been permanent pasture, we set up four 7 × 7-metre 'exclosures' to see how herbivores, including roe deer and rabbits, affected the landscape. In three of them, the rate of natural regeneration has been remarkable. One of these – in a field that, before the release of our free-roaming herds, self-sowed with thousands of oak saplings – is stuffed with oaks that are now more than 2.5 metres tall. In the absence of thorny-scrub regeneration in the grassland outside, the herbivores have grazed off all the saplings in the rest of the field. The fourth exclosure, in an area of high nutrients near the old dairy buildings, contains no trees or shrubs at all, just nettles.

This 7 × 7-metre 'exclosure' in the Southern Block at Knepp was built in a field of permanent pasture. Before large herbivores were introduced, the whole field had self-seeded with oak saplings. If our intention had been to establish trees and nothing else for, say, carbon storage, we could simply have fenced the animals out of this area, and the whole field would now be covered in trees without a single spade having entered the ground.

Another option would be to introduce a few pigs for a short period to rootle up the sward and expose bare soil, thereby speeding up colonisation by plants. In 2019 we began to experiment with this idea in a 2.8-hectare former horse paddock in the Middle Block. After several months of pig disturbance over successive years, hundreds of oak saplings and some good patches of thorny scrub emerged.

Most of the pasture in the Northern Block – apart from the 0.4 hectares of Crabtree Platt – was far from being ecologically valuable. Most of the fields were tired old medium-term rye-grass leys that had received quantities of fertiliser and occasionally broadleaf herbicides. We had no need to be particularly precious about them. Had we known how long it would take for diverse native plants to colonise we might have taken a more interventionist approach to kick-start the process.

The short-term rye-grass leys have been particularly resistant to change. Rye grass itself disappears quickly (developed as a fast-growing crop, it relies on large amounts of fertiliser, and when this is withdrawn it cannot compete and survive). But its immediate successors are a few aggressive species of perennial grass, taking advantage of residual nitrogen in the soil and continuing to suppress other plants. It has taken ten years for more diverse native grasses and broadleaf plants to start moving into these fields.

Rye grass for silage, normally a two- or three-year regime, is just another rotational monocrop. It certainly shouldn't be considered, in conservation terms, 'grassland'. With hindsight, we might have treated these short-term leys – and perhaps some of the medium-term leys – differently, to open them up for vegetation succession. We could have cut the rye grass very short and carted it off for silage in May or June for two or three years, to reduce the nutrient load in the soil. Then we could have gone in with heavy discs or power harrows, churning up the first inch of the surface layer, keeping as much of the soil structure as possible while leaving an open surface with very little grass. This, we reckon, would have resulted in the immediate colonisation of native plants, including thorny scrub and saplings, achieving a transformation similar to that of the arable fields of the Southern Block. To have even greater effect, we could have taken this approach in phases, returning each year to harrow another area, in the same way that we recommend a phased approach to rewilding arable land. This would have allowed different plants to colonise, according to the way each year differed. Or we could have used pigs.

Natural regeneration on moors and grassland in the highlands and uplands

Wet, peaty soils, characteristic of highlands (the low mountainous regions of Scotland and Wales) and uplands (ranges of hills up to about 600 metres) are devastated by conventional methods of tree planting.

On moorland above Megget Reservoir near Broad Law in the southern uplands of Scotland, a sheep fence has protected the vegetation on the right from grazing. In striking contrast to the sheep-grazed area, this part of the hillside is lush with heather, bilberry and a variety of moorland plants.

In the barren Highlands of Scotland, the biggest obstacle to natural regeneration is large populations of red deer and/or sheep. In most areas, seeds are available from trees clinging on in inaccessible gullies, gills or cliff faces or on small islands in lochs or along the coast. Numerous Scottish estates have begun to encourage natural regeneration simply by drastically reducing the number of deer. In a matter of ten years or so, Scots pine, rowan, sallow and birch are growing in areas that were previously heavily grazed, without any preparation of the ground. In a single year, from November 2019, some 1,400 hectares of new woodland were created in the Cairngorms National Park through natural regeneration alone. By comparison, that same area of woodland of *all* types was created in the whole of England in 2019.

Fencing off areas is effective, especially where deer numbers are still relatively high. Cost of installation and long-term maintenance must be a consideration, though, particularly if it's over a large area with challenging terrain. There's also the cost of removing the fence once its purpose is served. The fence posts may last twenty-five years before they need replacing, but it may be longer than that before trees cover the enclosed area. Also, fences can be easily breached in periods of snow or storm, and conventional fences can kill capercaillie and black grouse, which collide with the fencing when flying fast and low through the trees. Generally, fencing disrupts ecological processes and the movement of wildlife. A better system of fencing is two low, parallel, offset electric fences that are too wide for deer to jump over, won't kill black grouse, and will allow the movement of other creatures.[59]

However, avoiding fencing altogether by culling deer is best of all. At the National Trust's Mar Lodge Estate in Aberdeenshire, the magic number for the density of deer that enables natural regeneration is three per square kilometre.

Trees invariably establish in the most conducive spots first; in valley floors and at the bottom of slopes where nutrients have collected, where they are less exposed and there are milder temperatures. It may take decades for the trees to march up the hill to exposed areas at higher altitudes on poorer soil.

Natural regeneration at Mar Lodge, Aberdeenshire. With red deer numbers reduced, Caledonian forest species have self-seeded on even the barest hills. At 600 metres, rare mountainous scrub – thought to be functionally extinct in Scotland – has begun to emerge. The spontaneous return of this habitat, which includes the extremely rare mountain willow, as well as downy willow and juniper, has been one of the project's biggest surprises. As Mar Lodge's ecologist Shaila Rao says, 'thank heavens we didn't rush in and plant'.[60]

In upland areas where red deer are absent or rare, lack of tree regeneration is generally caused by the overgrazing of sheep. Again, the simple solution is to reduce the number of sheep and/or fence off areas for natural regeneration. The exclusion of livestock need not last for ever. Once trees are established, deer, ponies, pigs, cattle and perhaps even a few sheep could be allowed back in. Indeed, for the continuation of natural regeneration in these new woods, the disturbance of cattle, in particular, will be key. Without grazing animals, grassy thatch develops under the trees, preventing seeds from reaching the ground. Unlike deer and ponies, cattle do not target saplings specifically. They prefer long grasses, which they rip off with their tongues, and their trampling crushes grasses into the soil, where they can rot down. The rootling of pigs or wild boar provides added stimulus for regeneration by opening up bare soil for seed germination.

On heathland and in open-grown pinewoods, the volume of natural regeneration will depend on the height and density of the heather and other ground vegetation. Areas that were formerly heavily grazed will probably see a pulse of dense regeneration followed by a slower increase as the ground vegetation recovers. By contrast, areas with tall heather, where seeds are prevented from reaching the soil or seedlings are shaded out before they establish, produce a much lower density of natural regeneration. Some land managers consider lower-density regeneration a problem, and sometimes consider scarification to increase the germination of trees. But this view essentially seems to be influenced by the forestry mindset, which is all about stems per hectare. A more open woodscape will be much more beneficial for wildlife.

Bracken: an opportunity

There is estimated to be between 2,500 and 9,750 square kilometres of bracken cover in the UK, most of it in the uplands, with the largest areas in Wales and Scotland. These are areas that, generally through moor burning and over-grazing by sheep, have experienced a catastrophic shift to a monoculture of bracken. It is notoriously difficult to eradicate. Bracken roots descend a metre in some soils, and the plants can grow to over a metre in height. Burning when the bracken is dry can be effective but only if the roots are not very deep, and more burning further damages the ecosystem. Cutting is effective only if it is carried out three times a year

– an expensive and time-consuming exercise, and often impossible at scale or on steep slopes and rocky ground. Crushing with horse-drawn crimping rollers, trampling by foot or pulling by hand can be effective, although this is obviously labour-intensive and possible only on a very small scale. A more common alternative is using a herbicide, typically Asulox. In 2011 Asulox was banned by the EU, but conservationists and farmers have successfully campaigned for its continued use in the UK to control bracken. Of course, its impact on soil and water sources is devastating.

As ever, the best solution is to work with nature. Large herbivores – preferably a mix of native species (so not sheep) – will do the management. Cattle trample bracken, breaking the stems; horses browse it, especially later in the summer, when it is less toxic; and pigs dig up the roots.

Even more interestingly, bracken is a nursery for natural regeneration. Much like thorny scrub or gorse, it allows self-sown saplings to grow up within it. It provides a microclimate and a degree of protection from scorching sun, wind and excessive rain. Because it is unpalatable to sheep and deer, it affords some protection from browsing. Saplings may grow more slowly in bracken, but eventually the trees will shade out the bracken beneath them. Although bracken is unlikely to disappear completely through natural tree regeneration, what evolves is a patchwork of bracken in the broken and low distribution that is usual in mature woodland, creating the conditions for a far wider array of plants and insects.

When to put a spade in the ground

The landscape historian Oliver Rackham once wrote, 'The easiest way to create a wood is to do nothing.'[61] The benefits of natural regeneration over planting are clearly manifold, but there are circumstances in which natural regeneration is unlikely to happen or will take an excessively long time.

In temperate-zone Europe, without herbivory even the most barren landscapes will eventually be colonised by trees, as was demonstrated in southeastern Norway after the collapse of sheep farming in the mid-nineteenth century. Tiny pockets of shrubs and trees in inaccessible gullies and ravines, and the odd wind-blasted tree surviving in the open, were enough to seed a forest across the wider landscape. But this can take time. In the case of Norway, it took 50–100 years for a landscape once as desolate as the Highlands of Scotland to regenerate naturally into woods.

A rewilder keen for swifter results in an area denuded of trees, such as the highlands, uplands or arable belts without hedgerows, could plant a nucleus of trees and shrubs as a seed source from which these species can expand through natural regeneration. The Wildwood project at Carrifran in the Moffat Hills of the Scottish Borders, for example, has adopted this approach.[62]

A heavy grass sward without disturbance also tends to suppress natural regeneration. Seed 'rain' – the falling to the ground of wind-dispersed seeds – may be held up by thick, grassy thatch, preventing it from reaching the soil; if it does penetrate the thatch, it may be eaten by the small mammals that make their home there. Even if a seed manages to reach the soil and germinate, the thatch overlay may smother the seedling before it is able to reach the light. While trees and shrubs will win through in time, a keen rewilder may wish to accelerate the process by using animal disturbance or scarification to open up the sward and allow seeds to reach the soil.

In lowland areas where seed sources are particularly scarce, it may be worth considering temperate taungya.[63] This practice, which is used by ethnic minorities in China and was developed by the British growing teak trees in Myanmar during the nineteenth century, involves seeding trees directly into crops.[64] An arable crop acts as a nursery to help the tree seeds germinate and establish. In highland areas, direct-drilling native tree seeds into pasture or heather is an alternative method of tree establishment that is both more sustainable and more productive than planting saplings.[65] Both techniques are used primarily to establish woods for commercial production, but they can be used effectively for conservation to establish primary woodland. This can be allowed to expand through natural regeneration, and the disturbance of free-roaming herbivores and/or human intervention can then be used to make the woodland more complex.

In some areas, there may be a case for establishing native trees and shrubs that may be regionally scarce. It's easy to assume, particularly in the lowlands, that all native trees will be present somewhere in the vicinity, but they may not be. Trees of no commercial value, such as aspen, field maple, small-leaved lime, wild service, wild pear, crab apple, spindle, dogwood, alder, purging buckthorn and the wayfaring tree, have been slowly disappearing, losing out to the commercial species favoured by humans. Indeed, black poplar is now on the verge of extinction in the UK. In 2020 Plantlife launched a campaign to re-establish juniper, one of only three native conifers in the UK (alongside yew and Scots pine).[66] Once found everywhere in Britain and supporting more than a hundred specialist fungi and invertebrate species, as well as providing feeding and nesting habitat for birds such as goldcrests, firecrests and song thrushes, juniper is predicted to become extinct in lowland Britain within the next fifty years. Planting will help to prevent this from happening, until extensive rewilding on lowland chalk or limestone allows it to recover.

In other cases, it may be desirable to plant specific trees as landmarks or to replace lost specimens in a cultural landscape or park, to restore a vista or avenue, or to hide unsightly features, such as pylons, buildings or roads.

All these scenarios, while perfectly justified, should however be viewed as exceptions to the rule of natural regeneration. We must not underestimate nature's ability to regenerate even where it seems unlikely or impossible.

Engaging people with wilder trees and shrubs

There's no doubt that physically planting trees can be inspiring. The Woodland Trust and other UK conservation NGOs see tree planting as a powerful way to engage people with their work and connect children with nature. The satisfaction of watching a tree grow from your own labours confers a sense of agency, pride and optimism, and this can garner huge support for conservation projects.

But it is worth considering ways in which the public can engage with natural regeneration instead: by helping to establish thorny scrub; erecting fences to exclude deer and livestock from areas earmarked for natural regeneration; and collecting seeds from local trees, then casting them into the 'straggling bushes' as medieval foresters once did. The Wildlife Trust for Bedfordshire, Cambridgeshire and Northamptonshire, for example, extended the famous Gamlingay Wood near Sandy, to the north of London, on to an adjacent former arable field by exhorting visitors to collect seed from the shrubs and bushes in the wood in the autumn and broadcast them by hand into the open field on their walks.

Tapping into the public interest in establishing trees, and doing it in imaginative ways that promote natural regeneration, can help rewilding projects hugely, and help to shift the plantation mindset. So, too, can people engaging in 'rewilding' plantations and managing existing woodland to promote biodiversity.

Essentially, hedgerows are linear strips of thorny scrub. Rewilding them enables them to act as wildlife corridors. It also improves nesting and feeding habitat and creates seed banks for natural regeneration. Planting and restoring hedgerows, and allowing them to double in width or grow out even further, is something many smallholders can do. Studies at Knepp have shown that the relatively simple step of allowing hedgerows, with a preponderance of blackthorn, to billow out to a depth of 8–14 metres provides breeding habitat for nightingales, which – in common with many species – benefit from the protection of thorny shrubs. The cathedral-like interiors of deep hedges provide a protected place for insects to forage in the leaf litter.

This 170-metre stretch of hedge in the Southern Block at Knepp has billowed out to a depth of about 14 metres and now holds three nightingale territories. Dormice have also taken up residence.

Should we rewild existing woodland?

Buying a wood, particularly an ancient wood, as a private or community conservation project is becoming increasingly popular. Often people talk about 'rewilding' a wood, meaning either leaving it to its own devices entirely or opening it up to disturbance from free-roaming herbivores, such as pigs and cattle. Whether this is the right approach, however, must be carefully considered. Much depends on whether the wood is part of a much larger rewilding project and surrounded by recovering habitat, or a project undertaken in isolation.

In England, Wales and Northern Ireland woodland is classed as ancient if it has existed continuously since 1600 (in Scotland, that date is 1750) – before the era when planting became common. Centuries of undisturbed soils and accumulated decaying wood in these continuously wooded areas have created complex communities of plants, fungi, insects and other micro-organisms that cannot now be found elsewhere.

Just 2.5 per cent of UK land – around 600,000 hectares – is categorised as ancient woodland.[67]

Although ancient woodland may have evolved naturally, it is a mistake to consider it a functioning remnant of some primal forest, particularly if the idea of original 'forest' is closed-canopy trees. As we've seen, the idea of dense woodland as the dominant character of the landscape before human impact has been revised in favour of a more complex, dynamic, open and shifting mosaic of habitats ceaselessly influenced by wild herds of herbivores, apex predators and other natural processes. Only 20–25 per cent of the original 'wild wood' landscape is likely to have been characterised by closed-canopy trees.[68] These groves would have been constantly on the move, too, expanding at the edges and imploding over time as trees aged and collapsed. The glades created by these openings would have attracted large herbivores, expanding them further. Eventually, when they were big enough and herbivory dispersed enough, these open areas would have allowed thorny scrub to emerge, which would have acted as a nursery for a new generation of trees. The mosaic or cyclical nature of this type of succession in large areas of open woodland or wood pasture allows all species of tree and shrub – both shade-tolerant and light-demanding – to regenerate. It is the kind of woodscape that survives in the UK in the fluid landscapes of the New Forest and Hatfield, Sherwood, Epping and other forests, remnants of the medieval royal hunting forests that were analogous to the original 'wild wood'.

An 'ancient wood' today, alas, is contained within linear boundaries on a map. It is most likely small, surrounded by intensive agriculture, and disconnected from the dynamics of the greater treescape mosaic and the natural processes that would once have influenced it. It is, as the nature writer David Quammen might put it, a tiny, cut-off segment, fraying at the edges, of a once glorious Persian carpet.[69] As an isolated habitat, it is particularly vulnerable. With no soft, broad margins to act as a buffer, a wood may suffer from 'edge effect', whereby its perimeter is exposed to wind, heat and frost, chemical sprays, nitrogen drift, exhaust fumes, tyre particulates and other pollution. A wood often has a different microclimate and soil conditions at its centre from at the edges. The smaller the woodland, and the larger the relative area of its edges, the greater this edge effect.

Islands of woods, especially if surrounded by intensive agriculture, are also magnets for native and feral non-native deer. Arable and vegetable crops can support large numbers of deer for much of the year. These creatures use the woods as protective cover during the day, and when crops are harvested they move into the woods to browse the understorey,

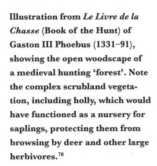

Illustration from *Le Livre de la Chasse* (Book of the Hunt) of Gaston III Phoebus (1331–91), showing the open woodscape of a medieval hunting 'forest'. Note the complex scrubland vegetation, including holly, which would have functioned as a nursery for saplings, protecting them from browsing by deer and other large herbivores.[70]

depriving ground-nesting birds and small mammals of habitat. Predators such as foxes, badgers and feral cats, moving easily through managed landscapes, zone in on these isolated pockets of woods for prey for whom the wood has become a refuge. In this way, a woodland can become a wildlife 'sink' – a patch of protective habitat that attracts species only to hasten their demise.

Also, an old or ancient wood is not an immutable or 'pristine' thing. It will almost invariably have been managed in some way, often intensively, over the centuries. It can be surprising how few trees within an ancient wood are very old. The wood's character will have changed, sometimes dramatically, according to the uses history has assigned it. The species of tree within it were selectively favoured for utility and commerce. In medieval times, as part of a wider wood-pasture landscape, domestic livestock are likely to have browsed beneath the trees, with pigs released to feed on acorns and beechmast. The trees themselves may have been pollarded or coppiced over centuries, as well as harvested for shipbuilding, roof timbers and coal-mine props. After the Second World War, in what Oliver Rackham called the 'locust years' between 1950 and 1975, fast-growing non-native conifers or beech may have been planted within it. Around half of all ancient woodland has been cleared and replanted as dense, single-generation forestry plantations, often a single species of non-native conifer, or very shade-tolerant trees, such as beech.[71]

Even if an ancient woodland has never been planted in this way, if it has been left to its own devices with no intervention to open up the canopy it will have grown darker, denser and colder, and the trees within it will have begun to shift to a more shade-tolerant community. Oak, rowan, birch and blackthorn, unable to regenerate in the deepening shadow, will give way to such species as sycamore, beech and hornbeam, which regenerate readily without direct or dappled sunlight. The darkening conditions will deprive the understorey shrubs and woodland flowers and plants of the light they need to thrive, too. There may be rare and interesting communities of saproxylic insects and fungi, but for most creatures the loss of light makes closed-canopy woodland poor habitat for feeding and breeding.

In these circumstances, a conventional conservation 'hands on' approach is likely to be best, especially if the wood is the main or only part of a restoration project and remains within a hostile landscape, such as intensive agriculture. The Woodland Trust's guide to ancient woodland restoration gives advice on how to halt the decline of biodiversity in ancient woodland and instigate gradual, long-term recovery.[72] Shade-tolerant trees that may be crowding out veteran oaks can be removed, in a process known as 'halo-thinning'. Plantation trees can be thinned or removed, glades and rides opened up, native trees promoted and recruited, and silvicultural techniques such as coppicing used to promote biodiversity. Such practices are supported by the Forestry Commission and DEFRA. But bear in mind that a licence is required from the Forestry Commission for any felling or thinning operations within woods of any description, if the tree is greater than 8 centimetres in diameter.

Many of these recommendations are about gradually and sensitively letting in the light – essentially using human management to mimic at small scale the natural processes of wind, disease and felling by bison, beavers and, in the distant past, elephants. As government guidance emphasises, 'open space, both temporary and permanent, is an important component of ancient woodlands'.[73] It is also helpful to keep fallen trees, resisting the urge to tidy them up, and leave dying and dead trees standing, provided there is no danger to rides or paths with public access.

While a wood is being restored, deer may need to be fenced out of it for a time or controlled by culling, so that the understorey – including protective brambles and thorny shrubs – has a chance to recover. Roe deer, being territorial and living in small family groups, tend to be less of a problem than fallow deer, which tend to live in woods in large numbers, and muntjac, which are much smaller and eat many native plants that are not palatable to native herbivores.

Where a wood is part of a much larger restoration project, and especially if thorny scrub has been allowed to establish over a substantial area around it, it may be possible to integrate the wood into a rewilding project. Bear in mind, however, that an ancient wood may contain very rare species and even archaeological and cultural features, such as Bronze Age hill forts, charcoal-making shelters, medieval woodbanks, green lanes and old boundary features, such as dry-stone walls. These may be sensitive to disturbance from large free-roaming animals.

At Knepp, we deer-fenced most small patches of ancient woodland to conserve these unique habitats. Where we allowed our free-roaming animals access, in younger plantations, the woodland understorey is now mostly browsed out. The bluebells that once carpeted the ground have been thinned by rootling pigs. Only poisonous wild daffodils, wood spurge and honeysuckle still thrive. But the ground flora normally associated with woodland – such as bluebells, daffodils and wood anemones – are now marching into the open areas surrounding the woods, into the tangled protection of thorny scrub.

In essence, woodland can be regarded as a reservoir of flora that, in time, will start to colonise the land around it. If or when to introduce a small number of free-roaming herbivores and/or pigs into the mix requires careful judgement. We recommend applying the rewilding approach (allowing in free-roaming large herbivores and pigs or wild boar) to existing woodland only if:

1. Careful consideration has been given to the quality of the habitat in the wood and surveys have been carried out to identify any rare or interesting species. If in doubt, consult the Woodland Trust. If the wood might be harmed by increased herbivory, large grazing animals must not be allowed in.

2. There is plenty of high-quality grazing and browsing habitat surrounding the wood – an area of, say, three to four times the size of the wood itself.

3. Stocking levels of large herbivores are based on the surrounding area, excluding the woodland.

4. The large herbivores are native or domestic analogues of native species. Avoid muntjac, sika deer, sheep and goats.

5. The soils and overall hydrology of the woodland are robust enough to cope with poaching by heavy animals, bearing in mind that, in shady conditions, the regrowth of flora may be slower in the wood than outside it.

Rewilding plantations

Timber plantations are, of course, a legitimate use of land, providing jobs and wood products that we all use (and should use more of, as we wean ourselves off plastic). Arguably, the UK should be producing more of its own timber and using it in construction for long-term carbon storage. The UK construction industry imports 80 per cent of its timber, and Britain remains the world's third-largest net importer of forest products, after China and Japan.[74]

Smaller plantations that are likely to be uneconomic, however, could be incorporated into a rewilding project. Small patches of plantation can be removed in one clear-fell operation. Such work is again subject to the appropriate licence from the Forestry Commission, which will want to ensure that there is no net loss of woodland. Permission to remove trees from a site entirely – in order to restore heathland, for example – is more challenging, and might require compensatory planting elsewhere if it is done at scale.

After felling a stand of plantation trees, native trees and shrubs (carried in by birds, small mammals and wind) can be allowed to regenerate naturally. At the same time, any regenerating seedlings of Sitka spruce or other non-native trees must be carefully removed. As with natural regeneration in ancient woodland, deer may need to be culled or fenced out until the young, self-willed trees have had a chance to establish.

It is possible to spice up the removal of plantation trees by being messier about it. Rather than removing the lot, consider leaving some trees standing – especially if they have been shattered by the trees being felled around them. Broken trunks and branches, even of Sitka spruce, provide niches for insects and birds, and, as they rot, fungi. Leaving the occasional fallen tree with its root ball exposed provides similar opportunities for wildlife, as well as the potential for a small pond to emerge in the hollow left by the roots. As always, the more uneven the surface of the ground, and the more broken and fractured the vegetation, the greater the potential for biodiversity. Go into your plantation thinking like a straight-tusked elephant!

5

Rewilding with Animals

How to balance numbers, scale and timing

Free-roaming herbivores, like these Exmoor ponies at Knepp, interact with vegetation in a way that produces complex, dynamic habitats. But it's all a question of numbers. Too many, and their impact can turn an area into species-poor grassland, like a pony paddock. Too few, and they will not be able to hold back the scrub from evolving into ubiquitous, closed-canopy woodland.

In all but the very largest landscape-scale rewilding projects, controlling the numbers of herbivores will be key to both stimulating and sustaining biodiversity.

The UK's landscape has been transformed by humans, to the detriment of wild animals. Once, vast numbers of aurochs, tarpan and elk wandered the countryside, being harried by packs of wolves. As we saw in the previous chapter, their feeding habits and other behaviours create varied and dynamic habitats – a huge stimulus for biodiversity. But much of the landscape is now agricultural or built on. Roads and railways cut through every pocket. Almost everywhere, there's a pressing need to keep free-roaming animals off agricultural land, out of forestry plantations and away from infrastructure, roads, built-up areas and private property. Even areas suitable for conservation grazing, such as national parks, heathlands and nature reserves, have to be fenced and cattle-gridded.

Perhaps, one day, when we have carved out more space for nature and been able to connect these areas using wildlife corridors – when we have moved towards the wilder end of the rewilding spectrum – megafauna will be able to roam the landscape once again in a more natural and dynamic way. But for the moment, introductions of herbivores to stimulate new habitat are limited to fixed areas, and that inevitably means a greater degree of hands-on management.

In particular, numbers must be controlled; in larger-scale rewilding projects this may be the only intervention that is required.

The ratio between herbivores and vegetation is key. On land that has been denuded of vegetation, it is best to wait until shrubs, trees and other plants have had a chance to establish before considering introducing animals. The ideal situation is a continuing battle between vegetation succession and animal disturbance – a kind of push-me-pull-you process – where neither side wins outright. The battle is dynamic and often unpredictable, resulting in losses and gains in vegetation cover over time. Vegetation may be heavily impacted in one area, yet gain ground in another. And this process will continue ad infinitum, at times more dramatically than others. It is this ever-shifting mosaic of vegetation generated, primarily, by interaction with large herbivores that results in habitat complexity – the messy margins in which life thrives. The greater the variety of large herbivores, the greater the opportunities for complexity.

Finding a balance between animals and vegetation

It is possible only through trial and error to recognise when this vegetation succession/animal disturbance dynamic is performing well, so it is wise to introduce animals in low numbers at first and allow them to build up over time, so that their impact can be gauged. Over a long period, if there are too few herbivores, they will fail to disrupt the vegetation which, in areas other than at high altitudes, will eventually progress to ubiquitous closed-canopy woodland. Too many, on the other hand, especially if they are introduced before woody vegetation is mature enough to fight back, can prevent the natural regeneration of trees and shrubs altogether, resulting in ubiquitous grassland. Both scenarios are less dynamic and biodiverse than the kaleidoscope of habitats produced by a more even battle between herbivores and vegetation.

The species and numbers of herbivores appropriate for a rewilding project will depend on the size of the landholding, its soil type(s), climate, altitude, the amount of existing grazing and vegetation ('available fodder'), and whether wild herbivores (deer, hares and rabbits) already have access. Some species – such as wild boar or pigs and beavers – have greater impact than others. If the land contains sensitive ecological sites, rewilding may not be appropriate, although such areas could be fenced out of the rewilding project, and/or virtual fencing used to manage seasonal or periodic access by herbivores.

At some point, an area of land becomes too small to sustain large, free-roaming herbivores all year round, particularly as animal welfare requires a minimum number of animals of any one species for social reasons. Having one horse, cow or pig should not be an option, and in many European countries it is illegal. The tables on pages 218–25 give a very rough sense of how scale relates to stocking density, and the tipping point where land may not be viable for large herbivores. On smaller areas animals could be brought on to the land for one season each year, or even every few years. On smaller areas still, human or mechanical intervention will be needed to mimic the impact of animals on the vegetation – in effect, the human custodian 'becomes' the large herbivore.

It's easy to feel nervous about getting the animal numbers right. Micro-managing is the default approach, particularly where livestock is concerned, and the mindset of conventional conservation is often fixated on a rather static notion of results. But rewilding is much more forgiving. Indeed, mistakes can be a good thing, mimicking the boom-and-bust cycles of herbivore die-offs and population explosions – something that benefits biodiversity. There is always a chance to adjust or change tack after a mistake.

It's important, too, to recognise that maintaining a constant number of animals over a long period is, essentially, unnatural. Varying the stocking density, allowing a vegetation 'pulse' to happen by reducing herbivore numbers for a while, then allowing them to increase again so they can hit back hard on the vegetation is a good long-term strategy. Of course, this can be more difficult for those who are selling meat from a project and depend on continuity of supply.

Understanding the behavioural traits, ecological benefits and husbandry requirements of the various herbivores available in the UK, in both wild animals and domesticated livestock as proxies for extinct species, will inform decisions about which species to include in a rewilding project (see Chapter 6). But first, it is useful to review how populations of herbivores are regulated in the wild, in order to appreciate the extent to which the rewilder can replicate these natural processes.

Population dynamics in the wild

It is often assumed that predators keep herbivore numbers down in the wild, and that this is the missing ingredient for regulating populations in the modern landscape. But in reality, predators account for only a small percentage of deaths. They exert pressure on population sizes in other ways (see page 195), and this, by increasing stress in herds, can have a lowering effect on the birth rate, as well as releasing pressure on vegetation in susceptible areas, thereby creating a more varied landscape. The presence of predators, however, is not by itself enough to regulate herbivore numbers.

What controls numbers in nature, primarily, is a mix of competition and the availability of food. In times of plenty, with good rains and sunshine at the right time and lots of vegetation growth, populations explode. In seasons when there is less to eat – notably during the dry season or drought (such as in south or east Africa), or over long, harsh winters (such as in Europe and North America) – they fall. Undernourished females will not ovulate; they may ovulate but not conceive; or they may abort, or even absorb, the foetus. Older animals – males in particular – weaken and die. In particularly harsh conditions, the whole population may be affected and a general die-off occur. A decline in herbivores releases the pressure of grazing and browsing on the vegetation, allowing a burst of regrowth when the conditions are right, and that stimulates another population spurt in the herbivores. It's an endless cycle of fluctuation.

Large herbivores often migrate in response to the seasonal availability of food, but cyclical die-offs happen even in migrating populations. And in places where animals cannot migrate – such as the Ngorongoro Crater in northern Tanzania, which has the highest density of predators in Africa – the dynamic is the same. Starvation, rather than predation, is the primary regulating factor.

Another, less common factor that affects populations is disease, such as the bacterial infection that killed off 200,000 saiga antelopes – 88 per cent of the entire global population – on the steppes of Central Asia in 2015.[1] Although much less frequent, outbreaks of disease release the pressure on food resources, often resulting in a dramatic vegetation pulse. The myxomatosis outbreak in southern England in the 1950s, which devastated rabbit populations, brought about a widespread regeneration of juniper and hawthorn. An outbreak of disease even once a century can have long-term effects. In southern Africa in the 1890s, an outbreak of the viral disease rinderpest wiped out 80–90 per cent of cattle, buffalo, eland, giraffe, wildebeest, kudu and antelope. The resulting impact on vegetation can still be seen today, in mature stands of mopane forest a hundred years old.

Managing populations in rewilding projects

People living in Europe are unused to witnessing starvation and disease in large populations of wild animals, and are unwilling to countenance it because of concern about animal welfare, especially in nature reserves. This is demonstrated by the public outcry against the Oostvaardersplassen reserve in Holland in 2018, when animals were allowed to die off in response to a seasonal shortage of grazing.[2]

Contraception has been used in some places, such as Dartmoor, to keep semi-feral herds of ponies down. But this is an expensive intervention. It restricts the normal social interaction of the herd, and therefore might also be considered an animal-rights issue. It also releases hormones into nature, which is unlikely to be wise.

The humane way to manage stocking densities in semi-natural areas, including rewilding projects, is by culling, pre-empting the onset of population stress and starvation. In Norway, for example, a country where half a million people (9.5 per cent of the population) are registered hunters, deer populations are managed to achieve optimum weight. The dressed carcass of a stag two and a half years old is expected to weigh 80 kilograms (around 20 kilograms more than in Scotland). When carcass weights go down, indicating excessive competition for food, the number of shooting licences issued rises until a population of deer of optimum weight is regained.

For rewilding projects, because it is – as yet – forbidden in Britain to leave carcasses on the land, culling also provides an opportunity to produce meat, and therefore an income, from those excess animals. However, meat production relies on regularity of numbers in order to ensure a continuous supply, and this is a consideration that runs somewhat counter to the dynamic principles of rewilding. Of course, rewilding projects must be pragmatic, but sometimes there are ways of involving randomness in a management plan to mimic the oscillating graph of wild populations. One could prioritise continuity of beef supply, for example, while allowing fluctuations in deer numbers so as to produce periodic variations in pressure on vegetation.

A little more difficult to replicate is the influence of apex predators in an ecosystem – the so-called landscape of fear. In a landscape devoid of predators (and the introduction of bear and wolves in the UK is unlikely to be countenanced any time soon), herds of herbivores relax, loosely grazing and browsing over an extended area. In the presence of predators – as the reintroduction of wolves, grizzlies and mountain lions into Yellowstone National Park in the USA has shown – they form tighter herds, for security in numbers.[3] Although they still want to graze and browse in the prime spots, such as meadows and riverbanks, they become more flighty, retreating into trees or up steep slopes at the slightest concern. They will even avoid some areas altogether, such as narrow canyons and bottlenecks or a peninsula on a lake, where they could be easily trapped by a predator. This has a significant effect on vegetation, releasing the grazing and browsing pressure in some areas, while concentrating it in others.

It may be possible to replicate some of these effects by using the new technology of virtual fencing to exclude herbivores from certain areas for a period (see page 208). There may also be some sensitive areas, such as ancient woodland or a Site of Special Scientific Interest (SSSI), that will make it worth the expense of physically fencing the animals out entirely. However, virtual fencing is, as yet, licensed only for cattle, sheep and goats. It would be impossible – and inappropriate – to put virtual fencing collars on wild deer, and the system has not yet been trialled for horses or pigs.

Deer in the landscape: the potential for rewilding

Roe and red deer are the only native large herbivores to have successfully remained at large in the British countryside since the last ice age. They have been joined in the last couple of centuries by populations of escaped exotics – sika, muntjac and Chinese water deer – as well as fallow deer, which have been present in the countryside since Norman times. Some regions, such as Norfolk, now have populations of all six species.

As deer have become more populous, their 'destructive' rather than their 'creative' impact is noticed most. Numbers of roe deer, in particular, have risen in recent decades. Having almost disappeared in the nineteenth century and again after the Second World War, they now number half a million: 350,000 in Scotland and 150,000 in England – the highest for 1,000 years.[4] These reclusive creatures use woodland for cover and, unlike fallow and red deer, live in families of two or three individuals, rather than in herds; that is, no doubt, one of the reasons for their success.

The understorey of the woods where deer shelter – mostly isolated pockets surrounded by farmland – are very quickly browsed out, denying cover to small mammals and ground-nesting birds, such as native woodcock and nightingales.[5] With food readily available from farmland crops, forestry plantations and gardens, nature's boom-and-bust cycle of starvation is unlikely to function forcefully enough to regulate deer in the wild in the UK, except when populations are extremely high. And, unlike in many countries in Europe, the British public – with the exception of sporting estates – seem to have lost their enthusiasm for hunting for the pot. With such large numbers of deer to contend with, conservationists argue for targeted culling regimes to release pressure on vegetation in key conservation areas, at least for long enough for the plants to recover.

This is being demonstrated to spectacular effect in Scotland. In the vast, denuded landscapes of the Highlands, with high and ever-growing populations of free-roaming red deer, NatureScot (formerly Scottish Natural Heritage) has encouraged landowner and deer-management groups to collaborate to control deer numbers across large areas, in order to encourage land recovery and natural regeneration.[6] This has been hugely successful, with Caledonian forest and rare montane scrub re-establishing across extensive areas, even at surprisingly high altitude. Red deer hinds (females) are generally hefted, or loyal, to a particular area or glen, so it is relatively straightforward to reduce the reproduction rate in that area; but stags will roam and, inevitably, be attracted by an area of recovering vegetation, so a regional culling approach is needed for successful land restoration. As we saw earlier, at the National Trust for Scotland's Mar Lodge Estate in Aberdeenshire, a stocking density of three red deer per square kilometre has proved to be the magic number for enabling spontaneous natural regeneration in this region.[7]

Natural regeneration is dramatic within the central deer-fenced enclosure, but it is happening outside the enclosure, too, thanks to a drop in the deer population. Scotland, June 2016.

While this type of deer management will probably always be necessary in the interests of conservation, we must also recognise that suitable deer habitat itself is lacking. With more space given over to nature, widespread reforestation and vegetation recovery, and connection between natural areas, the countryside will be able to sustain larger numbers of deer without them having such a devastating impact. Once again, food becomes the key factor. The Southern Block at Knepp, for example, an area characterised by thorny and woody scrub, sustains a much higher density of deer (as well as all the other large herbivores) than the wider countryside does, with no overall loss of vegetation cover.

'Over-grazing' and 'under-grazing'

As we have seen, assessing the stocking density that benefits the dynamism and complexity of an ecosystem is a question of trial and error, and should, wherever possible, take into account fluctuations in populations over time. In an effort to strike the right balance, to find the elusive golden rule, it is tempting to fall in with the kind of value judgement often used in conventional conservation and forestry, about the impact of herbivores on the vegetation. In particular, the terms 'over-grazing' and 'under-grazing' should be used with caution. These judgements are often associated with aesthetic sensibilities or cultural conditioning, for a particular 'look' is considered desirable or 'right'. They can also be derived from an ecologist or land manager's preference or concern for a particular species or suite of species.

'Over-grazed', for example, almost always has negative connotations. But what may be over-grazed for some species will not be so for others. On chalk grassland, an invertebrate specialist might consider the sward 'over-grazed' if it does not contain thick grassy tussocks as cover for grasshoppers and spiders, while a botanist would see paradise in closely grazed hillsides, where you can find up to fifty flower species per square metre. Even among a single group, such as the butterflies, habitat preferences vary wildly. A very short sward would be perfectly grazed for silver-spotted skipper and Adonis blue but horribly 'over-grazed' for the Duke of Burgundy and dingy skipper. Dung beetles would probably argue that there is no such thing as over-grazing. Even in conventional conservation, the terms can be meaningfully applied only when they are qualified: 'over-grazed' or 'under-grazed' for what?

In larger areas where a mosaic of habitats with complex vegetation is already established, and given the right mix of herbivores and considerably lower numbers, grazing and browsing may be distributed quite widely over the site during the course of a year. This is partly caused by seasonal variations in grass and vegetation growth, and partly because of the animals' changing requirements for nutrients, forage and shelter. But there will always be areas that animals favour over others.

'Over-grazing' or 'under-grazing' is not the biggest concern; it's *even* grazing, where the entire landscape is grazed at roughly the same level. It is vegetation patchiness – caused by a mixture of grazing and browsing levels, changing seasonally and over time – that is key to habitat complexity and hence biodiversity.

Grazing 'lawns': lessons of the New Forest

The New Forest in Hampshire, southern England, a royal forest dating back to the Domesday Book, is probably the closest analogue to the remaining wild wood and a useful illustration of the impact of free-roaming herbivores on habitat. Over the centuries herbivore numbers there have fluctuated considerably. When numbers drop, as they did in the Second World War or during the foot-and-mouth disease outbreak in 2001, a vegetation pulse results that produces the next generation of trees.

Today about 3,000 New Forest ponies (a distinct native breed), 3,000 cattle, 200–600 pigs (during the 'pannage' or acorn season each autumn), 1,300 fallow deer, 90 red deer, 350–400 roe deer and 100 sika deer graze and browse in the 375 square kilometres of Crown Lands of the New Forest.[8] Some of the areas richest in rare plants and beetles are the grazing 'launds' or lawns created and sustained by the ponies and cattle. Grazing intensity in these areas becomes a feedback loop of rising productivity. Because the animals spend more time grazing there, their dung and urine increase the fertility of the soil, enabling faster growth and recovery of plants and grasses, which, in turn, attracts the grazers. The drawback is that these lawns suffer more severely than other areas from drought during dry summers, and can be exposed to cold winds in winter because they are not buffered by the shrubs and taller vegetation that would have been present in the absence of grazers. Yet these conditions favour numerous stress-tolerant flora that cannot compete with perennial grasses and sedges: exquisite rare plants, such as yellow centaury and coral necklace, which thrive on areas that dry out regularly; pillwort and bog pimpernel on the wetter, trampled clays; and mousetail on areas that are continually trampled.

While the cattle, ponies and deer favour these grazing lawns, they do not spend all their time there. The ponies gather most of their food from rougher sources in the surrounding dry grassland, heaths and glades, including nutritious but challenging (because thorny) gorse, and the cattle browse from trees when they can. Both cattle and ponies graze on heather and the grass *Molinia*, too, although these are of poor forage value. In times of drought, both animals will venture deeper into the wet mires for fodder and browse. In the cold and wet, ponies feeding under cover of gorse seem to stay fatter and healthier than those that choose to remain on the more nutritious but more exposed grazing lawns.

It is clear, however, that the lawns are crucial to the annual economy of the animals. Without these high-quality lawns, which make up about 8 per cent of the area of the New Forest, the wider forest and heathland would not be able to sustain the herbivores throughout the year. Heathland is one of Britain's most biodiverse habitats, supporting many thousands of invertebrate species, all British reptiles and amphibians, and numerous endangered birds and plants. But only about 15 per cent of the heathland that existed in 1800 remains. The New Forest heathland is one of the last areas to be managed by naturalistic grazing. Small remnants of heathland in parts of southern England, once part of much larger, functioning ecosystems that would have included grazing lawns, are unable to support herbivores for long enough on their own. Consequently many small patches of heathland, abandoned as livestock grazing, have become overgrown with trees and simply disappeared.

The value of grazing lawns to the New Forest herds becomes even clearer when stocking levels fall, such as during the Second World War, when much of the area was requisitioned for military training, airfields and camps. Over time, across the forest, natural regeneration surges, with trees and shrubs expanding into grassland, and bracken spreading. With plenty to eat, the ponies that are left don't push so deeply into the thorn brake (areas of dense, spiny shrubs) to browse a few leaves, so oak seedlings, bramble and hawthorn take off. But the lawns receive almost the same grazing as before. Lowering grazing levels does not lead to an even drop in grazing pressure across the forest. The animals show clear preferences. When food is plentiful they ignore firstly beech, then oak, then bramble, and the lawns are the very last resource to be under-exploited. The reverse happens when stocking levels rise, until, with every other resource including the lawns depleted, the animals resort to chewing on holly bark and thorny twigs.

Some ecologists argue that grazing lawns such as those in the New Forest are possible only because of the absence of predators. In the past, they say, wolves would have released grazing pressure on the lawns by keeping the herbivores alert and cautious, even targeting preferential grazing areas for kills. But this missing fear factor can be overestimated. Herbivores will always be attracted to honeypots like grazing lawns or watering-holes, even in the presence of predators, because the benefit is worth the risk. The predators themselves, with huge territories, would not be present all the time, perhaps allowing the herbivores time to grow confident and making them easier prey. And the predators will not always pose a threat. Herbivores in Africa, for example, often graze in close proximity to lions and cheetahs that have eaten their fill. The herbivores can read the body language of sated, uninterested predators and become wary only when the predators are alert, using the opportunity, meanwhile, to graze in the prime areas.

Three different grazing scenarios at Knepp

The three different blocks of the rewilding project at Knepp evolved in slightly different ways and still have different management approaches in terms of the number and species of herbivores. Until we can create green bridges over or under the roads that separate them (an ambitious part of our original plan), these separate areas allow an interesting comparison that can inform other rewilders' decisions about how many and which animals to use, and how fast vegetation succession might be encouraged to happen. In the Northern Block, for example, which was either permanent or 'improved' pasture from the start, the colonisation of the grassy sward by woody species has been much, much slower than in the open arable fields of the Southern Block. In an effort to give the vegetation a chance to get away in the Northern Block, we introduced only cattle. Still, after twenty years, the grass has been so effective at suppressing colonisation that we're only just beginning to see hawthorn and blackthorn pulsing out from the hedgerows into the grassland, and oak saplings protected by tiny fists of bramble or gorse beginning to establish.

Land use on the fields sampled at Knepp from 1994–2019. Location of animal symbols on the timeline indicates the time of their introduction.

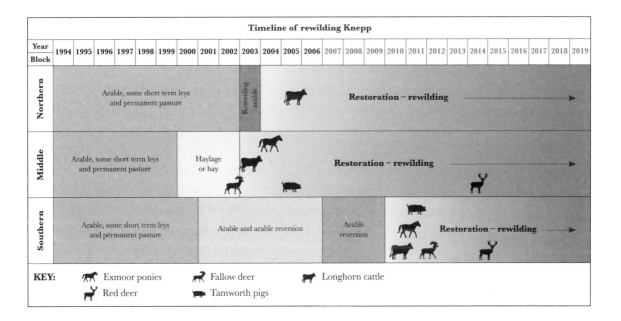

The Middle Block at Knepp began as the restoration of a nineteenth-century landscaped deer park around the castle, and is therefore particularly treated as a cultural landscape. It contains free-roaming cattle, ponies and red deer, as well as large numbers of fallow deer. This parkland is considered by some ecologists to be 'over-grazed'. Because we sowed it with grasses after it came out of arable, and introduced the herbivores immediately, there has been no chance for thorny scrub to take off, and any jay- or mouse-planted oak seedlings are browsed down almost immediately. This area has the distinctive horizontal browse lines on the trees and shrubs and, for much of the year, the tightly cropped grazing-lawn look of land-scaped parkland. Certainly, the volume of birdsong is far lower there in the spring than the deafening cacophony of the scrubland in the Southern Block. And the thorny scrub of the Southern Block is where we find our headline species – nesting nightingales and turtle doves – and our highest populations of small mammals.

Yet all three blocks have ecological values of their own. The parkland of the Middle Block is where the estate's first ravens, red kites and buzzards nested. The open sward is prime hunting ground for little owls, sparrowhawks, kestrels, peregrine falcons (which nest in the Northern Block) and bats. Canadian, greylag and Egyptian geese love the interface between the lake and the surrounding tight grassland. It's where we find green woodpeckers feasting on anthills, and hedgehogs – back after a thirty-year absence – excavating earthworms. Jackdaws and rooks hunt for grubs and worms, and in late summer swallows, house martins and swifts, storing up calories before migrating, swoop low over the grasslands to catch emerging insects. Leatherjackets – the larvae of the crane fly (and the bane of lawn enthusiasts) – feed on plant roots in short grassland, increasing microbial activity in the soil and providing a rich food source for corvids and white storks. The adult crane flies, which emerge in huge numbers for just a week or two in the autumn, feed everything from insects, spiders, amphibians and fish to birds and small mammals. Dung beetles abound. And ultimately, the condition of the herbivores themselves indicates that, for them at least, these areas are far from 'over-grazed'.

The stages of rewilding the Northern Block at Knepp (235 hectares)

Year 1

In 2004, improved grassland was allowed to revert to scrubland in the presence of a small existing population of roe deer. The few arable fields in this area were re-seeded with eight species of native grasses. In year 5, a small herd of 23 longhorn cattle was introduced.

Year 10

Cattle numbers were allowed to increase through natural breeding. In general, the grasses held back the emergence of woody shrubs. Without pig rooting or other disturbance to expose the soil, seeds of other plants found it difficult to germinate. Some blackthorn suckers moved out from hedgerows, and hawthorn, dog rose and a smattering of gorse began to appear in the grassland, and were lightly browsed by roe deer.

Year 20

The overall look is still of grassland, but here and there, woody plants have started to establish. They are kept low by the browsing of the cattle and a small population of roe deer and rabbits. The topsoil has begun to deepen as worms, dung beetles and soil biota recover.

Year 50

Speculation: by 2054, the landscape is still relatively open, but looks more like open wood pasture. Scrub and colonising saplings battle for dominance with the herbivores. Some of the early oak saplings are now strong young trees. Thorny species provide protection from browsing, allowing further saplings to grow. Dead wood – naturally fallen trees and branches – is now a conspicuous feature.

Year 100

Speculation: by the early 2100s, this landscape is mature wood pasture, with a strong cohort of oaks around 100 years old adding to the existing ancient trees. Dead wood is much more in evidence. Thorny scrub is very much part of the picture, shifting across areas in response to browsing pressure. The soil is now rich, deep and fully functioning.

The stages of rewilding the Middle Block at Knepp (277 hectares)

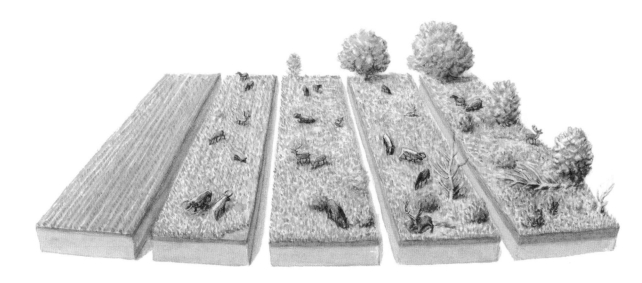

Year 1

In 2001, the Middle Block's arable fields were re-seeded with locally sourced wildflower and native grasses. For three years hay was cut and carted to lower the soil fertility. To align with the look of a historic park, fallow deer, longhorn cattle and Exmoor ponies were introduced in year 3. This meant any colonising shrubs and tree saplings were immediately browsed off.

Year 10

The Middle Block was predominantly a grazing 'lawn', like a traditional country house park, with a high browse-line on trees, hedges and understorey vegetation. Creeping thistle encouraged anthills, and was a breeding ground for common lizards and nectar source for butterflies. In 2009 a mass migration of painted lady butterflies from Morocco descended on the creeping thistle (food plant of their caterpillars) to lay their eggs, and the thistles then disappeared.

Year 20

Not much has changed in appearance above ground, apart from a few oak saplings that have somehow managed to establish. The grassland has lost most of its flowering plants to grazing, but the grasses are more complex and diverse. There are anthills everywhere. Below ground the soil is recovering, organic matter is building and more fungi are appearing. Our newly introduced white storks love stalking the parkland for grasshoppers and earthworms.

Year 50

Speculation: by 2051, more confident in our approach, we have collapsed the deer population, mimicking population dynamics in the wild, to allow colonisation of the park by woody plants. With just cattle and ponies and a few fallow and red deer, roundels of bramble and patches of scrub are appearing. The topsoil continues to deepen. A new generation of oaks emerges. Fallen branches and standing dead trees are a conspicuous feature.

Year 100

Speculation: the park now looks like mature wood pasture. The ancient oaks have been given a new lease of life and a new generation of open-grown fifty-year-old oaks have established. With lower deer numbers, a thick understorey has bounced back in the oak woodlands. Ubiquitous anthills provide alternative soil conditions for different plant communities.

The stages of rewilding the Southern Block at Knepp (450 hectares)

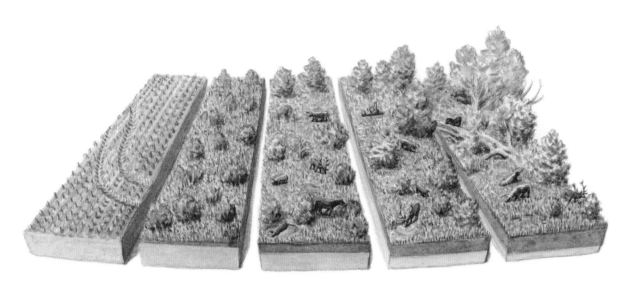

Year 1

Arable fields were left as stubble after their last harvest. The first fields to come out of arable had nine years without large herbivores. Others were cropped for six years before lying fallow for three. Without large grazers and browsers, woody shrubs, oak saplings, bramble, thistles, docks and ragwort were swift to colonise the open fields. The change is dramatic and, to the conventional eye, messy and challenging.

Year 10

Distinct differences in vegetation appeared between fields, probably relating to the year the field was left fallow. Some were colonised with dense thorny scrub; others had stands of sallow, brambles or just grassland. Our random retreat from farming created greater habitat diversity. Woodland plants began to colonise the former fields. In the ninth year, we introduced small numbers of fallow deer, longhorn cattle, Exmoor ponies, pigs and red deer.

Year 20

Herbivore numbers have increased and the battle between animal disturbance and vegetation growth has begun. Where thorny scrub dominates, jay-planted oaks are 20 feet high and bird numbers are breaking records. Fast-growing sallow has formed closed-canopy woods. The sward, though dominated by flea-bane, is more complex overall. The soil has become deep, rich and crumbly. Beavers are now free to roam, creating substantial wetland habitat.

Year 50

Speculation: the landscape is still very dynamic – an ever-shifting kaleidoscope of habitats, driven by the free-roaming animals. Patches of scrub continue to emerge and then fall away as the canopy of young self-sown trees begins to shade them out. The beavers have created hectares of new open water and channel complexes. Deadwood is ubiquitous. Topsoil continues to grow, and fungi proliferate.

Year 100

Speculation: the landscape now feels like a mature wood pasture system. The self-sown trees from a hundred years ago – some open-grown individuals; others in groves – are now gigantic. Thorny scrub has become less dominant, but is helping saplings to establish in the margins. Free-roaming animals, including introduced bison and elk, are driving the system, preventing it from turning into closed-canopy woodland.

In the Northern Block, where thorny scrub – including gorse – and oak saplings are just beginning to establish, there has been a higher uptake of flowering plants in the grassland, presumably largely owing to the absence of deer. One particular field has proved the best of the whole project for invertebrate diversity, with 200 species found in 2015.[9] Explosions of small skipper and Essex skipper butterflies have been recorded in the parkland.[10] Marbled white butterflies – a species of unimproved grassland – are common, and silver-washed fritillaries flutter in glades in the woods. As well as peregrine falcons, hobbies have successfully nested there over several years.

At Knepp, then, we consider the three different grazing regimes as having habitat distinctions of their own – the Serengeti of the Middle Block, the Okavango of the Southern Block and the ancient wood pasture of the Northern Block – which contributes to the overall heterogeneity. The important thing for us, is the mix of habitats across the project as a whole.

Equally important is the idea that these habitats will – if natural processes are allowed to play out – change over time. While conventional conservation is often focused on fixing a habitat in stasis for the benefit of certain species or suites of species, rewilding is about releasing that fixation and allowing dynamic flux. Variable management strategies can be part of that. The Advisory Board at Knepp was keen to give our stocking-density plans and species selections in the three different blocks ten years to show the results and general trajectory. Now that the project has, in a sense, proved itself as a system for nature recovery, we are feeling a little bolder about adding population dynamics into the more conventional parkland of the Middle Block, lowering the number of animals dramatically over the next ten years by culling the fallow and red deer – in effect, mimicking a natural population die-off – to see if we can stimulate the regrowth of the understorey of existing woodland, in particular. We still expect to have large expanses of closely grazed areas there, but there will be a rougher look overall, with some emergence of thorny scrub to provide greater complexity of vegetation structure within the Repton park.

Counter-intuitive though it may seem, from the point of view of nature restoration, the effects of over-grazing are easier to contend with than under-grazing. Once trees take hold, following a release in grazing pressure and in the continued absence of herbivores, the advance to closed-canopy woodland can be remarkably swift and another, very different 'catastrophic shift' takes place. It is much harder to rewild mature or even early succession closed-canopy woodland – to create heterogeneity within it – than it is to generate complexity of habitat on over-grazed land.

Seasonal and mob grazing

Another way to vary the grazing pressure over an area, or even concentrate it in others, is to manage the herbivores' access to the land, physically moving them on from one place to another using real or virtual fences. This is what happens in the wild in seasonal migrations. Grazing and browsing pressure, intensified by herds bunching up for security in the presence of predators, is released as the animals move on to pastures new. By the time the herds return, a season or so later, the vegetation has bounced back and is ready for herbivory again. Indeed, if the vegetation doesn't receive this grazing and browsing pressure, it will start to revert to domination by thuggish grasses and other plants, and/or evolve into closed-canopy woodland, and biodiversity and soil function will diminish. Vegetation needs animals as much as animals need vegetation.

This phenomenon of herd migration is echoed in the transhumance pastoral systems (whereby livestock is moved over the year from one grazing ground to another) of eastern Europe and summer grazing in the Alps, and in many other parts of the world. Inevitably, though, as traditional pasturing ranges shrink and cultures begin to change (often by force), the ability of the land to recover between grazing seasons begins to be lost.

Seasonal migration is the inspiration behind Allan Savory's livestock-management system, known as 'holistic management' or 'mob grazing' – a technique that can dramatically increase the carrying capacity of the land (in other words, the number of animals the land can support) and that has been shown to have a regenerative effect on even barren and desertified areas.[11] It has been successfully implemented on nearly 9 million hectares around the world, including in the UK.[12] At Knepp, we have recently begun mob grazing a herd of Sussex red cattle on 140 hectares of land next to the rewilding project. The herd is kept tightly bunched on a relatively small area of pasture, then moved on after a day or so to the next, and so on. After this brief spell of intensive grazing the land is left for months, even up to a year, depending on soil and climate conditions, giving it time to recover fully. The new technology of 'vence' (virtual fencing) has transformed the logistics of seasonal and mob grazing. Developed in Norway and now being used the world over, it works as an alternative to labour-intensive and unsightly electric fencing (which can also be a hazard to wildlife), and can be moved simply by resetting the boundaries via GPS on a mobile phone.[13] Animals are fitted with 'no fence' GPS collars, which emit an audio sound as they approach the virtual fence. If they get too close, the collar applies a mild electrical pulse, not even as strong as an electric fence. Mob-grazed cattle on the regenerative farm at Knepp learned within a day or so to be steered only by sound.

Mob grazing is a clever adaptation of a natural process used in holistic or regenerative farming that could also prove useful for rewilding. On dysfunctional, compacted soils depleted of nutrients, little will happen if the land is simply left to its own devices. A few thuggish plant species will tend to dominate, leaving no opportunity for the vegetation to increase in dynamism and complexity. Herbivores could kick-start the natural processes in these depleted areas, but what would attract them if there are more compelling nutrient-rich areas to feast on elsewhere?

A typical scenario would be a landholding covering both rich lowland and depleted upland areas. Given the choice, herbivores will avoid the poorer upland areas, where nutrients have been washed away over time and under poor management. They will instead graze lower down, where nutrients have concentrated. A system modelled on mob grazing, while relatively complex to manage, could, in the early stages of rewilding, ensure that the herbivores cover all the ground and don't linger too long in any one area. This would begin a process of nutrient transfer from lowland to upland via the animals' dung and urine, thanks to microbes in the dung – as well as dung beetles – stimulating soil function, and seed transfer in the gut and on the hooves and fur of the animals aiding plant colonisation in the poorer areas. The process could take several years, but would redress the balance of soil and nutrient loss, and kick-start natural processes in areas that would not, in their present state, attract herbivory.

One specific concern with this system, however, is the challenge of providing shelter for animals – particularly during hot, dry summers – in barren areas, where vegetation and naturally regenerated trees are yet to emerge. Corralling livestock in this way may also affect natural social dynamics and the animals' ability to forage naturally. For that reason, mob grazing is worth considering as an initial, short-term measure in order to restore the soil, before a more dynamic, free-roaming system is adopted.

Deer cannot be mob grazed in this way, being too wild to be fitted with collars. So far, horses and pigs have not been trialled for licence under this system.

Floral abundance and diversity under grazing

As with the grasslands of the Serengeti, only when the stocking density falls dramatically do the New Forest's grazing lawns burst unfettered into flower. During the outbreak of foot-and-mouth disease in 2001, with a drop of 20–30 per cent in cattle numbers and only the grazing ponies to contend with, carpets of camomile erupted into flower, and bluebells, anemones and other wildflowers bloomed in the bracken brakes. Many who witnessed this lovely scene argued for the removal of cattle or, at least, a dramatic reduction so that these spectacular flowerings would happen regularly.

But therein lies the paradox. Dramatic flowerings are an occasional event, triggered by a momentary pause or irregularity in grazing. Flowering plants in temperate regions depend on the long-term efficiency of herbivores to prevent thatchy grasses, scrub and trees from changing their habitat and wiping them out. Enough plants will flower in a year under intense grazing to allow the continuation of the species. If the grazing stops, the flowers may flourish spectacularly for a season or two – setting and distributing even more seed in the process – but if the grazers do not return, the plants will vanish altogether, lost to thuggish perennial grasses and sedges, and eventually closed-canopy woodland.

Visitors to Knepp are sometimes disappointed not to see carpets of wildflowers reminiscent of the glorious hay meadows of the past (Britain lost 97 per cent of its wildflower meadows between the 1930s and 1984[14]). But wildflower meadows are, fundamentally, a human construct. In Romania, where, in May, you can sit in one spot and count more than forty species of flower in just 0.1 square metre, the meadows have been managed for hundreds of years. The flower-rich grass is scythed in June or July, after the flowers have bloomed, and the seeds drop out of the hay as it dries in the field before collection.

Weather often dictates the timing of haymaking and, since different plants flower at different times, just a little annual variation in the cutting regime enhances the diversity of flora, allowing different plants to set seed every year.

Wonderful as they are as habitats in their own right – and well worth considering as a conservation project when there is just a field or two to play with – spectacular wildflower meadows are not characteristic of wild or naturalistic grazing systems in temperate regions. In rewilding terms, it can be very misleading to look at natural processes through a purely botanical lens. Carpets of flowers may be present at altitude or in barren, rocky areas. But this is where conditions are so extreme that perennial grasses, shrubs and other vegetation cannot compete, and grazing pressure is low (perhaps a few mouflon and chamois, and smaller herbivores, such as alpine marmot) or even non-existent. Blankets of bluebells in the spring – a phenomenon associated with ancient woodland – are in a sense 'unnatural'. They occur in the absence of free-roaming herbivores; wild boar (or pigs) and field voles, particularly, love eating the bulbs. Their very profusion is a strategy, used by other palatable plants such as wild garlic, to overwhelm herbivory.

At Knepp, the ancient woods and hedgerows provided a resource of wildflowers that have broadly colonised the former arable fields and flower where they can evade herbivory. However, species that weren't already present in some forgotten corner inevitably have a greater challenge to reach us. The absence of free-roaming ungulates (other than roe deer) and larger animals travelling in and out of the project is also relevant here. We are missing the vectors that would have carried seeds in their gut, coat and hooves/paws on to our land from elsewhere. Floral diversity at Knepp has therefore been relatively slow to increase, from a total number of 67 taxa (types) in 2007 to 72 in 2017. We can, however, now boast some rare species, such as water violet transported from pond to pond in the mud on herbivores' hooves or the feet and digestive tracts of wetland birds. For the first time in 2017 we spotted sharp-leaved fluellen, round-leaved fluellen and fumitory – native wildflowers with tiny, protein-rich seeds favoured by turtle doves and other 'farmland' birds. The wetland plant yellow loosestrife, which seems to have increased at Knepp since rewilding began, provides one unusual solitary bee (appropriately known as the yellow loosestrife bee) with the floral oils with which it waterproofs its nest in damp soils.

Wildflowers under herbivory cannot put on a show anywhere near as spectacular as a traditional hay meadow (largely because the flowers are eaten), and the understorey in closed-canopy woodland is also likely to be affected by herbivores. But we expect floral diversity to continue to rise at Knepp, thanks to the actions of the herbivores, particularly in the scrubland and wetland areas, and our dynamic, shifting habitats will continue to attract rare plants.

The flowers that do make themselves conspicuous at Knepp are the ones that are unpalatable to animals, such as fleabane, which currently covers a large part of the Southern Block. There, too, soil disturbance caused by the herbivores' trampling favours spectacular outbreaks of ragwort, and the Middle and Northern Blocks have swathes of creeping thistle. While hugely beneficial to insects – and birds such as linnets and goldfinches, which feast on thistle seeds and use thistledown to line their nests – these pioneer species are, conventionally, not considered worthy to be known as wildflowers. We have grown to love them.

We are used to thinking that fauna needs flora; after all, animals need plants to eat. But understanding the reciprocal relationship – that flora needs fauna – is crucial to rewilding. The way animals eat plants, stimulating the plants' defences and changing their structure and density; the way animals move seeds and nutrients from one place to another; the way they disturb the ground to provide opportunities for seeds to germinate; and the way their dung, urine and carcasses restore the soil – all this creates a complex mosaic of vegetation that generates life. But, as we have seen, it is also a question of numbers and timing. Deciding when to introduce animals into a rewilding project, which species to choose, and how many individuals of each, will affect the way a project evolves. Varying the numbers and species over time, mimicking nature's boom-and-bust scenarios, adds even more variety to the system. More than any other aspect of rewilding, large free-roaming animals, and the wild, dynamic habitats they create, will challenge the prevailing mindset. They will also bring immeasurable joy.

Putting it into practice

To fence or not to fence

Because deer are able to leap or penetrate hedges and most conventional livestock fences, and even swim across water, they can easily access most areas of land. This can be an advantage or disadvantage for rewilding, depending on the density and species of the local deer population and the condition of the land in question.

Having a porous boundary will save the cost of a deer fence. The minimum fence height for protection against red, sika or fallow deer is 1.8 metres; for roe deer it's 1.5 metres – so it ends up being very expensive, and corner posts will push the cost up further. However, you may still want to consider supplementing an existing population of deer with other herbivores, such as cattle, ponies and pigs, for different disturbances on soil and vegetation, and that will involve some sort of livestock fencing, which is less expensive.

But even if you plan to have only deer in your project – and especially if you need time to establish vegetation on the land and are in an area with high populations of deer – you may need to try to control the ingress of deer for a time, either by culling or by excluding them with a deer fence. On grassland, even low numbers of deer can suppress natural regeneration, so control or exclusion may be necessary, at least to begin with.

The smaller deer species, however – roe, muntjac and Chinese water deer – are almost impossible to fence out. Like badgers, they will eventually find their way under the fencing, or slip between fence and gatepost. If your rewilding project receives Countryside Stewardship or other government funding, you will be required to try to eradicate non-native sika, muntjac and Chinese water deer on your land. In practice, however, if your neighbours are not so assiduous, and particularly when your rewilding project starts presenting enticing habitat for them, they may continue to sneak in. Muntjac and Chinese water deer, being reclusive and tiny, can easily vanish into thorny thickets.

At some point, when the vegetation appears to be 'getting away', you will want to introduce some serious herbivory: time for battle to commence. At Knepp, in order to have as broad a spectrum of herbivore disturbance as possible, we decided to introduce fallow and red deer, as well as free-roaming cattle, ponies and pigs. This meant investing in a deer fence to secure our boundaries and protect surrounding roads, farmland and neighbours' gardens from wandering herds of deer. If we had been in an area with significant numbers of fallow and/or red deer, we would have simply used a conventional fence to contain the livestock and allowed the wild deer to come and go, leaping the fence, as they pleased.

However, this last approach introduces another variable that can make gauging the appropriate stocking levels difficult. Without knowing how many deer are accessing your land, it can be difficult to predict the fodder that will be necessary to support other species you may be managing, such as cattle and ponies. Ultimately, a deer fence allows more control over the numbers of deer on your property.

As always, having as wide a variety of herbivore species as possible provides the greatest opportunities for habitat complexity. Given that the non-natives – sika, muntjac and Chinese water deer – are not to be encouraged, and roe do not lend themselves well to capture and release, red and fallow are the two deer species to consider as introductions in a fenced rewilding project.

Clearly, deer fences – as well as sheep and pig netting – are less permeable for such creatures as badgers and foxes than more open livestock fences are, although these wily mammals often find their own way through. In the Highlands of Scotland, deer fences around regeneration blocks have proved lethal for capercaillie and black grouse, which collide with them as they fly low and fast through the trees. Deer fences are also unsightly and can present a hostile message, although the visual impact can be mitigated by using hedges, the edges of woodland and such natural topography as dips and hollows to direct them. The managers of the Mar Lodge Estate in Aberdeenshire prefer to use culling to control the impact of deer whenever possible, but where total exclusion is required, they use two parallel, low fences offset from each other.[15] Red deer, with their eyes on the sides of their head, find it difficult to judge this distance to jump. A fence of this kind doesn't kill woodland grouse, and allows the free movement of other creatures. It is also much easier than a conventional deer fence to take down and relocate. The approach has helped to improve Mar Lodge's relations with other estates. In his book *Regeneration: The Rescue of a Wild Land* (2021), the estate's ecologist, Andrew Painting, quotes the poet Robert Frost: 'good fences make good neighbours'.[16] The offset-fence system, however, is effective only for red deer.

Another consideration is the opportunity provided by fence posts as perches for raptors and corvids. Particularly over open, flat land with ground-nesting birds, this can increase predation. It is sometimes suggested that, as a deterrent, nails can be driven into the top of fence posts, with the points protruding upwards, but there is no clear evidence that this works.[17] Most important for improving the chances of prey species avoiding predation is habitat. Where lapwing productivity and numbers are low on RSPB reserves, for example, the charity concentrates on improving the habitat, as well as limiting access by foxes through the use of electric fences.[18]

If you're planning to introduce cattle and/or horses, and are happy to allow local deer in, a simple cattle fence will be all you need and the deer will come and go as they please. If you're including pigs, the cattle fence can be fortified with a couple of additional lower strands of barbed wire or pig netting. Adventurous piglets may squeeze through the lower wires but they'll always want to wriggle back in to get to their mother.

Obviously, a fence that is intended to keep deer and/or livestock out of an area to allow time for natural regeneration can be used to keep animals in later, when the emerging vegetation could do with some browsing and grazing. If this is a possibility, it is worth designing the fence for longevity.

Stocking density

The question potential rewilders often ask us is 'How many free-roaming animals should I have on my land?' It's almost impossible to answer, because every situation will be governed by a different set of variables, and rewilding should, anyway – if possible – involve fluctuations in populations over time. But we have come up with a very rough formula based on our experience at Knepp. This gives some idea of possible stocking levels, and can help identify some parameters for a rewilding project.

We used the Southern Block at Knepp, an area of roughly 440 hectares, as our model because that is where we have the most dynamic battle between vegetation and animal disturbance and, hence, the greatest biodiversity. It is where, at the outset, we allowed a vegetation pulse of thorny scrub, trees and other plant communities to evolve on ex-arable fields over a period of between four and nine years before introducing free-roaming herbivores. The vegetation was well established and able to produce defensive mechanisms, such as thorns and tannins, by the time the animals arrived. The dynamic, complex habitat of the Southern Block is able to sustain a suite of six species (roe, red and fallow deer, cattle, ponies and pigs) roaming freely in socially cohesive herds all year round. By comparison, in the parkland restoration of the Middle Block, there are no roe deer at all. We have removed the pigs there – for the time being, at least – because their rooting was damaging the parkland aesthetic and depriving other herbivores of pasture. The extensive grassland in the Middle Block, however, can sustain much higher numbers of fallow deer.

The number of each species in the Southern Block has grown over time. At the outset we introduced our three large herbivore species in low numbers – 53 cattle, 23 ponies and 42 fallow deer – to supplement the small existing population of roe deer (we also started off with 20 pigs, but have since reduced the number to 6 or 7 breeding sows). However, as the established woody scrub began to get away, our Advisory Board recommended increasing the fallow population and introducing red deer for greater browsing impact, so as to prevent the scrub from taking over completely and, eventually, turning into closed-canopy woodland.

We intended to leave the deer to fend for themselves. However, we wanted to ensure that our semi-feral Exmoor ponies, old English long-horn cattle and Tamworth pigs had enough grass and vegetation to survive all year round without supplementary feeding. That allows us to adopt a relatively 'hands-off' approach and allow the animals to drive natural processes.

Body Condition Scoring

The only major intervention at Knepp is culling the animals to keep the stocking density to a level that ensures there is enough food for them. We do this, principally, according to the condition of the cattle. Of all the available free-roaming herbivores for rewilding, cattle are the least hardy. Having been domesticated for so long and bred so intensively, they have lost some of their ancestral skills and toughness. Many cattle breeds, for example, have lost the ability to dig in the snow to find grass – something the Exmoor ponies, deer and pigs still do. As ruminants (animals that chew cud), they generally need less food than, say, horses, but their complex digestive system needs softer forage and more water.[19] For that reason, we base our overall stocking rate on the number of cattle we reckon can come through the harshest, longest, wettest possible winter without supplementary feeding, and retain satisfactory body condition scores.

Body Condition Scoring (BCS) is a veterinary management tool, based on observation, which uses a numeric score between 1 and 5 (in some systems, 1–9) to estimate the relative fatness or body condition, 5 being too fat and 1 too thin. Body condition varies over the year, particularly in a naturalistic grazing system where the cattle overwinter outside and receive no supplementary feeding. Individuals in a herd can have different scores, too: sometimes to do with genetics (thinner animals may be less suited to rewilding, in which case you would identify these for culling); sometimes because of a temporary affliction, such as intestinal worms. An average score is therefore worked out for the herd as a whole. At Knepp we hope for a score of 4–5 after a good summer's grazing, so that the animals enter winter with good fat levels; by the end of a long, harsh winter the score may have dropped to 2. This is significantly lower than intensive feedlot farming, but it is far more natural. Studies show that a cycle of weight loss and gain may actually be beneficial to the animals' health. Evolved over millennia to cope with the boom and bust of the seasons, the metabolism of herbivores may be unsuited to high calorific intake all year round. The test is in the animals' overall health at the end of the winter, and how quickly they put on weight with the first flush of spring grass and new foliage. In this naturalistic system, we cull the cattle at the beginning of autumn, after they have benefited from the spring and summer growth of grass and plants, and before they go into the lean months of winter.

Assessing the stocking density is always a question of trial and error, as well as personal judgement, but because we built up our cattle herds gradually from very small numbers it was relatively easy to identify the tipping point. One year it became clear that the numbers had increased too far: the BCS of the cattle had fallen and their trampling impact over the winter had increased beyond what we thought the land could sustain in the long term. We therefore drew back a little, dropping the cattle numbers by 20 per cent in the Northern Block (where only cattle are present) and 10 per cent in the Middle and Southern blocks, where we also dropped the number of deer to lower the competition for forage.

A further important indicator is the carcass condition of the deer, which, unlike the livestock, we cull on site. A qualified marksman does this from high seats, fixed to trees. He randomises the sites he shoots from, so the deer do not begin to avoid a particular area. Because we cull the deer from August until the end of February, this gives us a very accurate health check and body score for both fallow and reds right through the winter.

Livestock units

The stocking density that we now reckon is the right level for year-round grazing in the Southern Block (on heavy clay) is 0.27 livestock unit (LU) per hectare. This reckoning appears, perhaps not unsurprisingly, to tally with the ratio of herbivory that seems to work well for biodiversity, ensuring the impact on vegetation *overall* remains beneficial (stimulating vegetation complexity) rather than detrimental (almost or entirely preventing vegetation succession). Of course, some areas are naturally more heavily targeted by the animals for grazing and browsing, either through ease of access or for the plants or minerals they contain, and other areas are not so popular (disturbance from walkers, particularly dog walkers, on the footpaths, for example, has an effect on the deer and ponies but not on the pigs or cattle), and this variation of grazing and browsing impacts, which may also be seasonal, is important for the habitat mosaic.

The schematic of livestock units is a little technical but it is extremely useful as a guide for comparing different grazing regimes and stocking densities. A livestock unit is based on a single animal's feed requirements and/or its weight, and different schemes across the world use different measurements based on their dominant livestock, generally either a cow or sheep. The simplest UK system is the government's 2006 formula, which equates one livestock unit (LU) to the liveweight (LW) of the average dairy cow (650 kilograms) – which is why livestock units are sometimes referred to as 'cow equivalents' (ce). All other livestock are then calculated against the dairy cow yardstick. So, for example, the average ewe, weighing 98 kilograms, equates to 0.15 LU. Using livestock units in this way shows that you can have many more sheep grazing a given area than cows. Put simply, in a low-input farming system, a 10-hectare pasture might be able to support 15 cows or 99 ewes – both scenarios have an equivalent of 15 LUs, or 1.5 LUs per hectare.

The tables that follow show some different scenarios and potential LUs per hectare in low-intensity farming, intensive farming and Knepp-style rewilding.

Using sheep and cows as an example, in a low-input farming system, a 10-hectare field might support 14 cows or 93 ewes, equating to 1.4 LU/ hectare.

Low-input farming: a 10-hectare field with 1.4 LUs per hectare

	Weight			LU value per animal species	Total LU of animals in 1 hectare		Total
Ewe	**98 kg**	98/650kg	=	**0.15** LU per ha / LU value per species	**9.3** head per ha × number of ha	=	**93** ewes
Cow	**650 kg**	650/650kg	=	**1.00** LU per ha / LU value per species	**1.4** head per ha × number of ha	=	**14** cows

Under an intensive farming system, where you have more sheep and cows in the 10-hectare field, the LU is higher. Part of that intensive system would be taking the animals indoors over winter and feeding them silage from the field. The field would be modern rye grasses with lots of artificial fertiliser.

Intensive farming: a 10-hectare field with 4.2 LUs per hectare

	Weight			LU value per animal species	Total LU of animals in 1 hectare		Total
Ewe	**98 kg**	98/650kg	=	**0.15** LU per ha / LU value per species	**27.9** head per ha × number of ha	=	**279** ewes
Cow	**650 kg**	650/650kg	=	**1.00** LU per ha / LU value per species	**4.2** head per ha × number of ha	=	**42** cows

Under the Knepp rewilding model, the average LU for the whole project is around 0.3 per hectare – less than one tenth of intensive farming. Note that we are not actually using sheep in the rewilding project at Knepp.

The Knepp rewilding model: a 10-hectare field with 0.3 LUs per hectare

	Weight			LU value per animal species	Total LU of animals in 1 hectare		Total
Ewe	**98 kg**	98/650kg	=	**0.15** LU per ha / LU value per species	**2.0** head per ha × number of ha	=	**20** ewes
Cow	**650 kg**	650/650kg	=	**1.00** LU per ha / LU value per species	**0.3** head per ha × number of ha	=	**3** cows

LUs become really useful in calculating the stocking density for several species in a given area. Natural England, for example, stipulates that, for conservation grazing with cows, ponies, sheep or deer, or all four, there should be no more than 1.4 LU per hectare. That figure is much lower than for farming systems.

To calculate the stocking density at Knepp we have given all our animals an LU measured against the 650-kilogram dairy-cow unit. Our longhorn cows are slightly smaller than a dairy cow, at about 600 kilograms, so they equate to 0.92 LU; longhorn bulls are heavier, at 1,000 kilograms, so they equate to 1.54 LU; and so on. Because we are required by law to manage the cattle as domestic livestock, all the data (births, deaths, slaughter weights, movements in and out of the project, and so on) is entered into a software programme.

The Exmoor ponies, too, are closely monitored. A pony is about half the weight of a dairy cow, equating to 0.46 LU for a mare, 0.51 for a stallion, 0.4 for a filly, and so on.

Because the deer are essentially wild animals, we can't monitor them from birth in the same way as we do the cattle or ponies. We can, however, be accurate about the weight of the adult fallow and red deer, because they are culled on site. Our deer LUs are therefore based on adult weight only. Multiplying the weight of the 'dressed' deer carcass (without guts, feet and head, but with skin) by 1.26 gives the liveweight equivalent, from which the LU is calculated: 0.05 LU for a fallow doe; 0.08 for a fallow buck; 0.12 for a red deer hind; and 0.14 for a red deer stag.

Tamworth pigs are more difficult to quantify. Because they are livestock, we're able to monitor their weights in a similar way to the cattle and ponies, but because their disturbance in pasture is so much greater than the purely herbivorous species, the ratio of their weight to environmental impact is higher. They are also a relatively new component in conservation grazing, so there are few studies to go by. Pigs graze throughout the spring and summer, but, being omnivores, they also rootle all year round – whenever the ground is soft enough – nosing up the turf in search of insects, roots and worms. These rootled areas effectively remove pasture grazing from the system until the land has recovered, which can take four or five months. To factor in their additional impact we've added 0.6 to the LU calculated by weight. This is an estimate, but it gives us a yardstick. The number of pigs will always be much lower than that of other animals because of this impact.

The system we've evolved from our experience at Knepp is far from perfect, but the important thing is that we stick to a particular method of LU calculation. That will allow us to identify patterns over time.

Livestock units (1 LU = 650kg LW)		Average live-weight (LW) in kg	Livestock unit value
Longhorns	Cows	600	0.92
	Bulls	1000	1.54
	Heifers (25–32m)	550	0.85
	Steers (25–32m)	600	0.92
	Heifers (13–24m)	350	0.54
	Steers (13–24m)	400	0.62
	Calves (7–12m)	225	0.35
	Calves (0–6m)	50	0.08
Fallow deer	Does	34	0.05
	Bucks	54	0.08
Red deer	Hinds	81	0.12
	Stags	92	0.14
Exmoors	Fillies	260	0.40
	Colts	280	0.43
	Mares	300	0.46
	Geldings	330	0.51
	Stallions	330	0.51
Tamworths	Sows	200	0.91
	Boars	250	0.98
	Weaners/porkers	50	0.38

Totalling all the LUs of the five species of animal in the Southern Block and dividing them by the area (440 hectares), gives an LU of 0.27 per hectare. This is the average over the whole year, taking into account offspring arriving in early spring/summer, and the autumn cull. This gives a very basic idea of the stocking density that works for rewilding on heavy clay soil.

This is significantly lower than the 1.4 LU stipulated by Natural England for conservation grazing (where animals are often supplementary fed and/or removed from the land in the winter). It is also, of course, far lower than conventional farming systems, which can be as much as 5–7 LU per hectare.

Stocking density on heavy clay in the Southern Block at Knepp (large scale)

Area of land	**440** hectares		**1087** acres	
Average livestock units	**0.27** LUs		**119** total LUs	
Type of soil	**Heavy clay**			

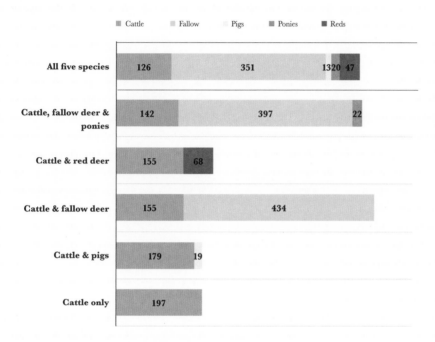

Legend: ■ Cattle ■ Fallow ■ Pigs ■ Ponies ■ Reds

- **All five species:** Cattle 126 | Fallow 351 | Ponies 13 | 20 | Reds 47
- **Cattle, fallow deer & ponies:** Cattle 142 | Fallow 397 | 22
- **Cattle & red deer:** Cattle 155 | Reds 68
- **Cattle & fallow deer:** Cattle 155 | Fallow 434
- **Cattle & pigs:** Cattle 179 | Pigs 19
- **Cattle only:** Cattle 197

The top bar shows the animals we actually have in the Southern Block at Knepp.

The bars underneath show some of the options you could have if you decreased the number of species.

This bar chart shows some of the many permutations that will maintain the same stocking density (0.27 LU per hectare) with different combinations of species. But there are a few other observations to make. In the cattle, fallow and ponies scenario we have kept the number of ponies roughly the same (20 in the first bar; 22 in the second). One could, in theory, have more ponies than this, but (as we discuss in Chapter 6) managing a population of feral ponies can be difficult, because there is such a low market for them. Have too few ponies, however – especially if you take out the stallion – and you lose the social dynamics of the herd.

In the cattle and pigs scenario, with 19 pigs (6 more than we currently have in the Southern Block), the increased ground disturbance from rootling could affect the grazing availability for the cattle. According to our rough reckoning, in a wet winter on heavy clay, a single sow can rootle about 20 hectares.

Clearly, it is not just the area of land but also the available fodder that affects stocking density. Most herbivores, including cattle and horses, will turn more towards browsing – even nibbling bark on fallen and living trees – as grasses and herb plants diminish over the winter, so the amount of complex woody vegetation is important. Bark, leaves and twigs will also have medicinal benefits for the animals.[20] Sallow, for example, contains salicylic acid, the main component of aspirin. It provides relief from aches and inflammation, as well as relieving animals of worms. Eating bark and twigs has also been proven to ease digestion and reduce the amount of methane the animals produce.[21] But shrubs and trees take up grazing space, so animals that predominantly feed on grass may lose out if there is less pasture.

Stocking density on free-draining soil (large scale)

Area of land	**440**	hectares	**1087**	acres
Average livestock units	**0.43**	LUs	**189**	total LUs
Type of soil	**Free-draining**			

The numbers of animals you can have on free-draining soil are likely to be higher than on heavy clay. Even with a slight rise in LU per hectare, the numbers change considerably, broadening your options for numbers of species and dynamic herds.

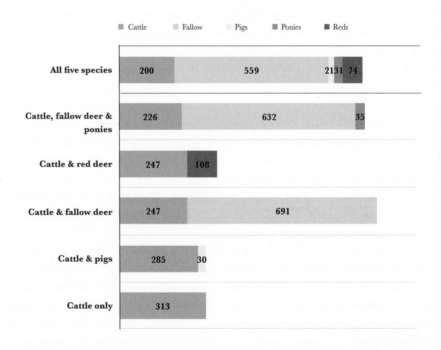

Soil conditions will also determine the stocking density. Heavy Sussex clay is very easily 'poached' – turned into mud by trampling – during wet periods, not just in the winter. On free-draining soils, trampling is likely to have much less impact on grass and grassland plants. This means more year-round grazing will be available, allowing the possibility of a higher stocking density. Again, trial and error is the only way to find out what works on your land. If you increase the herd sizes slowly and carefully while watching the impact on the land and vegetation, you may find you can sustain a higher LU per hectare while maintaining the benefits you want for habitat creation.

Stocking density on heavy clay soil (smaller scale)

Area of land	**40** hectares	**99** acres		
Average livestock units	**0.27** LUs	**11** total LUs		
Type of soil	**Heavy clay**			

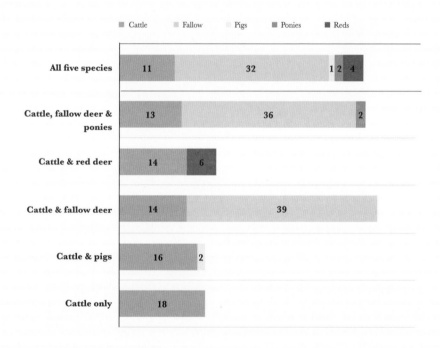

As the area of land diminishes, so do the options in terms of species mix. On the same soil as Knepp, with the same stocking density, having five species on 40 ha starts to become problematic. For animal welfare reasons you would never want to have fewer than 2 pigs, 3 horses and 4 red deer but, at these low levels, you have lost natural herd structure and dynamics.

The smaller the area, obviously, the more restricted the options in terms of number of species and herd size. On heavy clay with plenty of scrub habitat, with stocking levels at around 0.27 LU per hectare, a 40-hectare site would support a very small number of animals all year round. Even if you have only cattle, the herd would amount to just eighteen head.

This is where herd dynamics suffer. Horses are a good example: a herd any smaller than, say, about fifteen ponies – especially if you remove the stallions, so that there is no breeding – can become passive and sedentary. Instead of charging around, play-fighting and contesting dominance, with the mares using energy to feed their foals, the herd can become slower and lazier. Without natural social stress the time the ponies spend grazing can rise, leading to a risk of laminitis, an extremely painful disease whereby the whole body, and particularly the feet, becomes inflamed. It is caused by carbohydrate overload. All this has an effect on ground and vegetation disturbance.

In terms of welfare, no herd animal should ever be on its own, but they can still feel vulnerable in very low numbers. As a rule of thumb, we recommend the following.

Cattle: No fewer than four cows and their young; bring in a bull when needed.
Horses: No fewer than four mares and their young; bring in a stallion when needed.
Red deer: No fewer than four hinds and their young; no fewer than one mature stag.
Fallow deer: No fewer than four does and their young; no fewer than one mature buck.
Pigs: No fewer than two sows and their young; bring in a boar when needed.

With livestock you can bring in a bull, stallion or boar when you want to breed. Deer, being wild, cannot be handled in this way; in a closed system surrounded by a deer fence, you'll need to run at least one red stag or fallow buck with the females.

Clearly, however, stocking animals in such low numbers is a big step down from the dynamics of a natural herd system. At around the 40-hectare mark there comes the question of whether to have more species but fewer of each, to take advantage of the different impact each species has on the environment, or to have fewer species in higher numbers, for a more natural herd structure – which, again, will have a different effect on the environment.

At this size, soil type can become a more crucial factor. On 10 hectares, on free-draining soil, for example, you could have – just about – three relatively large and dynamic herds of cattle, fallow deer and ponies, for a good compromise between number of species and number of animals.

You could add dynamism by bringing in two or three pigs every now and again for a bit of rootling.

Stocking density on heavy clay soil (small scale)

Area of land	**10** hectares	**25** acres		
Average livestock units	**0.27** LUs	**3** total LUs		
Type of soil	**Heavy clay**			

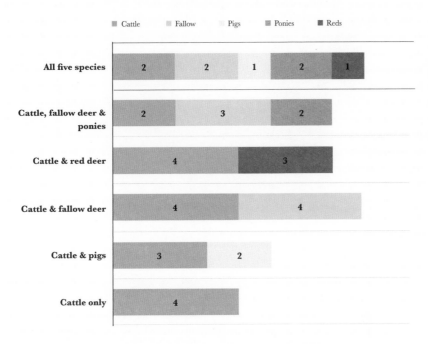

At 10 hectares, having free-roaming animals all year round starts to become unviable. The numbers are too low for natural herds, unless you have just a small herd of fallow deer. This is the point at which seasonal grazing should be considered.

At around 10 hectares, where it is not really viable to have free-roaming animals all year round, you could consider bringing in animals for some seasonal disturbance for just a few weeks or months each year, or even every few years. At this size, the potential for wild populations of deer to access the land also becomes more significant, since this may be the only herbivore disturbance you can handle. The potential for joining forces with neighbouring land also becomes important, and could determine whether you can be part of a larger rewilding project with free-roaming herbivores, or if you go it alone with seasonal grazing and browsing on your own land, and/or mimic their disturbance yourself by mechanical or other means.

6

Types of Herbivore

Traits, ecological benefits and management

Large herbivores are ecosystem engineers. The way they disturb the ground, their different eating preferences and techniques, their dynamic social and territorial behaviour, the way they transport seeds and nutrients from one place to another, the way their dung and urine replenishes the soil; all these things help create new habitats and niches for other creatures.

Their impact can kickstart natural processes again on even the most depleted and static land. And the more herbivore species there are, the greater complexity they generate in the ecosystem.

'Where species are extinct, consequent changes in the ecosystem can indicate a need to restore the ecological function provided by the lost species; this would constitute justification for exploring an ecological replacement.'
Guidelines for Reintroductions and Other Conservation Translocations,
IUCN Species Survival Commission, 2013

To some extent, all large herbivores are keystone species. The disturbance caused by large animals is rocket fuel for biodiversity, principally because it generates open, sunlit conditions by limiting competitive perennial plants such as trees, bracken, brambles and grasses. This creates vegetation complexity and movement, and the kind of messy margin where life thrives. Many species, including metapopulations of small mammals and invertebrates with complex life cycles, require different habitats and food sources at different stages of their lives to survive and reproduce. They benefit from the random niches and fringes that large animals create. Conventional conservation finds it difficult to replicate such natural dynamism and complexity. In our quest to stabilise shrinking, deteriorating habitats and safeguard particular endangered species, we have tried – at enormous expense and effort – to restrict the variables, and lock nature in stasis. This approach has been vital for preserving snatches of habitat and species on the brink of extinction. In many ways, nature reserves (often small and isolated) are our Noah's arks, but they are also, generally, enterprises of superhuman endeavour, crazy cost, over-simplification and control.

Introductions of free-roaming animals can indeed drive and sustain complex habitats within nature-conservation areas, shifting the burden and expense of management from human shoulders. But they can also kick-start dynamic natural processes on land that, at the start, has little or no value for nature – like that at Knepp.

The role of large herbivores as keystone species is very differ-ent from that of apex predators. Apex predators are critical to the dynamics of trophic cascades, whereas large herbivores are ecosystem engineers. Much of the land on Earth has, thanks to human interven-tion, undergone what scientists call a 'catastrophic shift'. Hydrology, soil types and communities of plants, animals and invertebrates have moved, over time, to a depleted state of equilibrium. Left to its own devices, this land could take tens if not hundreds of thousands of years to re-establish its former biodiversity. Apex predators on their own cannot change this. Put a pack of wolves or a lynx into a planta-tion or on to depleted farmland and – provided they can survive there – they will be relatively ineffectual at altering the habitat. Put bison, elk, beavers, wild boar, cattle, water buffalo and ponies there, on the other hand, and they will immediately shake things up and start generating different habitats: opening up closed-canopy trees by ring-barking (gnawing the bark) and felling them, rootling up the ground, trampling, puddling, tearing off branches, building dams, browsing and stimulating vegetation regrowth, and transporting and evacuating seeds across the landscape in soil-enriching dung.

The original native animals of a landscape are the ideal ones to use for rewilding, but often these species are severely depleted or even extinct. 'Proxies' (domesticated descendants or close relatives of the original species) are successfully used in rewilding projects across the world. In some places, quite foreign species – such as camels and water buffalo in Australia – benefit the ecosystem, in this case replac-ing extinct marsupials.[1]

One species that divides rewilders and conservationists in the UK is sheep. Sheep have been present as domesticated livestock in Britain for thousands of years, but they are non-natives (being descendants of mouflon from the stony, arid mountains of Mesopotamia), so the flora has not evolved defences against their specific browsing and grazing techniques in the same way that it has for native species. Sheep are, however, used to good effect in conservation grazing – on chalk grassland, for example – as well as low-intensity grazing in traditional parkland and churchyards. As always, it's a question of impact: how many animals in a given area, and for how long. There may well be a role for sheep in rewilding projects in the UK, but for our experiment at Knepp, with heavy clay soil that is unsuitable for the year-round grazing of sheep, we decided to stick with native species and their domesticated descendants.

Large herbivores range from specialist grazers, such as antelope, which feed predominantly on grass and wildflowers, to browsers such as roe deer, which feed almost exclusively on woody vegetation. Many are somewhere in the middle, habitually browsing and grazing. But even beavers, famous for gnawing trees, eat grass in summer; and wild horses and bovines take advantage of tree fodder in winter.

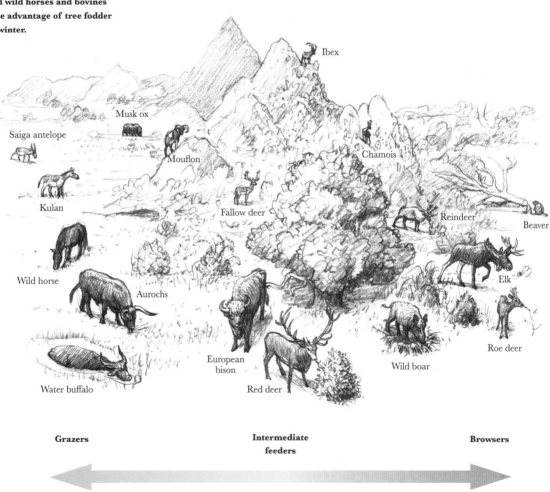

Ibex

Musk ox

Saiga antelope

Chamois

Mouflon

Kulan

Fallow deer

Reindeer

Beaver

Wild horse

Elk

Aurochs

Roe deer

European bison

Wild boar

Water buffalo

Red deer

Grazers

Intermediate feeders

Browsers

Horses and cattle

After the last ice age, in many original ecosystems in temperate-zone Europe where there were no saiga antelope, wild ass or water buffalo, the only predominantly grazing herbivores were aurochs and tarpan. Other species that might have been present, such as musk ox, fallow deer, red deer, mouflon, ibex, European bison, chamois, wild boar and reindeer, are 'intermediate feeders' – that is, they browse as much as graze – or are purely browsers, such as elk and roe deer.

The old English longhorn (bottom left), an old breed of cattle, looks remarkably similar to its ancestor, the aurochs, depicted in cave paintings in Lascaux, France, around 17,300 years ago (top left). The Exmoor pony (bottom right) is also strikingly reminiscent of its ancestor, the tarpan (top right), immortalised in the same caves.

Habitat creation

Grazing by aurochs and wild horses consequently played a crucial role in shaping the grasslands and semi-open forest ecosystems of post-Ice Age Europe. It is no coincidence that it was these two species among the larger herbivores that were domesticated, and that subsequently became extinct in the wild. They lived in precisely the open grassy areas where humans wanted to keep their livestock or grow their crops.

The different grazing techniques of the two species – the aurochs with its long tongue wrapping around longer grasses and its ruminant digestive system developed for digesting cellulose; the tarpan with its nipping, forward-pointing teeth for cropping shorter grasses, forbs and thistles, and its ability to digest dead and nutrient-poor grassy thatch – complemented each other. We are only just beginning to understand this interdependent equine/bovine relationship. Studies in Kenya show that cattle put on more weight and condition if donkeys and zebra graze alongside them than if they graze the same area on their own. Worm infestations, and hence the incidence of disease, have proved to be lower in both cattle and horses if they graze together.[2]

Horses use their forward-pointing front teeth for precision nibbling; while a cow, with no top front teeth, uses its long, wrapping tongue to rip and tear grass and vegetation

In terms of rewilding and the potential for habitat creation, the disturbance and grazing effects of cattle and horses are fundamental, and the impact they have together – if both can be accommodated – will be exponential.

Selection criteria

Whether the plan is to introduce cattle and/or horses at key stages in the rewilding process, to allow them to roam free for just a few months every year, or to have them outside all year round, hardy, primitive breeds will be required, with the old survival traits.[3]

These include:

- Being well adapted to the terrain and local climate, preferably a regional breed
- Taking easily to foraging, without the need for supplementary feeding
- The ability to store fat reserves to draw on in hard times
- No need for artificial shelter
- The ability to grow thick winter coats and shed them efficiently for the summer (where appropriate)
- Being robust enough to require minimum veterinary attention, and preferably resilience to the more common diseases
- The ability to calve or foal without human assistance
- Well-developed maternal instincts
- Alert and confident behaviour

It is also worth considering whether they will look the part – that is to say, 'read' well in a wilder landscape, and have less of a 'farmyard' appearance. Belted Galloway cattle, for example, with their designer white belts around the belly, do well as a species for rewilding but may look a little contrived. At Knepp we've chosen old English long-horns partly because of the wild look of their sweeping aurochs-like horns – also a tool for disturbance – and their mottled, woolly coats. 'Dolly mixtures' of different horse breeds may function just as well for rewilding, but visually you may feel they don't hang together as a wild, integrated herd. This is all subjective, of course, but it feeds into the public perception about rewilding. In the wild, any animal that stands out from the crowd can be picked off more easily by predators, so herd animals tend to have the same or similar coloration.

A breed that has historic associations with the region will add to the rewilding story, conjuring up a spirit of place and exciting local interest and pride.[4] Government funding may also be available under the 'native breeds at risk' programme.[5] The Grazing Animals Project, run by the Rare Breeds Survival Trust, offers courses, information and advice on conservation grazing.[6]

Domesticated cattle

Descended from the extinct aurochs, cattle – particularly old or primitive breeds – retain many of the traits of their extinct ancestors. They are proving to be an important catalyst for rewilding in both Europe and the UK.

Habitat creation

Old breeds of domesticated cattle, allowed to roam at will, perform the role of their extinct ancestor, the aurochs, by generating an open wood-pasture landscape. Their hoofprints and trails provide micro-habitats for insects; their fur, nesting material for birds; their dung, food for beetles, as well as introducing micro-organisms that increase the fertility of the soil.

Like all bovines (a group that includes bison, yak, musk ox and water buffalo), cattle are ruminants. They specialise in consuming grasses, although they will supplement their diet with foliage and other plants, often for self-medication (instinctively eating plants with medicinal properties in order to keep themselves healthy).[7] During the growing season they focus mainly on the most nutritious grasses, a circumstance that opens up opportunities for flowering plants. In the winter and during droughts they will eat dried plants and foliage, evergreens (such as bramble), and twigs and other woody materials.

Cattle are vectors for a huge range of plant species, transporting seeds in their gut, hooves and fur, and their considerable weight contributes significantly to ground disturbance: in particular, the treading in of thatchy, unproductive swards. On clay soils like ours in Sussex, heavy animal traffic creates important micro-habitats through ground compaction. In dry weather, a pinch point in an animal trail creates compacted soil in which solitary sweat bees make their nests, and the compacted hollow beneath the low branch of a tree favoured for back-scratching becomes an ephemeral pond, habitat for such aquatic species as the endangered fairy shrimp (see page 119). Cattle hoofprints provide specialised invertebrates, such as the Adonis blue butterfly, with exactly the right hot microclimate conditions they need to lay their eggs, and the cattle's grazing and trampling are thought to have provided the habitat at Knepp where a species of fly, new to Britain, was found in 2019.[8] Cattle are also particularly good at treading down bracken, breaking up the bracken thatch and bruising young growing fronds.

Cattle dung, being particularly loose in texture and with high levels of lignin and cellulose, is excellent for dung beetles. In most modern farming systems cattle are routinely treated with antibiotics, wormers and other pesticides, which filter into the dung and kill any insects (including dung beetles) and microbes that come into contact with it. Organic dung is one of the vital elements missing from the landscape today. Dung beetles are considered a keystone species for their impact on the soil (see page 258). The process of a dung beetle tunnelling down through the cowpat, and eating and digesting the dung to feed their larvae in underground chambers, increases soil fertility, aeration and structure. Cow-dung micro-organisms, too, have been shown to have a natural ability to increase the fertility of the soil through their actions on phosphates.[9] Numerous invertebrates in addition to beetles benefit from cattle dung as habitat and/or food. In turn, they become important food for birds and bats. Such birds as little owls and night-jars particularly love eating dung beetles.

In a natural herd system, bulls generate ground disturbance, pawing up the soil and rolling in the dust. It is thought they do this partly to rid themselves of parasites and to mark territory, but more importantly as a kind of fitness gym: training without danger of disturbance from an opponent. In this way they create 'bull pits', bare patches in the vegetation, often several per hectare, providing habitat for pioneer plants, sand bees, wasps, beetles, snakes and lizards.[10]

Heck cattle

In the early years of the Weimar Republic the German brothers
Heinz and Lutz Heck attempted to breed back from eight domestic
breeds, including White Park and Highland cattle, to an approxima-
tion of their wild ancestor, the aurochs.[11] The resulting 'Heck' cattle
have been used effectively in various conservation-grazing projects,
including the Oostvaardersplassen in the Netherlands, and they
are hardy enough to withstand low temperatures and nutrient-poor
forage. But as a breed they are controversial. Many ecologists consider the Heck brothers' programme to have been deeply flawed,
inspired by Nazi ideology and mythology. Heck cattle appear to be
no closer to the aurochs physiologically than many other existing
older breeds, such as the Spanish fighting bull. While their aggressive
temperament is sometimes overplayed, they are challenging to handle
and are particularly problematic with dogs. They are unsuited to all
but the largest and most ambitious wilding projects.

The Tauros Programme

The Tauros Programme, which was started in the Netherlands in
2009 by the Taurus Foundation, is a more scientifically based initiative to create a modern equivalent of the aurochs, primarily as a
driver for rewilding.[12] Animals are selected according to genetic closeness to the aurochs (whose genome was first sequenced in 2011[13]), as
well as on appearance and behaviour. Now in their fifth breed-back
generation, Tauros are much larger than most cattle (a Tauros bull
stands up to 180 centimetres at the withers – the highest part of the
back – compared with 140 centimetres for a Heck bull). They are
also relatively slender, generally more athletic, and have long legs
and strongly developed shoulder muscles, like the aurochs. This gives
them the ability to defend themselves better against wolves and bears.
Like their ancestor, they sway their long, sharp, forward-pointing
horns from side to side to ward off a predator and, if engaged, can
toss the wolf or bear aside. They have strong herd instincts, surrounding and protecting their young within the nucleus of the herd. Horses
also seek their protection at night, as observed through night vision
cameras in Croatia. Experienced cows guard the young calves in a

créche while the rest of the herd is away feeding. Herds of free-roaming Tauros cattle have been introduced to nature restoration projects in Croatia, Romania, the Czech Republic, Spain, Portugal, Germany and the Netherlands.

According to Ronald Goderie, director of the Taurus Foundation, Tauros herds would work well in the UK and could be licensed in the same way as any other domestic cattle. They would certainly add excitement, and are worth considering for any enterprise that will include ecotourism. The Tauros Programme insists on being involved in all breeding and selection decisions in the herd, to continue developing the breed. While some Tauros animals – generally those considered detrimental to the breeding programme – are sold for meat from rewilding projects in Europe, the emphasis is on the creation of truly wild herds. A Tauros herd would suit a system where minimal management is possible, and preferably in a low-risk TB area, where mandatory testing is once in four years.

Territorial bull and stallion pits are swiftly colonised by insects and wildflowers. Here in Europe, where this Tauros bull – bred in the Netherlands to resemble an aurochs – is kicking up earth, nearly 100 species of tunnelling bees and wasps find vital habitat. Among them are the four-spotted furrow bee, specialist of steep edges, and the black-headed mason wasp (very rare in the UK), which makes burrows with tiny chimneys.

Temperament and traits

Apart from the ancient Chillingham herd in Northumberland, which has special exemptions given its ancient history, all cattle in the UK are categorised as domesticated livestock. They are therefore subject to the same regulations and welfare checks as cattle in conventional systems. This makes cattle, in many ways, the most challenging species to manage within a rewilding project.

Until we have a scenario that allows wild herds of cattle in the landscape, docility and ease of handling will be the most important considerations when choosing a breed for rewilding, especially when it comes to cattle with horns, when one unwarranted or accidental toss of the head in close proximity can be extremely dangerous. It's also important to use cattle that fall in easily with being rounded up, put through handling systems, physically handled for veterinary treatments (including periodic vaccinations for common diseases) and – if the surplus are being sold for meat – transported to the abattoir. A conventional hands-on system, more like ranching, is also an option, whereby the bulls are kept separate from the herds, allowing control of the breeding season and protecting young heifers from being covered too young (see page 268).

There are, as yet, no dispensations for free-roaming cattle in rewilding scenarios in the UK or elsewhere in Europe. One of Rewilding Europe's GrazeLIFE objectives is to create a 'Status Wild' designation for de-domesticated herbivores.[14] At the moment, in the UK, if calves are to go into the food chain, they must be ear-tagged within twenty days of birth and registered for a passport within twenty-seven days of birth. Cattle should, in theory, be checked once a day when they're not calving, and several times a day when they are – quite a challenge if you're having to search for animals dispersed in challenging terrain or dense scrub over a large area, although tracking devices, such as GPS collars and radio transmitters, can make this easier. *A Guide to Animal Welfare in Nature Conservation Grazing* (2001), published by the Rare Breeds Survival Trust, contains useful stocking guidance and information on the legal requirements for free-roaming pigs, horses and cattle.[15]

It is to be hoped that some of these restrictions are relaxed – in line with dispensations for semi-wild horses in the UK – as more rewilding and conservation-grazing projects begin. Having a system (as there is for domesticated pigs) whereby ear-tagging is required only if the animal leaves the farm would save enormous effort searching for calves being born out in the wider landscape, and remove the unsightly, conspicuous yellow ear labels, which detract from the re-wilded look of the herd. This is a particular consideration if ecotourism, which brings with it a high demand for nature photography, is part of the plan. Ear tags do, however, make it easier to keep track of breeding and family lines, and identify an injured animal.

As the herd becomes established it will be possible to start breeding for specific traits, bearing in mind that selecting for one trait might involve losing another of benefit. It is an easier decision to select *against* a trait, such as mastitis (inflammation of the udder caused by infection) or excessively large calves in utero (which threatens the health of the cow herself, as well as her calf). Traits that are beneficial for rewilding should be selected positively: strong skeletal structure, hard hooves, good condition at the end of winter and swift weight gain with the spring flush of grass and new foliage. In many ways, this is simply breeding back to natural traits that have fallen away with intensive modern breeding for feedlots.

Resistance and resilience to common diseases can also be selected for. Many of the drugs sanctioned even by organic certification standards are, we believe, undesirable in a nature conservation project, so it makes sense to treat animals only when absolutely necessary. That is our approach at Knepp. But even better than this is to bite the bullet and stop medically treating the livestock, instead breeding only from those animals that show themselves able to cope with a particular disease.

At Knepp, our cattle are particularly prone to liver fluke, because our land is predominantly wet. As well as targeted breeding, our most important ally would seem to be nature itself. Wildfowl eat mud snails – the other host in the fluke's life cycle – so we hope this will reduce the pathogen. In a similar way, we have seen a complete collapse in the population of ticks. When we were farming, ticks were common on our livestock, our dogs and occasionally ourselves. It was one of the reasons we dipped our sheep. We haven't noticed any ticks for more than fifteen years. We don't yet know why, although we presume something in the food chain is interrupting the tick's life cycle. This would be a valuable study. If we could prove that restoring an ecosystem can reduce ticks, there would be an incentive to improve the management of areas with the tick-borne bacterial infection Lyme disease, which can be devastating to human health.

As a herd grows and gains experience, the animals naturally become more adept at living out in the wild: knowing where to go for the best grass, where to find herbs, nettles and sallow for self-medication, the best watering and wallowing holes, where to lie up in a storm. In natural, multi-generation herds, dominant females play a vital role. Most herd animals, including deer, bison and horses, are governed by a matriarchy, led by experienced old battle-axes who keep even the boisterous young males in check. This makes a natural free-roaming herd safer for the public, particularly dog-walkers, than most encounters on conventional farms. Almost all accidents on farms involving cattle and the public are caused by young, single-generation, usually single-sex groups of animals penned together in a field, and often provoked by the sight of a dog. Deprived of natural herd dynamics, steers or young heifers are like bored teen-agers lacking parental control.

The longhorns at Knepp barely raise an eyebrow at walkers or their dogs. Only if someone gets between a mother and her calf does the cow raise her head, strained and alert, her eyes beginning to flicker. However docile longhorns – or indeed any cattle – are, maternal instincts win the day. Sometimes it is difficult to know where a newborn calf is laid up in undergrowth, and the mother may be some distance away, grazing with the rest of the herd. We've often been caught unawares, walking between the two, and had to back away from a protective cow approaching at a brisk trot. Occasionally, newborns are secreted right beside the footpaths. Sometimes the herd establishes a crèche system, leaving the young calves resting under the watchful eye of several older nanny cows so that the rest of the herd can wander longer distances.

One trait that domestic cattle seem to have lost is the ability to rec-ognise yew as lethal. The Exmoor ponies at Knepp steer clear of yew, and red and fallow deer only nibble at it, presumably in medicinal quantities. When we introduced our first herd of longhorns into the Middle Block, the few yew trees in the park already had a browse line from the fallow deer, so the branches were beyond the cattle's reach. But disaster struck one winter after a prolonged snowstorm. Eight cows browsed on the branches weighed down by snow and were dead within hours. It is crucial to remove or fence out any yew trees that are accessible to cattle.

Old English longhorns and other breeds for rewilding

Old English longhorns – an eighteenth-century breed improved for beef and saved from oblivion by the Rare Breeds Survival Trust in 1980 – have proved perfect rewilders for us in lowland Sussex.[16] Renowned for their longevity and ease of calving, they're hardy and resilient enough to withstand wet winters on our heavy clay. Docile as they are, their intimidating horns encourage most people to give them a wide berth, which is a bonus in a rewilding system where the intention is to encourage natural behaviour and a healthy flight distance in the animals. They also have a distinctive white line or 'finching' down the spine, which makes them easier to spot when they disperse into the scrub. And, from an economic point of view, the quality of their meat is a great advantage, being rated by chefs as some of the best in the world.

While horns can be a hazard for stock handlers, they are important for natural disturbance. Cattle use them for pulling down branches and rubbing against trees. Bulls dig out bull pits with them, helping them assert their dominance. They are also handy for scratching an itchy flank. Cattle with horns, rather than polled cattle breeds, are, therefore, desirable for rewilding. Short-horned breeds may be more suitable for projects that will involve meat production, however, since some abattoirs have width restrictions in their races (see page 274).

Some cattle breeds, having been bred principally for dairy, produce milk that is far in excess of the needs of the calf. They therefore tend to have problems with mastitis, a condition that, in intensive dairy systems, requires the routine use of antibiotics. While old English longhorns are originally a dual-purpose (beef and dairy) breed, we had several cases of mastitis every year in the early days of our project. This is a trait that can be selected against, and indeed that tends to resolve itself. Eventually, rewilded or de-domesticated cows will naturally revert to producing less milk and having much smaller, more natural-looking udders. We now have hardly any cases of mastitis.

Other UK breeds suitable for rewilding and renowned for their meat are Dexter, Sussex, British White, South Devon, Short Horn and Shetland for lowland conditions, and Galloway, Highland, Welsh black and Devon for the uplands.[17] Galloways, although polled (without horns), are less reliable in temperament than the very docile Highland cattle with their fearsome-looking horns. This is probably understandable since, throughout history, any trait of aggression in domesticated horned cattle would not have been tolerated because of the danger. The Welsh black and the Devon are normally horned – and reliably docile – although they do have polled strains.

Water provision

In intensive systems cattle require large amounts of drinking water to help them digest grains and high-performance feed. Pastured cattle require far less because they're eating what they're designed to eat, the water content in grass and foliage is higher than in dry feed, and they can take advantage of natural moisture, such as dew, hoar frost and even snow. But they still require more water than sheep, for example. In large rewilding scenarios of hundreds or thousands of hectares, with restored ponds, rivers, streams and floodplains, and where water is allowed to lie on the land in puddles and ephemeral ponds, there should be plenty of drinking water and the cattle will alternate between options.

If there is no natural water source, the area may not be suitable for grazing at all. In smaller areas fenced away from natural water sources or where water regularly runs dry, an artificial source must be provided. This could be an excavated water pit lined with clay or a tank connected to a piped water system. Creating artificial water sources will inevitably change the behaviour of the animals and affect areas around the water source, so they must be positioned carefully.[18] Mains-fed troughs without additional on-site storage capacity also require checking daily.

Many river restoration projects, where banks have been badly eroded by livestock, are designed specifically to exclude cattle. But in low numbers the disturbance of heavy herbivores, such as red deer, cattle, horses and water buffalo, can have a beneficial effect on water margins. Their hooves create watery hollows in the boggy ground – which provide habitat for the larvae of dragonflies, and spawning spaces for newts and frogs – and their browsing of thuggish reeds and rushes gives space for other aquatic plants.

The importance of browse

Trees and shrubs are important browse for herbivores, especially in the winter, when grazing provision is low, and also for self-medication.[19] In old pasture systems, browsing from species-rich hedgerows, shelter belts and 'tree-hay' in winter was considered important for their health.[20] There is also evidence to suggest that the tannins in leaves, twigs and bark, as well as plants such as ribwort plantain and wildflowers – including angelica, common fumitory, shepherd's purse and bird's-foot trefoil – increase the animals' growth, improve digestion and reduce methane emissions.[21]

We have lost much of this ancient wisdom about natural medication, but agro-ecology is beginning to rediscover the health benefits for herbivores of different trees, shrubs and plants.[22] At Knepp, the cattle browse regularly in stands of sallow, which is an important source of cobalt and zinc, as well as salicylic acid.

Horses

The original tarpan – the European horse – was hunted to extinction. The last tarpan is said to have died in 1887, in Moscow's zoo, although it is now thought to have been anything but a pure original. Its genes live on in numerous regional herds of semi-wild ponies, however. Small herds of wild horses have always been tolerated in remote areas, often as a resource for cross-breeding with domesticated horses for resilience, vigour and genetic variation.[23]

These herds have developed distinct regional characteristics. In Europe, they are now widely used in rewilding projects according to their specialist topography and conditions: the Yakut pony for harsh climates in northernmost Europe and Siberia; Norwegian Fjord and Swedish Gotland ponies in Scandinavia; the Konik in wet, lowland Europe; Hucul in the Carpathian mountains; Przewalski's horses on the continental steppes of eastern Europe; Camargues in the Mediterranean deltas; the Giara/Achetta in Sardinia and other Mediterranean islands; and several sturdy mountain pony breeds with wolf and bear awareness in the Balkans.[24]

There is a surprising number of native pony breeds in the UK, including Exmoor, Dartmoor, New Forest, Welsh mountain, Shetland,

Fell, Dales, Highland and Eriskay. All are suitable for rewilding and can cope with both uplands and wet lowland pasture. Konik ponies (originally from Poland) have been used effectively in various wetland sites in the UK, but it makes sense, where possible, to use native breeds rather than importing horses from the continent: for spirit of place, to spread the genetic load, and for rare-breed funding.

Habitat creation

Hardy breeds of pony are especially useful in reclaiming neglected grasslands, and their impact complements that of cattle.[25] They trample and eat coarse, abrasive grasses that cows cannot stomach – including thuggish tor-grass and bushgrass – which helps to increase plant diversity and establish a mosaic of low and high vegetation.[26] Their forward-pointing teeth enable them to nibble the tops of thistles and the tips of gorse, holly and sloe, helping to apply the brakes to scrub encroachment. Their flexible lips, meanwhile, allow them to graze closer to the ground than cattle. They can also eat a surprising amount of live bracken, especially in late summer and early autumn when grass is low and bracken toxicity has declined. Indeed, horses may use bitter plants, such as bracken, ivy and ferns, to increase the metabolic bacteria in their gut and help them digest.[27]

In winter, when grazing is minimal, horses take to browsing, even scraping the bark off poplars, willows, spruce, beech and woody shrubs with their teeth. This helps to open up patches of closed-canopy woodland to the light.[28] Their ability to gain nutrition from dead and desiccated material means they are less likely than cattle to experience nutrient stress in the winter. Also unlike cattle, which scatter their dung randomly in individual cowpats, stallions mark their territory with strategically located dung latrines, and this localised nutrient-loading effect encourages patches of nutrient-demanding flora. They also tend to travel further than cattle, resulting in a more diffuse impact.

Horse dung is particularly attractive to insects. Its rounded lumps create a more complex surface than the thick pancakes of cowpats, providing niches for moth flies, lesser dung flies and fungus midges as well as a wide range of dung beetles. House sparrows, yellowhammers, chaffinches and linnets love to peck seeds from horse manure.

Even the seemingly casual actions of herbivores have a knock-on effect. At Wild Ken Hill in Norfolk, a lovely, scarce plant called shepherd's cress grows prolifically where the Exmoor ponies have hoofed away at anthills on the heathland. No one knows why the ponies do this, although it's possible they find it an easy way to lick at minerals in the soil.

Herd dynamics

Horses in the wild live in herds that can number thousands, but the megaherd is, essentially, an amalgam of small groups, or 'natal bands', living side by side.[29] This smaller social group, consisting of about seven mature mares and a stallion and their followers – say, between ten and twenty individuals in all – is the most likely size for a rewilding project in the UK. As with most herbivores, it's the leading females who call the shots, deciding what the group does, when and where. The main task of the stallion is to keep the herd together and protect it from other stallions or predators. Sometimes he joins forces with a lower-ranking stallion, to whom he grants limited mating rights to lower-ranked mares. Young mares and stallions are expelled when they reach sexual maturity, and join other herds or form bachelor groups, which avoids inbreeding.

The colts and young stallions often play-fight, preparing for the day they'll be stallion of their own social group. Mock fights, just like bullfights, disturb the soil, and in sandy soil horses will create territorial sand baths just as bulls dig up bull pits. All this generates opportunities for other species.

Wild boar and pigs

Wild boar, a native species that was once widespread in the UK, became extinct here in the seventeenth century, although there are now around 2,000 in the Forest of Dean and scattered populations roaming Kent, East Sussex, Devon, Dorset, Bedfordshire and Scotland. These are descendants of escapees and illegal releases from boar farms. In open countryside, the legal status of wild boar is that of any other non-notifiable wild animal, such as deer, badgers and foxes. Wild boar are subject to the Dangerous Wild Animals Act and require a licence to introduce them into managed systems, so it's well-nigh impossible to release them into a rewilding project.

Hardy breeds of domesticated pigs, however, make excellent proxies for wild boar. At Knepp we chose the Tamworth, a lovely ginger pig closely related to European forest swine, with long legs and snout, fringed, 'prick' ears, narrow back and long bristles. Of all the UK's native breeds, it was the least influenced by imports of Asian pigs in the eighteenth and nineteenth centuries.[30] It's another rare breed back from the brink. In the 1970s there were only seventeen surviving boars. Today, there are ninety or more registered Tamworth boars in the UK – still not a huge number.[31] Tamworths are muscular, strong-boned, active and alert, but also extremely docile and excellent mothers. They're renowned for their meat, and are hardy enough to cope with the cold climates of Scotland and Canada. All this makes them ideal candidates for rewilding. We do, however, live in hope that wild boar from East Sussex – tempted, perhaps, by the captivating scent of our lovely sows – will one day break through the deer fence and become self-naturalising denizens of Knepp, cross-breeding with our pigs. After all, a boar can weigh 125 kilograms fully grown, jump 2 metres and reach a speed of nearly 50 kilometres an hour. Our registered Tamworths would lose their pedigree status, but we would gain an Iron Age pig. This cross between a pedigree Tamworth sow and a Eurasian wild boar, first developed nearly fifty years ago, is thought broadly to represent how early European domesticated pigs looked and tasted 4,000 years ago.[32]

Other breeds of pig that are good for rewilding are Berkshire, Gloucester Old Spot and Middle White. All are hardy, ready foragers and generally docile.[33]

Habitat creation

Wild boar and free-ranging pigs are a keystone species, primarily because of their rootling. They are nature's plough, opening up the thick, grassy sward or dense bracken brake for colonisation by other plants and invertebrates, such as solitary bees. Ants use the clods of earth turned over by the pigs to kick-start anthills, which become egg-laying habitat for common field grasshoppers, basking spots for small copper butterflies and common lizards, and feeding sites for mistle thrushes, wheatears and green woodpeckers. In the winter wrens, dunnocks and robins trail the pigs, picking for insects in the

Domesticated pigs can act as proxies for wild boar, turning over clods of turf with their snouts, exposing bare, damp soil where shrubs such as sallow (top right), food plant of purple emperor butterflies, take root. Wildflower seeds, blown in by the wind or deposited by birds, also germinate in these rootled patches. The tiny, protein-rich seeds of these wildflowers (often known as 'arable weeds') feed birds such as the rare turtle dove. The earth exposed by the pigs heats in the sun, providing basking spots for insects and reptiles.

furrows. Pigs have a penchant for plants that other grazers cannot find or stomach, such as the stubborn, subterranean roots of docks and spear thistles. Unlike other ungulates (hoofed mammals), they can also eat bracken and its rhizomes, even in its most toxic stage, neutralising the toxins and carcinogens in their gut. They can't tackle poisonous rhododendron as a mature shrub, but they are an effective ally in its eradication, suppressing growth by eating the new shoots.

It's thanks to the pigs that new stands of sallow (naturally hybridising willow) are regenerating at Knepp. Sallow seed needs to find open, damp, bare soil in a two-week window in April in order to germinate. And it's thanks to the sallow – food plant of the purple emperor butterfly – that Knepp now has the largest breeding population of one of the UK's rarest butterflies. We also think it's the rootling of the pigs that is partly responsible for the return of turtle doves to Knepp, by encouraging colonisation opportunities for native wildflowers such as scarlet pimpernel, fluellen, fumitory, black medick, red clover, vetches and vetchlings with their small, protein-rich seeds.

Pigs also love water. Having a natural dive reflex to hold their breath, just like humans, they can submerge their faces to seek out acorns and rhizomes. At Knepp they have even learned to forage for freshwater mussels. This provides opportunity for plants, particularly at the margins of water systems. The wallows they create in soft mud and floodplains fill with water, creating micro-aquatic habitats.

We were surprised at Knepp to find our free-roaming Tamworths behaving like hippos. Pigs, like humans, can hold their breath underwater. They search for rhizomes, submerged acorns and even freshwater mussels to eat.

Wild boar/pigs are omnivores. In the summer, with rich, short new grass and clover, and when the ground is too hard to dig, they spend most of the time grazing. When the ground is soft, they rootle for rhizomes and insects. But they also eat carrion. They are the hyenas of Europe. If regulations were relaxed regarding fallen stock in rewilding scenarios, pigs would be one of the many species that would contribute to the swift processing of carcasses.

They may even help in the battle against climate change. A recent study shows that rootling, or 'bioturbation', by wild boar/pigs in closed-canopy woodland – mixing up leaf litter and organic matter on the forest floor with the mineral soil – increases the stability and storage of carbon in the soil.[34]

Fallow deer

Fallow deer are present in most of England and Wales below a line drawn from the Wash to the River Mersey. Thought to number at least 130,000 in the wild, they are descendants of animals imported from the eastern Mediterranean by the Normans for their deer parks and deer 'forests'. However, fossil remains show that fallow – or, at least, a very close relative – were present in huge numbers in Britain in the last interglacial period, about 130,000–115,000 years ago. They are therefore, in evolutionary terms, closely connected with the UK's ecology and are a good choice for an introduction to a rewilding project.

They also look spectacular. Unlike roe deer, they congregate in large herds of up to 150 animals and are happy to be seen out in the open, which gives them a visual similarity to antelope in Africa. The bucks, which live together for most of the year in bachelor herds, have palmate antlers, which are shed every year and can grow to 1 metre across and weigh 4 kilograms a pair. A buck can weigh 150 kilograms when mature, and lives for between twelve and sixteen years.

Fallow bucks at Knepp spend most of the year in male groups. Predominantly grazers, they help maintain open grassland. They have particular impact during the rut, when they create 'lekking' sites: gladiatorial arenas where they fight for mating territory by smashing branches with their antlers, and turfing up the ground, spraying it with pungent urine.

Habitat creation

Unlike roe, fallow are predominantly grazers, creators of grassland, benefiting a host of grassland-specific insects and biota. They browse shrubs and trees, but only at a relatively low level, since they generally do not rear on to their hind legs to reach taller leaves. The most active and dramatic season for fallow (as with red deer) is the rut in autumn, when the biggest bucks stake out 'lekking' sites, or gladiatorial arenas, where they compete in hard-fought head-to-head battles for dominance over the females. Pawing up the turf with their hooves, drenching themselves and the earth in urine, they create enriched, disturbed stamping grounds in the grassland. In the run-up to the rut, bucks will rear up and toss their antlers in the branches of trees for fighting practice, shredding foliage in the process.

Red deer

We waited years to introduce red deer at Knepp because we were advised that they might be a danger to the public, particularly dog-walkers, even though red deer live with minimal concern in popular dog-walking destinations like Richmond Park, Woburn Abbey, Chatsworth Park, Raby Castle and Wentworth Castle.[35] We have had no problems at Knepp since we introduced them in 2013, and feel we could easily have introduced them earlier. As long as people act sensibly and keep their distance (and do not approach hinds at calving time, or red deer stags for a selfie in the rut) they should be entirely safe.

Red deer at Knepp love the water, indicating they may originally have been a riverine species before humans drove them away from agricultural land into remote and mountainous areas.

The red deer is the fourth-largest deer species after elk (known as moose in the USA), wapiti (known as elk in the USA) and sambar, and is native to Britain. They are generally associated with the uplands, particularly sporting estates in the Highlands of Scotland. But observations of them at Knepp and elsewhere in the lowlands suggest that they may originally have been a riverine species, pushed into the uplands as humans took over their preferred habitat. Across Eurasia, red deer are still key grazers in reed beds and marshes, just like the sambar in Southeast Asia, Père David's deer in China and sitatunga antelope in central Africa. They tend to frequent the wettest areas at Knepp, and are often seen swimming in the water.

Red deer also exhibit remarkable growth rates in lowland conditions. The stags at Knepp are twice as heavy as Scottish stags – nearing the size of wapiti – with antlers up to three times the weight.

Habitat creation

Red deer help control the growth of rushes and reeds in the margins of watercourses, and their physical disturbance in marshy areas controls aggressive perennials such as reeds, reed mace, rushes, purple loosestrife and willows, opening up space for less dominating aquatic plants. They are also grazers in open grassland and will browse vegetation more than the fallow deer. They are often to be seen lifting themselves on their back legs to reach inaccessible foliage.

As with fallow, the most active season for red deer is the rut, which occurs slightly earlier in the autumn than for the fallow. They do not create lekking sites, but tear up turf with their antlers in territorial displays, their hooves kicking up the soil during their energetic head-to-head battles.

Just like fallow bucks, red deer stags enter the winter in relatively poor condition following their exertions in the rut. Many of the weaker die, contributing food reserves to carrion eaters. Similarly, their annual shedding of antlers and velvet in the spring returns minerals and amino acids into the ecosystem.

In the winter, when the nutrient value of vegetation falls, red deer strip off the bark of branches and trees, which can kill both young and mature trees. Ringbarking by red deer helps generate more open habitat.

Bison

Along with wild boar and beavers, introducing bison at Knepp was in our original plan when we wrote our 'letter of intent to establish a biodiverse wilderness area in the Low Weald of Sussex' to the UK government back in 2002, at the very start of the project. They are still high on our wish list.

Once, bison – more commonly known as 'wisent' in Europe – ranged in winter herds of many hundreds from France to Belarus and Ukraine, and across parts of central and northern Russia to the Urals

Snatched from the jaws of extinction, the European bison – or wisent – is proving a keystone species in rewilding projects in Europe. In particular, its penchant for de-barking trees in winter make it a powerful ally in the protection of rare heathland from tree encroachment.

and Siberia. All three subspecies of European bison were hunted to extinction relatively recently, in the nineteenth and early twentieth centuries. The European bison that survive today are descendants of just a dozen *Bison bonasus* that had been kept in zoos. Around 7,000 bison now roam in 50 or so free-ranging herds, from the relatively small 330-hectare Kraansvlak reserve in the Netherlands to much larger areas in Germany, Northern Spain, the Southern Carpathians, Belarus, Ukraine and Russia.

There is ongoing debate about whether bison were ever present in Britain after the last ice age. No bison bones from the Holocene – our current post-Ice Age epoch, which began around 11,700 years ago – have yet been discovered. But that's not to say they weren't here. Absence of evidence is not evidence of absence, and fossil evidence is notoriously difficult to come by. Bison bones from the Holocene have been found in Doggerland, the broad land bridge that connected Britain to Europe until rising seas separated us 8,200 years ago. It's almost inconceivable that bison, along with all the other fauna, did not continue along the land bridge into Britain.

But even if they didn't, we know they were here in the Pleistocene, before the last ice age, and they were prevalent until relatively recently in regions of Europe with the same climate and habitats as Britain.[36] They are an important part of the fauna of temperate northwest Europe today. Their ability to stimulate and sustain ecosystems, and improve biodiversity, is argument enough for the role they should play in nature restoration in the UK.

In 2022, the Kent Wildlife Trust, supported by the Dream Fund Lottery, introduced four adult European bison (three females and a male) into an enclosure, alongside a small herd of Exmoor ponies and some Iron Age pigs, in the Blean Woods near Herne Bay, Whitstable. It is hoped the animals and, in due course, their offspring, will open up non-native, species-poor conifer plantations to a more species-abundant mosaic of open habitats, principally by de-barking the trees in winter.

It was a landmark moment that we hope will pave the way for rewilders to introduce more large herbivores to help drive the recovery of landscapes across the UK.

Habitat creation

An adult bull bison consumes 30–60 kilograms of fresh food a day. Their influence on an ecosystem is consequently huge. They browse much more than cattle, are more selective in their choice of forage and can cope with much coarser grasses. Between 20 and 65 per cent of their diet consists of bark, branches, leaves and seeds such as acorns, with the proportion changing depending on the season. In winter, they strip the bark off trees. In nature reserves like Kraansvlak this has played a vital role in habitat restoration, reversing the encroachment of sycamores, white poplar, dogwood, hawthorn and other shrubs which were threatening to overwhelm the sensitive sand dune ecosystem.[37]

Bison would have shared the landscape with Europe's other large bovid, the aurochs, for millennia, so it's not surprising the two species evolved to command different niches. While cattle tend towards open, wet environments like grasslands, European bison seem to prefer open, dry areas such as heath, sand dunes and steppe. That said, the European bison possesses aurochs genes, indicating hybridisation between aurochs and steppe bison at some point in the Pleistocene.

A conspicuous impact of bison is the creation of sandy wallows. Unlike cattle, bison family groups create wallows together. Pawing with their front hooves and tossing away the turf with their horns, they shoulder into banks and roll in the exposed sand or dirt to rid themselves of itches, old fur and parasites. Also, unlike horses or cattle, which often use the same rolling places, bison continually rub out fresh wallows to avoid reinfestation by parasites. During the rut, creating wallows becomes part of the bison bulls' territorial displays of machismo, along with breaking young trees and hoofing the ground.

This dynamic, continuous disturbance opens up new habitat for invertebrates like sand wasps, mining bees and tiger beetles, lizards, snakes and small mammals, and creates a micro-climate for flora, fungi, lichens and mosses.

Like cattle, their urine and dung plays a vital role in soil restoration and the transfer of nutrients around the landscape. In the past, their carcasses, too, would have sustained many other species, including vultures.

Perhaps more than any other large herbivore, their ability to consume coarse, thuggish, thatchy grasses and other dry matter would have considerably reduced flammable materials in fire-prone regions. One of the main aims of the reintroduction of bison into northern Spain is to prevent fires.

In the UK, some of the most endangered and vulnerable habitats, including sand dunes, moors and heathland, could benefit immeasurably from the presence of bison. They could, at the very least, be used as seasonal browsers to reverse the encroachment of trees – now the largest threat to protected areas of UK heathland; 80 per cent of lowland heathland has been lost since 1800. A quarter of that acreage was lost in the last fifty years. The most extensive remaining tract of heathland (3,000 hectares) in southeast England is in the Ashdown Forest. But the collapse of commons grazing and tree cutting here, especially since the Second World War, has led to the encroachment of birch, sweet chestnut, hazel, alder, oak, Scots pine, gorse and other woody shrubs onto the heath. The heath supports astonishing biodiversity including 5,000 invertebrate species, more than 500 species of spider, 27 out of all 39 British dragonflies, all the British reptiles and amphibians, and endangered birds such as the nightjar, woodlark and Dartford warbler, and scarce plants such as the marsh gentian and bog asphodel. In the absence of herbivores, trees are cleared by mechanical and physical cutting. This is obviously hugely labour-intensive and expensive, and often fails to supply the beneficial impacts of animal disturbance. Bison could do this work for free and, unlike heavy machinery, in a way that creates rather than destroys habitat. Sites owned by the Ministry of Defence for army training, such as Pirbright and Sandhurst near Farnborough, Salisbury Plain, and Lodge Hill in Medway, Kent – former MOD land holding the UK's largest populations of nightingales – could be prime sites for bison disturbance.

Water buffalo

Europe was once home to a third indigenous bovine species – European water buffalo (*Bubalus murrensis*) – which became extinct, most likely through hunting, towards the end of the Pleistocene or early Holocene.[38] The wild Asian water buffalo (*Bubalus arnee*) is a close relative. Its domesticated descendants were introduced into Europe hundreds of years ago, possibly as early as the fifth century AD, by Eurasian nomads, after the fall of the Western Roman Empire. Subsequently, water buffalo became commonly used as livestock in Europe, particularly by pastoralists in eastern Europe and in Italy, home of buffalo mozzarella, where they developed regional characteristics such as a dense winter coat that allows them to survive in sub-zero temperatures, and resistance to European parasites.

Recently, domesticated water buffalo have been identified as an important animal for wetland conservation – a proxy for the native European species which used to thrive in the continent's river valleys.[39] They have been introduced into rewilding projects in the Netherlands, Germany, Austria, Spain and the Danube delta in Ukraine.[40] They are also used for nature conservation in the UK, in the Teifi Marshes in Pembrokeshire, Ham Fen Nature Reserve in Kent, and Kingfishers Bridge, a wetland nature reserve in Cambridgeshire.[41]

Habitat creation

Water buffalo have the same positive effects on the ecosystem as other large herbivores. Their grazing and trampling prevents a few thuggish species of plants from dominating, allowing less competitive plants a chance to thrive. Their dung provides habitat and food for insects. The way they rub their bodies against large trees and toss their horns in younger trees and shrubs helps beat back the understorey of woodlands and scrub. And they disperse seeds – of more than 200 plant species – via their dung and hair.

Asian water buffalo, used here as livestock in Sibiu, Romania, fulfil the role of an extinct species of European water buffalo. A keystone engineer of wetlands, they can also create wallows and ephemeral ponds in wood pasture, which is vital habitat for reptiles and aquatic insects.

But their great advantage over cattle, horses, bison and even wild boar or pigs is their affinity with wetlands. Other herbivores will wallow in water margins to varying degrees, and even graze on riverine and marshland plants. But wetlands are the water buffalo's kingdom. They can wade through the thickest mud and peat bog, which cattle and horses tend to avoid, using their horns to dig up rhizomes and roots. Unlike cattle, their digestive systems are adapted to digest rush, sedges and alder.

A water buffalo's skin is about six times thicker than that of cattle and they have about one-sixth as many sweat glands, so they need to keep cool either in the shade of trees or, preferably, in water. Their wallowing and digging deepens puddles into pools, creating water that remains in the landscape for longer. In the wood pasture and meadow systems of Europe, these wallows create habitat for tadpole shrimp, common newts, and red-bellied and yellow-bellied toads. In floodplains and deltas their wading routes and wallows open up dense reedbeds, creating feeding habitat for water rail, heron, snipe, great and little bittern, spoonbill, egrets, all species of crake, and white and black storks, as well as providing hunting grounds for marsh harriers. In the 150-acre Kingfishers Bridge Nature Reserve in Cambridge, the wallows of a herd of twelve individual water buffalos have encouraged the proliferation of the very rare fen violet.

Conservationists spend fortunes managing reedbeds and shrub encroachment in wetlands with diggers and mechanical reed-cutters and, when the land is too soft for heavy machinery, hand-held tools. In Germany, two areas of reedbed at Herter See in Lower Saxony were even blown up with dynamite to create pools for amphibians and birds. Water buffalo are clearly an easier and more attractive option, and we would love to have them at Knepp, to keep on top of reed encroachment in our historic 60-acre mill pond.

Dung and urine

The movement of herbivores, eating in one place and dunging and urinating in another, transfers nutrients around the landscape. Dung contains phosphorus, magnesium and calcium, and urine is rich in nitrogen, sulphur and potassium, all of which are made available for recycling when voided. Some 80 per cent of the nitrogen taken up by plants is recycled in urine. On dry sites, grass can be 'scorched' by urine, and this grass die-off in patches itself enables the colonisation of the sward by other plants. Cattle tend to avoid areas that have recently received dung and urine, as a behavioural defence to limit the spread of gastrointestinal parasites present in the faeces. In drier conditions, when dung remains on the surface for some time, this can be long enough to allow vegetation to grow taller, which benefits a range of invertebrates.

The sheer volume and variety of creatures that rely on dung signals its importance in ecosystems. Dung plays a vital role in the nutrient cycle and replenishes the soil. However, if livestock are dosed with insecticides, as they commonly are in conventional farming, their dung becomes toxic and can kill the creatures that try to digest it.

Birds are attracted to the dung eaters and their larvae

Whale dung is a driving force of life in the sea

Fox scat attracts purple emperor butterflies

Badgers search for insects under dung

Some seeds germinate directly in dung

Cow pat

Horse droppings

Seed eaters are drawn to undigested seeds

257

Herbivore dung is famously excellent fertiliser, but it requires a further stage of digestion before it can be successfully taken up by the soil. The role of fungi, bacteria and invertebrates as dung digesters in this process is key. On degraded soils, and especially if – as is routine in conventional farming – the dung contains anti-worming agents and antibiotics, which can kill invertebrates that eat it, a cowpat can sit on the surface for weeks, barely changing. Eventually it may be washed away by rain into the watercourses and be lost to the land.

Dung beetles and dung fungi

Most dung beetles in temperate-zone Europe are dung-dwellers, ranging in size from 3 to 15 millimetres, living in the dung itself or just beneath it. But some, like the three-horned minotaur beetle, and earth-boring *Onthophagini* and *Geotropidae* from the scarab beetle family, are tunnellers. Bigger in size (up to 25 millimetres), they dig vertical holes beneath the dung, creating nesting chambers deep in the soil. The actions of dung beetles, digesting dung and pulling it down into the soil, make them a keystone species, vital to the nutrient cycle and soil fertility.

Dung beetles, in particular, are masters of dung decomposition, and their exponential impact on the soil categorises them as a keystone species. The activity of different species of dung beetle varies according to seasonal temperatures, but as a group they are active all year round.

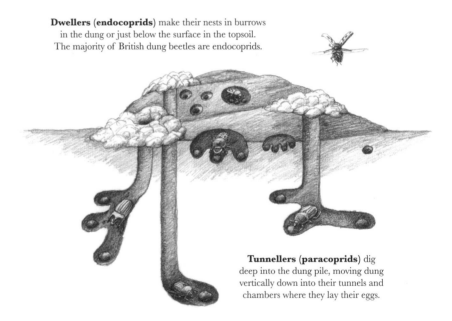

Dwellers (endocoprids) make their nests in burrows in the dung or just below the surface in the topsoil. The majority of British dung beetles are endocoprids.

Tunnellers (paracoprids) dig deep into the dung pile, moving dung vertically down into their tunnels and chambers where they lay their eggs.

Dung beetles have been on the planet for 30 million years and exist on every continent except Antarctica. In northern temperate regions, with historically high populations of large herbivores, most dung beetles specialise in herbivore dung, feeding on the dung 'soup' of waste products, body cells, microbes and small plant particles.[42] Some species show a clear preference for the waste of particular herbivores. The UK has about sixty native species of dung beetle, of which forty are 'dwellers', living in the dung, and twenty 'tunnellers', pulling the dung down into the soil to nest chambers that can be up to 60 centimetres deep, either near or directly beneath the dung site. The dung provides food for the beetle's larvae, allowing them to develop deep inside the nest, away from predators. The process of a dung beetle's tunnelling, eating and digesting adds organic matter, increases the fertility, aeration and structure of soil, and improves rainwater filtration and the quality of groundwater run-off.

Ironically, by breaking down, desiccating and burying the dung, thereby reducing its suitability for gastrointestinal parasites, dung beetles reduce parasite transmission between animals and hence the need for chemical livestock wormers.[43] Only now, when several of the UK's dung beetle species are on the verge of extinction, are farmers beginning to appreciate their value. In 2015 dung beetles were estimated to save the British cattle industry £367 million a year by improving soil function, encouraging the growth of healthy grass and preventing transmission of parasites.[44] That figure will have increased substantially since.

Dung beetles are active all year round. Most livestock, even in organic systems, are brought indoors during the winter, and this essentially starves the dung beetles of food. Rewilding, pasture-fed systems and conservation grazing where animals are pastured outdoors all year round throw a lifeline to these engineers of the soil system.[45]

And, of course, the dung beetles are food themselves. At Knepp, little owls – beetle specialists – are breeding for the first time, boosted by our burgeoning numbers of dung beetles. We now have twenty-one species of dung beetle on the estate. Their ability to scent fresh dung over long distances impels dung beetles to travel. We don't know where they've come from – presumably they were clinging on to a pocket of permanent pasture nearby – but when they find the right conditions again their populations rocket.

Dung, or coprophagous, fungi also play a role in breaking down animal dung so that it can be taken up by the soil. Some of these species, too, are becoming rare because of the widespread use of chemical wormers in livestock. The New Forest is home to what may be the rarest fungus in Europe, the nail fungus *Poronia punctata*.[46] Its spores are ingested by ponies via vegetation and germinate in weath-ered horse dung. Dung fungi themselves provide food for others, such as the lesser earwig, which feeds on the mycelia of moulds found on horse dung.[47]

Carrion

The absence of carcasses in the landscape is another lost aspect of natural processes. In the UK and Europe it is forbidden to leave fallen stock (sheep, cattle, pigs, goats and poultry) – including stillborns and afterbirth – on the land. The carcass must be collected by an approved transporter and taken away for rendering and incineration, or given to hunt kennels or maggot farms. In the meantime, animals and birds must not be allowed access to it. Wild animals, such as deer (unless they're farmed), are exempt from these regulations, and pets (oddly, in England, including horses whether they are pets or not) are allowed to be buried. The regulations are intended to prevent the spread of disease. The primary concern is for intensive farming systems, where soil biota, bacteria, fungi, necrophagous (carrion-eating) insects and other natural processes for breaking down carcasses are absent, as are large scavengers, such as birds of prey and wild boar. In near-natural conditions, carrion simply does not last long enough to cause a problem.

In today's farming systems the removal of grazing and browsing livestock from the land for human consumption is a one-way street of depletion of the soil's nutrients. In the past, when animals were slaughtered on the farm or the house-pig or house-cow killed at home, unwanted offal and bones were thrown out for the birds, rodents, dogs, foxes and pigs. Having consumed the meat, the humans produced manure – 'night soil' – which was collected by waste workers known as 'gong farmers' and spread on the fields. Waste carried out of London went to grow vegetables in fields in the Home Counties, and those vegetables were transported back to the city for Londoners to eat. It was essentially an enclosed food cycle. But now, thanks to legislation and the invention of the flushing loo (WC), all the minerals and nutrients taken up by livestock from the land are flushed out to sea or disposed of away from the land. We are endlessly depleting our soils of nutrients and trace elements, as well as depriving a whole host of necrophagous wildlife of food.

Now that there is greater understanding of the natural processes that dispose of carrion safely and swiftly, Rewilding Europe and Rewilding Britain would like to see the legislation relaxed for fallen stock in rewilded areas. Allowing carcasses to rot down on the land keeps nutrients in the food cycle. Without carcasses to feed on, entire

communities of insect species, such as clown beetles and blowflies, as well as fungi and bacteria, have collapsed. The now-extinct dead donkey fly, which gets its name from the site of its last British sighting, used to lay its eggs on decaying carcasses at the advanced skin-and-bone stage. Fallen stock provides carrion for corvids and buzzards as well as fat for songbirds, particularly in the winter – something we now provide for them artificially on bird tables. In Portugal and Spain, fallen stock is the primary resource for rising populations of vultures and eagles. Foxes, badgers, hedgehogs and wild boar – the hyena of Europe – all happily tuck into rotting carcasses. Indeed, in a healthy ecosystem it's astonishing how quickly a carcass can be rendered down by the myriad species it attracts in a matter of days.

Carcasses can have a dramatic effect on vegetation succession, too. A study in Norway in 2018 revealed how new vegetation growth can be triggered by fallen animals in the barren landscape of the Arctic tundra.[48] When a lightning storm struck Hardangervidda National Park, electrifying the ground and triggering instantaneous cardiac arrest among a herd of 323 wild reindeer huddled together (a not uncommon occurrence in the wild), the carcasses were left on the land and monitored by camera.[49] An array of scavengers, including ravens, crows, eagles, buzzards, smaller birds, foxes and wolverines,

Carcasses of large mammals return nutrients back into the ecosystem and sustain a vast array of creatures, from bacteria and necrophagous insects to scavenging birds and wild boar. Rewilding Europe, Rewilding Britain, Knepp and other rewilding projects would like to see a change in legislation to allow corpses of wild animals and fallen livestock to decompose naturally on the land, at the very least in areas designated for nature restoration.

261

visited the graveyard. Initially, the increase in ground acidity from the carcasses eliminated plant life in the vicinity. But two years later saplings from crowberry seeds dropped in the faeces of foxes and birds attracted to the site began to emerge. In the Arctic tundra, the crowberry plant is a keystone species. It requires bare, nutrient-dense soil – exactly what the carcasses created.[50]

Herbivores of the future

The introduction of beavers and bison (pages 112 and 251) brings a whole new dynamic to the suite of herbivores currently available for rewilding in the UK. But why stop there? Elk (known in the US as moose), a marshland animal, seem to be associated with habitats engineered by beavers and survived in Scotland until 4,000 years ago, possibly even later than that.[51] Successfully reintroduced to sites in Poland, Denmark and the Kamchatka Peninsula in Eastern Russia, they might well suit the fens of East Anglia and the Lincolnshire coast. Elk browse an astonishing variety of plants (as many as 355 in Russia) which generates vegetation complexity, helps keep wetlands open and increases the movement of seeds around the landscape. There is considerable potential for introducing elk within large-scale, fenced rewilding projects. At Lille Vildmose in Denmark they have been fitted with tracking collars. Russian experiments in elk domestication for haulage and as mounts for the cavalry prior to the Second World War and, latterly, for milk production at Kostroma Moose Farm, indicate their suitability for ranching-style management.[52]

Reindeer probably disappeared from Britain around the same time as elk, though a herd 150-strong roaming across thousands of hectares of mountain top and hillside in the Cairngorms National Park in the present day show what good candidates they make for northern rewilding projects. They, too, could be ranched in a similar way to de-domesticated livestock, with excess numbers providing premium meat.

And then there are the smaller herbivores – native keystone species such as water voles and wood ants whose populations have declined catastrophically over recent decades.[53] Their reintroduction to nature restoration projects would add greater habitat complexity still.

And we shouldn't lose sight of the brown bear, however distant it might seem on the horizon. Though its reputation today is primarily as a predator, nearly three-quarters of the brown bear's diet is herbivorous, consisting of roots, grasses and insects in spring, when they emerge from hibernation; berries and grasses throughout summer; and vast amounts of acorns, chestnuts and beechnuts in autumn to put down fat before returning to hibernation. They also feed on carrion, and catch migrating salmon in rivers, acting as important vectors, transferring nutrients around the landscape.

Numbers of brown bears have been rebounding in Europe with land abandonment. Europe now has around 17,000 brown bears (distributed between ten separate populations across twenty-two countries), compared with just 1,800 grizzly bears (the larger sub-species) in the lower forty-eight states of the USA – an area twice the size of Europe. A recent pan-European study suggests there is still plenty more land – about 380,000 square kilometres – for bears to expand into.[54] As a highly charismatic species, the draw for wild-life tourism is immense. In Croatia, Slovenia, Finland, the Southern Carpathians, and the Central Apennines in Italy, the construction of bear hides has proved enormously popular with tourists and wildlife photographers.[55] Locals are employed as guides for hiking, trekking and wildlife watching, and the local hospitality industry has boomed. This has been a significant factor in promoting human co-existence with an animal previously considered entirely undesirable.

In Britain, native brown bears were still present in the wild after the Roman conquest in the first century AD. The last populations of British bears may have lived in Yorkshire and the Pennines, and were most likely hunted to extinction a few centuries into the post-Roman period. They may have survived in Scotland into the later Middle Ages, when the remaining animals, wary of humans, would have retreated into the wildest, least populated areas of the Highlands and their ultimate demise would have gone unnoticed. The UK is not ready for bears yet. But perhaps one day, when landscapes are restored and reconnected, when wild areas can once again provide enough food for them, bears will be allowed to join the other herbivores driving ecosystem recovery.

Putting it into practice

Domesticated cattle

Buying cattle for rewilding

When buying cattle, it's best to find a cohesive, single-suckling herd of varying ages and with genetic diversity. A reputable seller will keep good records of bloodlines and if the cattle are pedigree they will be entered in the Herd Book kept by the breed's particular society. When we bought fifty-three longhorns for the Southern Block from a farmer in Cumbria, we thought we were buying a healthy and established herd. Our stock manager had gone to view them before purchase. But when we released the animals into the rewilding project, it was clear that some of them had never met before. Most probably some were recent purchases, 'flipped' on to us by the farmer. They were wary and unsettled and would scatter at any approach, a fact that was a concern particularly because of footpath users. They also tested positive for a number of diseases, which was probably why thirty-one of the forty-five cows failed to calve the following year. It took time to eradicate the diseases and two years before the animals settled down, established a social matriarchy and began to behave as a cohesive herd.

Breed societies can be a reliable source for purchasing stock, since they often know the breeders, particularly if a rare breed is involved. The Grazing Animals Project and Pasture for Life forums may be helpful, and the Rare Breeds Survival Trust website has a livestock advertisement page.[56]

If you have limited experience in buying cattle, you are best advised to employ a livestock agent. Setting up founding stock is an investment for the long term, and it will always pay to take trouble with this step. An agent will do the legwork and know the backstory of the farmers in their region, and they can give you guarantees. Reputation is everything in this game. Farmers also love to quibble over deals, and if you don't have an agent they can have you on the back foot. In our experience, it's worth securing the right herd even if it means paying a little over the odds.

Choosing a stock manager

If you are not intending to manage your cattle, horses or pigs yourself, you will need to employ a stock manager. It can be difficult to find someone with experience who is also open to the challenges and the different mindset of managing stock in the wild. Our own stock manager was originally an arborist and took up the post with no previous experience of animal husbandry. He is amazing at his job and clearly has an instinctive way with animals. We feel we also benefited hugely from his lack of conventional training and preconceived ideas.

A stock manager in any scenario must be able to turn their hand to anything (which in rewilding, could be fencing in beavers or catching wild piglets), calm under pressure, resourceful when things don't go to plan, willing to learn from others, happy to work on their own and physically fit. When it comes to rewilding, for example, this means being able to lift a two-week-old calf single-handedly into a safe box for ear-tagging.

They must also understand the soil and vegetation – native grasses and how they grow, the importance of woody and herbivorous species for browsing and self-medication, plants that could be harmful – as well as the overriding principles of stocking density and management for the benefit of biodiversity and stimulating natural processes. A readiness to read up on the subject and research imaginative solutions for problems as they arise is therefore important.

Livestock vets

A good livestock vet who is genuinely interested in the project is key. Vets who are mired in the conventional farming mindset are unlikely to be helpful in a rewilding scenario. Your vet must be experienced yet open-minded, and able to rise to the unforeseen challenges of managing free-roaming animals – something they are unlikely to have done before. A good dose of 'can-do' pragmatism is critical, as is the curiosity to research solutions. If you're registered as organic, you should choose a vet who is knowledgeable about organic systems, and able to communicate with the Soil Association or relevant organisation.

Many vets feel uncomfortable with rewilding. Rather than looking solely at the individual animal, a rewilding vet must consider the good of the herd and maintain social structure and 'wildness'. If an individual animal loses condition or becomes sick because it is not well equipped to deal with life in the wild, don't nurse it back to fitness and return it to the project. Instead, take it out of the system, treat the welfare problem, then fatten it up for culling as soon as possible. Isolating an individual from the social group for treatment (as is common in intensive farming) may in itself be an insensitive thing to do. Bulls are, perhaps, the exception, since they often live apart from the herd anyway.

A livestock vet encounters every livestock manager in the area. For that reason they can, in a low-key but very effective way, also be a useful ally, disseminating information about your rewilding project and myth-busting when the community bush telegraph gets ahead of itself.

TB testing and the cattle trade

Testing for bovine tuberculosis (TB) – an infectious respiratory disease of cattle caused by the bacterium *Mycobacterium bovis* – is an added stress for managing cattle, particularly in the rewilding scenario. The normal rate of testing in the UK is once every four years, but it will be more often in a high-risk area, and more often still if a herd is found to have a 'reactor' – a cow that produces a positive response to the test. Animals that test positive are likely to be compulsorily slaughtered, with statutory compensation paid, and the herd put under official restrictions.

The testing procedure involves catching the entire herd and injecting two types of tuberculin serum, avian and bovine, into the neck skin to elicit an inflammatory response in an infected animal. The reaction usually takes about 48 hours.

Bovine TB can infect humans, and used to be a common problem for farmers and their families in the early decades of the twentieth century. Now, though, according to DEFRA and the Health Protection Agency (now part of Public Health England), the risk to people of contracting TB from cattle in the UK is extremely low. The main reason for culling all TB-infected cattle, or even those suspected of having it, is to try to eradicate the disease entirely.

Unfortunately, eradication is proving elusive, and testing is itself problematic. Bovine TB can infect other mammals, including deer, goats, pigs, cats, dogs and, notoriously, badgers. Arguments rage around whether cattle in intensive farms spread the disease among each other, or whether TB-infected wildlife play a part.

Our experience with TB testing at Knepp has been painful. Our area of West Sussex has long been free of TB. Most dairy farms around us have closed herds, breeding their own replacements rather than bringing in fresh stock from outside. But with the spring flush of grass many farmers and landowners see an opportunity to import beef animals to graze their land, fattening them up – or 'finishing' them – for the meat market. They contact local dealers, who organise the transport of store cattle from markets that are often in high TB-risk areas, such as Salisbury, Exeter and Frome in the West Country.

Those defending the trade argue that animals leaving high-TB areas will have been compulsorily tested for the disease. But the test in current use is not infallible. Research estimates that 25–50 per cent of recurrent TB breakdowns are caused by infected cattle not being detected by the skin test. There is a cattle vaccine for TB (similar to the human one), but it is only partially effective, and at the moment it is impossible to distinguish vaccinated from infected animals. Vaccinating would not, therefore, prevent the spread of the disease. If Britain went ahead and vaccinated anyway, the EU would ban its meat and livestock. TB levels are higher in Britain and Ireland than anywhere else in Europe, so for Britain test and slaughter remains the best option, in the hope of eradicating the disease entirely.

Inevitably, though, while the current system is in place in Britain, TB-infected animals slip through the net and enter disease-free areas. Some traders are also notoriously unscrupulous and fail to adhere to bio-security regulations. The disease has also been known to spread from the wheel arches of livestock lorries and even traders' boots.

When a couple of our neighbour's cattle tested positive in 2017, all cattle within a 3-kilometre radius of their farm boundary had to undergo rigorous testing. Mustering dispersed, semi-feral herds of longhorn cattle across heavy clay terrain, often through thorny scrub, is not something to be done lightly. Unfortunately for us, two of our more than 400 cows were inconclusive reactors – that is, showed some form of reaction to the test. This is not unusual, and does not in itself prove TB. There followed a merry-go-round of further tests on these two cows, which proved both negative and inconclusive. Meanwhile, as the question marks continued to hover, we were compelled to test the rest of the herd every six months. The two suspect cows were quarantined, and when one finally tested positive, both were culled. Ultimately, however, both cows tested negative in all their autopsy tests. It is highly unlikely that we ever had TB.

In April 2020, to our intense relief, the entire herd at Knepp tested negative and we could return to the four-yearly cycle of testing. But our close brush with disaster was sobering. If we had had a positive reactor we would have had to cull the affected animals and continue testing every sixty days. We might also have had to embark on wildlife culls – which would have been soul-destroying.

We remain in a low-risk area. However, now that we're aware of the continued danger, we've been petitioning government to halt altogether the potentially devastating export of cattle from high-risk TB areas.

Starting up a cattle herd in a high-TB area is a brave choice for anyone embarking on rewilding. It is important to consider whether you are prepared to introduce free-roaming cattle on to your land with this sword of Damocles hanging over your head. Even in low-risk areas it is impor-tant to know the reputation of your cattle-farming neighbours. Do they run closed herds? Are they meticulous in their biosecurity? Might they be tempted to import cattle from areas of high TB to fatten up on their land in the spring? A good local livestock vet will inevitably have the inside track on local farmers, and is one of the best people to advise on this.

Selling grazing rights for rewilding

Had Knepp been in a high-TB area we might not have opted to have our own cattle in the project at all. One solution in this scenario is simply to sell the grazing rights, thereby allowing someone else to graze their cattle on your land, perhaps just for the summer. Your rewilding project then benefits from at least some cattle disturbance, but you are not responsible for anything to do with TB. The success of such an arrangement depends on finding someone with a single-suckling (that is to say, multi-genera-tion, cohesive) herd who is willing to entrust their valuable animals to a free-roaming scenario – a prospect that may be unfamiliar both to the owner and to the cattle themselves. You must also decide on the param-eters of your joint venture. Will you be responsible for the daily checking and general management of the herd while they're on your land; and will the herd owner send someone out if there's a problem? The logistics depend largely on your knowledge, the level of animal husbandry you're comfortable with, and your relationship with the grazier.

Cattle management in a rewilding system

Natural grazing throws up some intractable conflicts with British live-stock regulations. Of particular concern, if you're receiving funding for a rare breed, is pedigree status. Allowing a number of bulls to run with the herd makes it impossible to identify the sire of calves, which is a tenet of the pedigree rulebook. There may also be a weight-gain requirement to qualify the breed as pedigree. Our longhorn bulls, for example, were required to reach 310 kilograms in 300 days – more than 1 kilogram a day. This can be achieved only by feeding them intensively on grain, hence the importance to those in the know of the term 'slow-grown', as a matter of both taste and animal welfare. At Knepp the average daily weight gain for a steer is between 0.6 and 0.7 kilograms. We hope the stipulation will change as it becomes clear how useful rewilding projects can be in reviving and sustaining rare breeds.

Originally, we hoped to run our cattle herds on an entirely natural system. It took eight years for our first herd of cows in the Middle Block to recover their natural rhythm and synchronise their cycle to give birth in the spring. During those years, locating calves for ear-tagging within days of being born was a frustrating and time-consuming game of hide-and-seek. Even if our stock manager knew where a certain cow liked to give birth – and some of them, mercifully, were so loyal to a location that he could plot them on a map – the random calving pattern scattered across the year meant we never knew quite when she would. Checking just the Southern Block twice a day on nearly 40 kilometres of tracks could take hours.

In 2007 we invited a group of vets from DEFRA, the RSPCA, Natural England and several conservation projects to see our rewilding system in action. We were keen to dispel any concerns about animal welfare, specifically with the cattle. They were extremely happy with the condition of our animals but had one concern: that leaving the herd to its own devices meant that heifers sometimes as young as six or seven months old were being covered (mated with) by bulls. They were worried that this posed a risk to the young heifers because of the comparatively large size of the calves and the mothers' immaturity, which could lead to problems giving birth.

It is very hard to determine how significant this risk is. We had possibly two such calving problems in a herd of fifty cows and heifers over eight years, which is very little in terms of conventional farming. However, the rewilding project was still at an early stage and we felt we were not in a position to argue. We couldn't risk engendering public criticism by ignoring advice from leading vets. Had this happened later in our rewilding project, we would have defended our position more vigorously. The National Trust, for example, has run bulls with their free-roaming herd of Highland cattle at Wicken Fen in Cambridgeshire since 2003 with no problem; the wild cattle in the park at Chillingham Castle in Northumberland have been left to their own devices since the Middle Ages; and a considerable number of rewilding projects in Europe are now operating with entirely natural herds.

The system we now manage at Knepp is certainly a great deal easier for our stock manager. The bulls are segregated from the herd, grazing on 16 hectares of organic conservation land separated geographically from the project by the village, until the breeding season begins. When the bulls join the herd in June or July the young heifers are taken out, and they are swapped over again in September.

We now also pregnancy-test the cows and separate the hundred or so that are in calf into a fenced-off area within the project, so that they are easier to ear-tag and check for calving, suckling difficulties and general health. If there's no problem, they are released straightaway to return to the herd.

Breaking up the herd, though, has an inevitable impact on relationships. When the heifers are taken away, aged twelve months or more, even though they are now entirely dependent on grass, they can still occasionally be suckling from their mothers. When they return to the herd ten weeks later it is sad to see that mothers and daughters have lost their former bond. Periodically withdrawing bulls from the herd will also have a profound effect on herd dynamics.

There will, perhaps, inevitably be tension between the needs of a stockperson to make the job easier, more efficient and (in terms of meat production) more profitable, and the desirability of a natural herd system for rewilding. The form of management you decide on will depend largely on the scale and topography of the land, whether you intend to keep pedigree status for your breed of cattle, considerations of public access, and the importance of meat production as an income stream.

The Bud Williams technique of cattle mustering

Managing free-roaming cattle in a rewilding project is much like ranching. Conventionally, mustering on a ranch is done on horseback. In dry areas, such as Australia, cattle are often rounded up by quad bike and all-terrain vehicles, but this can be stressful for the animals and is often destructive to habitat and soil. In the early days of our project we bought three Camargue horses – an ancient breed designed for rounding up fighting bulls in the rough marshland terrain of the Rhône estuary in the South of France. Unlike many horses, they are undaunted by water and thick scrub, know instinctively how to behave with cattle, and are generally unafraid of pigs, habituated as they are to wild boar.

Our longhorns were beginning to respond to some trial Camargue mustering, and this might be a good option for a project with experienced horse riders. But neither our stock manager nor his assistant are natural cowboys, so we have instead adopted the Bud Williams technique.[57] Although requiring considerable fitness, this astonishingly effective system of mustering on foot has the bonus of dispensing with the added cost and effort of looking after domesticated horses.

The late Bud Williams, a cattle handler originally from Alberta, Canada, preached a method of moving cattle based on an empathetic understanding of herd mentality. He could gather up animals of any description – from cattle, sheep and 'hogs' to reindeer, elk and bison – on foot and over any terrain. He grazed his own cattle on rotation, without fences. The hours of Bud Williams videos, filmed on a shaky camera by his wife, Eunice, are amateurish compared with the quality of many online videos today, but highly instructive nonetheless.[58] Using just the right angles of approach, always in sight, never directly behind an animal,

Bud stirs the herd into pacific motion, using a system of gentle pressure and release. He gauges just the right pace – the pace of herd migration – to magnetise them. Stray animals, nervous about being left behind, are sucked into the direction of travel like drops of mercury. It took our stock manager a couple of years of trial and error to master the technique, but now the cattle respond swiftly and instinctively and round-ups have become relatively easy and stress-free.

Temple Grandin cattle-handling systems

The designs of the American scientist and animal behaviourist Temple Grandin have been adopted in more than half the handling and slaughter systems across the USA, transforming a stressful and dangerous process into an efficient, humane and ultimately cost-saving exercise. Her extraordinary life as both a spokesperson for autism and a proponent of the humane treatment of livestock was documented in the biopic *Temple Grandin* (2010).

Our Temple Grandin cattle-handling system has been revolutionary. Previously, it took five people a whole stressful day to process an adrenaline-fuelled herd of 100 cattle for vaccinations or TB testing. Now, two or three people can process the herd – which remains relaxed – in under two hours, with considerably less risk to themselves. Grandin's designs for handling systems to suit all sizes of herd are published in her book *Livestock Handling and Transport* (2019) and on her website.[59]

The systems are easy to build. We built ours in-house, using our existing mobile hurdles and oak posts from wood thinnings, erected on a concrete surface around our former agricultural buildings. The only specialised kit we had to buy was the cattle-crush: in our case, one tailored for longhorns.[60]

Rewilded beef

As the anti-feedlot movement gathers pace, so does the market value of sustainable, extensively reared, organic, pasture-fed meat. There is increasing interest in rewilding, too, of course, and educating the public in the relationship between free-roaming herbivores and nature restoration has become one of the core aspects of the work we do.

Regulations since the outbreak of bovine spongiform encephalopathy (BSE, 'mad cow disease') in the UK in the 1990s dictate that beef entering the food chain after thirty months of age must have the backbone removed. This means you can't hang the whole carcass to mature, or 'age', and this can have an effect on the quality of the meat. In natural grazing systems many steers reach prime weight only at thirty months or more, so producers who want to hang the meat for flavour and tenderness may need to cull the animals before they reach prime weight.

For years we sold our meat under the generic 'organic' label of a supermarket. This anonymity was always a source of regret, and in 2019 we began to take sales in hand, selling direct to the consumer, thus cutting out the middlemen, improving our profit margins and marketing our own brand and story. Shrinking the supply chain is also often greener. It pays, too, to know your customer and for them to know you. Not only does it build up a sense of trust and commitment on both sides, but it can also foster a deeper understanding in the consumer of the animals they are eating.

Selling direct to the consumer opens up other potential revenue streams, too, particularly if you're able to add value by processing your meat into sausages, bacon, burgers and charcuterie, and/or enter the realm of ready meals, catering and hospitality. The vogue for nose-to-tail eating and buying local has given rise to hugely popular 'pop-up' restaurants. From River Cottage in Devon to Balgove Larder in Fife, barns, outbuildings and even woods have become atmospheric venues for home-grown feasts, as well as butchery and charcuterie courses. Rewilding projects can add an extra dimension by showcasing the role of free-roaming animals in nature conservation, and making that vital connection between pasture and plate.

Selling directly also means that the meat can be sold frozen, which is crucial if it is to be supplied all year round. Because rewilding involves naturalistic grazing, the animals reach age and condition at the end of summer, meaning that we slaughter our steers and heifers over a relatively short period, between August and October. We hang the beef for five weeks before butchering. We can sell fresh cuts – if customers want it – in late summer or autumn, but almost all the meat is quick-frozen and stored in our own freezers, giving us continuity of supply. Having a direct connection with the customer means we're able to explain the seasonality of the meat and the rationale behind freezing, and people have been surprisingly accepting of that. Often, the meat goes straight into their freezers, anyway. We use high-quality, eco-friendly wool-insulated packaging that ensures it remains frozen to the point of delivery.

Freezing is convenient and efficient for storing meat, but it does come at an environmental cost. Studies have shown that the refrigerant hydro-fluorocarbon (HFC) commonly used in cooling systems, including fridges and freezers, is one of the greatest contributors to greenhouse gases, with a global-warming potential (GWP) of 600–4,000, compared to carbon dioxide, which has a GWP of 1.[61] A previous offender, the refrigerant CFC (chlorofluorocarbon), which depleted the ozone layer and so would have killed us all, was phased out of global production in 2010, although evidence suggests that CFC emissions are currently on the rise again in China.[62] The Kigali Amendment to the Montreal Protocol, which came into force on 1 January 2019, has scheduled a phased reduction of at least 80 per cent in HFC production and consumption by 2047 (although the USA has yet to ratify the Kigali Agreement).[63] So, while it is still possible in the UK to buy refrigerators that run on HFCs, it makes sense, if possible, to make the switch to carbon-dioxide refrigeration. Such fridges and freezers are, unfortunately, currently around four times as expensive as HFC options, since the government does not tax or ban environmental costs. However, the running costs are between eight and ten times lower, which is a considerable saving. We have a CUBO2 Smart carbon-dioxide refrigeration system at Knepp, and we also exploit its thermodynamic properties to provide all the hot water for our butchery.

Organic status

A rewilding project involving large herbivores will invariably involve going organic, if only for the organic dung and urine that play such a crucial role in the restoration of soil biota. There are several different organic associations, of which the Soil Association is perhaps the best known, and all have slightly different standards.[64] Grants are available for the conversion of land to organic, which generally takes two years.[65]

Organic accreditation is costly and complex, but it should guarantee a premium price for your meat. It is also worth acquiring Pasture for Life accreditation, which certifies meat from animals that are entirely pasture-fed. The label 'grass-fed', deceptively, can apply to animals that have spent only part of their time out grazing or have even lived indoors, eating conserved grass.[66] 'Grass-fed' beef animals are often 'finished' (fattened) on grain and concentrates.

Given that it takes two years for UK land to qualify as organic, it is worth applying for organic status as soon as you know you're going to rewild. You could buy a small number of non-organic animals at the outset, which would increase in number, converting to organic status at the same time as your land. Or, if you're allowing a few years for a vegetation pulse to happen without grazing animals – in which case the land has time to convert to organic – the animals you introduce later must be organic already. It may be easier to buy organic founder heifers from a variety of sources but it will take them a while to settle and form a cohesive herd; if you can get an established herd of your chosen breed from an organic farm, so much the better.

If, rather than managing your own herd, you have an arrangement with a local farmer to graze their livestock on your land and they are non-organic, your land will not lose its organic status if the animals are with you for less than 120 days. However, you'll need to ensure that the cattle have undergone the correct amount of withdrawal time from any wormer or other chemical treatment before they come on to your land.

Organic systems consider the welfare of the animal as primary. If the need arises, you can give conventional drugs to individual animals as required without them losing organic status. There must simply be double the legal withdrawal time – in other words, the time that must elapse between the last administration of medicine and the slaughter or pro-duction of food from that animal – before selling it as organic. However, many of these drugs are still very harmful to the environment and can enter the ecosystem through dung and urine, permeating watercourses and persisting for a long time in the soil.

Abattoirs

It's important to discuss logistics with local abattoirs before you intro-duce any livestock, in order to plan your system of herd management and slaughter. Unfortunately, there are very few small abattoirs left in the UK. As of 2020 there were only around sixty small red-meat abattoirs, a third having closed in the previous decade.[67] The Sustainable Food Trust, among others, is campaigning for a reduction in burdensome paperwork, improvement in cross-government coordination and help with the cost of waste collection – the issues that are currently driving small abattoirs out of business.[68]

Very few abattoirs are set up for organic processing, or have a market for it. There are even fewer purely organic abattoirs. Most larger abattoirs have specific days on which they process organic meat, and require you to sign a contract specifying the number of animals you will deliver to them for processing on each given date. Any animals above that number may have to be sold at non-organic prices. Crucially, the abattoir(s) must be able to cope with the seasonal flush of animals in late summer, since most farmers on pasture-fed systems want to kill their animals at this time.

The lairage (holding pens) of some abattoirs cannot accommodate cattle with horns, and some simply don't want to be bothered with them. In common with many abattoirs, ours has a width restriction in the race, or alley, so we are required to cut the tips off the wide, upsweeping horns of our longhorns before they go to slaughter. The horns have to be cauterised as they are sawn through, to stem the flow of blood. Although the animals are treated with anaesthetic and pain-relief drugs, which requires a withdrawal time for the drugs to leave the body if the meat is to be allowed to enter the food chain, the process adds stress for both animal and stock manager.

Organic regulations do not allow any 'goads', or prods (often electric), which are commonly used in conventional abattoirs. On a couple of occasions, on arrival at the abattoir, our docile longhorns have stayed quietly in the lorry; since we were unable to encourage them to leave, they held up the entire system (on one occasion for two hours), leaving a hundred abattoir workers standing around, waiting for the cattle to come out of their own volition.

Ultimately, we would much prefer to be able to slaughter on site, and indeed mobile slaughter and butchery systems are used very effectively elsewhere in Europe.[69] Killing on site is kinder to the animal, involving none of the stress of transportation and large industrial abattoirs, so the animal can remain relaxed and eating right up to the last moment; it is also reassuring for the stock manager, who can oversee the whole process. Mobile slaughterhouses are, invariably, small and handle only a few animals a day, providing time for kindness. Essentially, this approach closes the circle, allowing the animal to live and die in the same familiar place. We would love to see this industry develop in the UK to fulfil the needs of small livestock and rare-breed farmers and rewilding projects.

Horses

Laminitis

Most hardy pony breeds are very adaptable and their fur, hoof size and growth, muscles and tendons should all respond within a few years to the requirements of climate and terrain. They are also less susceptible than domestic horses to such ailments as equine influenza. But laminitis can be a problem, especially if animals are imported from marginal land with poor nutrition into more fertile areas of ex-arable and improved grassland.

At Knepp we had a couple of cases of laminitis in our starter herd of six Exmoor fillies, which we managed by rounding up the ponies during the spring flush of grass and feeding them just small amounts of hay in a paddock for four weeks before releasing them on to the rougher summer grass. After two years, however, none of the ponies showed any sign of the condition. With declining artificial nitrogen in the soil, the sugars and fructans (chains of fructose molecules) in the grass had finally dropped to a level that the ponies could metabolise. It is likely, too, that their digestive systems had begun producing the intestinal bacteria to help them adapt to the different diet.

However, herd dynamics also plays a vital role in susceptibility to laminitis. Our six fillies initially had little to distract them, but the introduction of a stallion in their third year at Knepp added natural stress and interaction. Breeding and lactating, in particular, significantly reduce the risk of laminitis. We've had cases of laminitis in the Middle Block herd since the stallion was introduced, but mainly in barren or older mares. A combination of wet winters alternating with hot, dry weather, and the shorter grasses of the parkland restoration, may also have resulted in higher fructose levels in the grass.

There is a genetic predisposition to laminitis, which can be selected against through culling or not breeding from affected mares. But it is important to balance against this the need to maintain genetic diversity. Most wild horse breeds have a very small genetic pool. Koniks originate from only twenty-two founders (six male, sixteen female), for example, and Exmoors from just fourteen. So, even in a relatively large rewilding project with a herd of thirty to fifty horses, it's important to bring in new bloodlines whenever possible. Some bloodlines of pure-bred registered Exmoors are more commonly bred from than others. The Exmoor Pony Society is working with breeders to try to ensure that rarer lines are bred from, too.[70]

Managing herd numbers

A problem arises, of course, when the herd grows beyond the number you feel is appropriate for your project. At Knepp, we judged this to be around thirty ponies.[71] There is a limited market for selling excess semi-feral, unbroken ponies. Most excess feral ponies, such as those from the New Forest and Dartmoor, are culled, with the meat going to zoos or packs of hounds. They can also be exported live to the continent, where there is a market for horsemeat.[72] The long journey is particularly stressful for ponies that have never known a trailer and are unused to being separated from the herd. None of these prospects appealed to us.

One option is castrating the stallions. We tried this exercise initially and for five years ran a non-breeding Exmoor herd. Castration was expensive − £250 to dart and operate on each stallion − and stressful for all involved. The stallions took more than double the amount of tranquilliser normally required for a domesticated horse, and half the amount of antidote.

Vasectomy is also possible. It is more expensive because more complex, but is considered healthier for the animal and less disruptive to the social hierarchy of the herd. However, research suggests that sterilising only dominant stallions results in a modest reduction in population growth, and ends up favouring the genes of the less-dominant males.[73] IUDs (intrauterine devices) and uterine marbles have been used in feral horse populations with varying results, and can cause injury and stress.[74] Another option is to dart the mares with synthetic birth-control hormones, but there remain serious concerns about the effect on the soil, water and other species of introducing artificial oestrogen and progesterone into an ecosystem. The Dartmoor Hill Pony Association use an immunocontraceptive administered to the mares to control the free-roaming herds on the uplands of Dartmoor. Being a 'dead vaccine', it does not leave the animal and enter the ecosystem, and has proved 100 per cent effective. The mares do not cycle, so the stallion believes them to be pregnant. This system means that hill-farmers can continue to run a breeding herd with intact stallions, allowing only certain mares to come into oestrus in order to maintain the rare Dartmoor hill pony genetics. They can also plan for the number of foals they can sell at market.

You could, of course, run just mares, with no stallions, as we did at the outset, and one could argue that a non-breeding herd is better for rewilding than no equines at all. This may certainly be a pragmatic option for smaller rewilding projects. The mares still contribute to habitat creation with grazing, dung and other natural disturbances. But, in the end, depriving these herd animals of a dynamic, natural social life, and increasing their risk of laminitis, felt wrong to us at Knepp. With no foals at foot, colts play-fighting or bossy matriarchs, the spark of interaction goes and, with no future generations forthcoming, the acquired wisdom of the herd is halted in its tracks. The impact on habitat creation is also significantly reduced. We felt an obligation, too, to support the continued existence of a rare and ancient breed. Free-roaming Exmoors, like all native British wild pony breeds, are rarer than the tiger, and Knepp now has one of the largest naturally breeding feral herds in the UK.

Ultimately, for larger rewilding projects, at least, we feel that contraceptives and removing the reproductive rights of animals should not be the preferred option. The aim should be to establish herds that are as dynamic and natural as possible. A strong stallion, for example, will hold a herd with numerous mares and keep to the most abundant grazing areas. A weaker stallion will hold fewer mares and graze less favoured areas.

The boundary between the two will hardly be grazed at all, helping create diversity of habitats.

While we do manage to sell some of our Exmoors to other rewilding projects, rising numbers outgrowing our space remains a problem if we want to continue with a breeding herd. The only remaining option for our excess ponies is to eat them, as we do our cattle, deer and pigs.

Rewilded horsemeat

The horsemeat taboo in the UK is difficult to explain. Horses are eaten in countries across Europe, South America and Asia. Every year 100,000 live horses are transported into and around the European Union for human consumption. In Britain, horse was eaten throughout the Middle Ages, despite a papal ban on horsemeat in 732 and strict taboos among Romany and Jewish communities. Quite when eating horse became culturally unacceptable is hard to say, although recently the British public has shown itself to be more open to the idea, with horsemeat being served in several 'nose-to-tail' restaurants. In 2007 a readers' poll in *Time Out* magazine showed that 82 per cent of respondents supported the chef Gordon Ramsay's decision to serve horsemeat in his restaurants. The flavour is very similar to that of beef, and the meat is tender, especially if it is aged in the same way.

Selling horsemeat has been instrumental in safeguarding the future of the Dartmoor hill pony. In the 1960s there were thousands of ponies grazing on Dartmoor. Adapted to survive the harsh moorland climate, they have grazed here for at least 4,000 years. Now there are only about 1,000 still roaming wild on the moor. As a breed, the Dartmoor hill pony is in crisis, and without their particular type of grazing, the moorland faces the prospect of being overcome by rough grasses and losing important habitat and diversity. Numbers need to increase. An enthusiastic market for pony meat in restaurants, pubs and farmers' markets around Dartmoor – encouraged in recent years, ironically, by horse lovers concerned about the welfare and future of the breed, and boosted by the tourist trade – has provided farmers with a source of income from the ponies and given the wild herd a new lease of life. New foals can now be guaranteed at least three years living wild and free on the moor.

Only three abattoirs in the UK are licensed to slaughter horses for human consumption. In theory abattoirs would need only minor alterations to accommodate horses, but they are generally reluctant to attract negative attention from British horse lovers and to deal with a mountain of extra paperwork and inspections. There remains the problem of transport to a horse-licensed abattoir, which may be very far from the site. In principle, any good butcher should be able to butcher horsemeat, since it is anatomically similar to beef, and generally has the same names for muscle groupings and cuts. There is increasing interest among butchers, although some are unwilling to do it because of anticipated criticism by customers.

For us at Knepp, our long-term goal is to maintain dynamic, breeding herds of Exmoors. Eventually we hope to develop a system using a mobile slaughterhouse on site for all our excess free-roaming animals, avoiding the need for transport to an abattoir. This will allow us to oversee our animals' welfare every step of the way, and we can process the horsemeat in our existing onsite butchery. For rewilding projects without these facilities, to be able to manage free-roaming herds of horses in this way and provide conservation-grade horsemeat will need greater co-operation from abattoirs and butchers, and that is likely to require a dramatic change in perception and support from the public.

Equine management

When it comes to finding the right person to look after semi-feral equines in a rewilding system, knowledge of horses in a traditional equestrian sense may not be the best attribute. It can take a lot of 'unlearning' to shed the habits of managing domestic horses, to let go of the idea of stabling, routine medical treatment, hard feed and turning horses out into individual paddocks.

The key is to find someone who has a deep appreciation for the value of equines in rewilding and conservation grazing. That could be a stockperson who is in charge of other free-roaming livestock. If they have no previous experience of free-roaming horses, they should be willing to research and learn the latest thinking and studies on the subject, including equine ailments. As with cattle and pigs, it's crucial to find a supportive vet who is enthusiastic about rewilding, open to different ways of doing things and likely to come up with imaginative solutions, as well as being able to judge when to intervene and when to give a nature a chance.

An equine manager or stockperson must also understand the importance of genetics and breeding strategies, and be able to think of the overall welfare of the herd as much as that of the individual horse or pony. To this end (as well, of course, as checking regularly for illness and accidents), it's vital that the manager is happy to spend time in the field, documenting what they see and learning the social grouping and interaction. Breed societies and rare-breed appreciation groups, such as the

Exmoor Pony Society, can be extremely helpful with leasing stallions and providing new colts to diversify bloodlines.

Penning horses for veterinary treatment can be more challenging than for cattle, owing to their flight response. But similar principles to the Bud Williams technique apply, and food, patience and body language are key. An understanding of equine senses, such as the incredibly wide-ranging field view with their large eyes (horses have the largest eyes of any terrestrial mammal) and monocular vision, and their ability to see in low light and to visualise distant objects clearly, as well as of their body language with each other, helps you gauge which individuals will initiate flight and whether you should press forward or drop back. At Knepp we train certain individuals to come to a bucket of food, which allows others to follow, particularly when luring them into a temporary pen that is not in their usual surroundings.

Horses do not need as much paperwork as cattle in terms of record-keeping and legislation. It is, however, important to record herd information so that you don't lose it with inevitable staff changes, particularly if you're keeping a registered pure-bred herd. You'll want to keep an eye on behaviour and aggressive traits if horse riders use the land (see Chapter 10), and traceability may become important if you intend to sell horsemeat for human consumption. In addition, it's important to be organised about microchipping (a legal requirement for all equines, even free-roaming ones), undertaking worm counts, registering youngsters and tracking DNA for breeding lineages.[75]

Pigs

Managing numbers

Important though pig disturbance may be, a little goes rather a long way. At the beginning of our project we envisaged a herd of seventy or more, from which we would make our own air-dried *jamón*. A fully grown sow weighing 230 kilograms, however, can plough 20 hectares over a wet winter. The impact of only two mature sows and eight offspring was so tremendous in the 280-hectare park restoration of the Middle Block in the first eight years that we have decided to give the area a break from rootling for the time being. However, we are considering introducing pigs for a period into the Northern Block, which is currently grazed only by cattle and roe deer, to open up the sward and speed up the succession of thorny scrub. In the wilder, scrubby 440 hectares of the Southern Block we now reckon on six breeding sows and their offspring – so, in summer, thirty or more pigs. This is not enough to sustain a separate pig-based income stream, but they contribute to our organic, wild-range meat venture.

We also came to realise, relatively early on, the wisdom of intervening in the breeding process. The rate of pig reproduction can increase very swiftly. Tamworths become sexually mature at six to seven months. Their average litter in a farmyard is seven to ten piglets, although in rewilding it seems they have fewer: generally four to six, sometimes only two or three, especially as they get older (their life expectancy is between fifteen and twenty years). The gestation period is three months, three weeks and three days, and piglets wean quickly, so sows can easily produce two litters a year. Of course, this is the result of domestication. Truly wild sows have a breeding season like any other species. We have no idea how long it would take for a de-domesticated herd of pigs to sync their breeding cycle, and we haven't wanted to risk the mayhem in terms of numbers to find out.

Bringing in a boar for just a couple of months ensures the sows have only one litter a year, and the piglets arrive at pretty much the same time, so we can keep eye out for them. We have rented three boars several times each over the years. They have all taken to wild living and foraging for themselves easily, although they had only ever been free-roaming on conventional farms before. Our stock manager looks for availability, docility and genetic vigour (the gene pool for Tamworths is not huge) in that order. They have all had tusks, although generally breeders keep tusks trimmed to avoid accidental damage. This has never seemed to be a problem, and we've noticed no wounds on the sows that might have been caused by them. Our only concern has been, on occasion, that the boars, being used to regular handling, can be over-friendly. This can encourage people to pet them and/or feed them, and to expect the other pigs to be as amen-able, which can cause problems.

However, this breeding strategy has not been entirely foolproof. It is much easier to lose pigs – not to mention piglets – than cattle or horses in scrubby, rewilded terrain. Sometimes it can take days of intensive search-ing to round up young pigs for the cull. One year we were missing a young male, presumed escaped or perished. Several months later one of the sows gave birth to six piglets. At six months this adventurous young boar had covered a sow three times his size (we hope it was not his mother).

Regulations for pigs

Classified as domestic stock, free-roaming pigs are governed by the same regulations as farmyard animals. This means having to check them daily, and twice daily if you're organic – which is virtually impossible in a rewilding scenario. This leaves us in a vulnerable position. The Soil Association inspector who came to look at our pigs claimed he had never seen such healthy, independent animals. He fully appreciated the challenging logistics of checking them regularly. The daily checks in an intensive farming system are, in large part, to make sure that food- and water-dispensing systems are operating correctly and that the animals continue to have access to them – something that clearly does not apply in rewilding circumstances. We are now working with the Soil Association to resolve the problem and secure dispensations from this onerous checking regime for semi-wild pigs.

We train the pigs to come to the shaking of a bucket of organic pig-nuts in order to make the round-up and general health checks easier. We do this from a quad bike, so that the pigs associate the vehicle – rather than humans on foot – with food.

Unlike cattle, pigs do not require ear-tagging at birth, although our stock manager does tag the sows to help with management.

Rewilded pork

Like deer, pigs gorge on acorns and beechnuts in the autumn, putting down an astonishing amount of fat in very little time in preparation for winter. Lack of oak or beech trees on your land to provide this vital protein boost may affect the number of pigs your land can support, although Iron Age pigs may be better at coping with a lack of mast (fruits and nuts). We cull all our piglets after the acorn mast, around December, by which time they weigh 60–80 kilograms, an astounding growth rate in just eight or nine months. This leaves plenty of rootling forage for the core group of breeding sows over the winter. If it has been a spectacular mast year, we cull in January or February, so that some of the fat reserves have been depleted and the meat is a little leaner. The fat, though, gives incredible flavour, and the meat – thanks to an active life and foraging – is much darker and denser than the pallid factory pork sold in most supermarkets.

Pigs and horses

Horses generally don't like pigs, we presume because of some atavistic memory of wild-boar predation on newborn foals. Our Exmoor ponies took a year or so to get used to the presence of the pigs and not see them as a threat, although even now if a piglet strays among the ponies it is likely to get kicked to death.

Even if you don't keep pigs and horses together on your land, it is important to make riders on bridleways aware that their horses could shy if there are pigs close by.

Deer

Fallow and red-deer management

Because fallow and red deer are categorised as wild, they can be shot and processed on site in an estate or mobile larder to provide meat for public consumption. This must be done by someone with a rifle licence (.245 or higher), a qualification to shoot deer and, preferably, a Deer Stalking Course Level 2 certificate.[76] If you are not going to cull in-house, there should be plenty of certified deer stalkers in your area who would be interested. The British Deer Society, the Deer Initiative and the British Association for Shooting and Conservation are good places to enquire.[77] The important thing is to make sure that anyone you employ for this job is genuinely interested in rewilding and understands your priorities for creating the right genetic herd for the site. We strongly recommend being businesslike and entering into a clearly stated contract, rather than a loose agreement, with your deer stalker. This should cover stalking rights, the areas of your land where they can stalk, insurance guarantees and shares (or not) in any profits generated from the sale of venison.

There are strong conventions in the stalking and parkland traditions for culling fallow deer to favour size and form of antler, as well as color-ation, in the case of fallow. Fallow have four main colourings: 'common' (classic chestnut coat with white mottles in the summer); 'menil' (very distinctive spots, which continue throughout the winter); 'melanistic' (almost black, with no spots); and 'leucistic' (the rarest, white with no markings – the 'white harts' of history). For us, at Knepp, coloration is not so important and, for both red and fallow, maintaining as natural a social structure as possible is the priority. We follow the Hoffman Pyramid system, culling across age groups and maintaining a roughly equal ratio of males to females.[78] The Deer Initiative website has a useful spreadsheet population model to help you calculate the numbers, sex and ages to cull.

To ensure the deer remain settled and visible – as much for human enjoyment as anything else – it's important to cull in a way that retains their group confidence. Culling at night, although challenging and requiring a special licence, will ensure the animals remain entirely relaxed during the day. It can also be safer if you have busy footpaths.

At Knepp, where we cull a large group in a single day to reduce the overall period of stress, we do not want the deer to associate vehicles with culling, so we're careful never to shoot from a vehicle. We have erected numerous high seats for the marksman in remoter areas, and we rotate their use randomly, so the deer do not learn to avoid an area because of its associations.

It is really important to use copper rather than lead bullets for deer culling. Between 50,000–100,000 wild birds die in the UK each year from ingesting lead pellets from used shotgun cartridges and fragments of rifle bullets that have shattered on impact.[79] A further 200,000–400,000 birds suffer from lead poisoning which can compromise their immune systems. Small mammals can also eat lead fragments, and animals that predate on wildfowl also suffer. In 2021 the UK government announced plans to phase out the use of lead ammunition entirely but so far has given no timeframe for the ban, and users of lead ammunition have been slow to shift to alternatives voluntarily.[80]

Unfortunately, none of the carcass of the deer – including the entrails (which are known as the 'gralloch') – can legally be left in the field (see page 260). Until regulations change, this – as described earlier – deprives a host of creatures of food, from foxes and carrion-eating birds to insects and bacteria.

Venison

If your land has been certified organic and you have a boundary deer fence so that you know no fallow or red deer are coming in from non-organic land, you can sell your venison as organic for a premium. If you can sell direct to consumers and/or restaurants with a particular interest in the rewilding provenance and the welfare of the animal, your venison can command an even higher price.

It is always best to hang venison to allow it to tenderise. The muscles go through rigor mortis and then relax, allowing the flavour to develop. It is also better to hang it in the fur to prevent the meat from drying out. We've found that seven to ten days is optimal; any more and the flavour can become too strong for most people's taste. Building a deer larder, including a butchery area and chiller unit to mature the meat, should be part of your plan if you're intending to sell venison yourself. Your local Environmental Health Authority is a good first base for advice. The Food Standards Agency requires records of all animals sold for meat, and will inspect your deer larder regularly.

There are strict regulations for disposing of all inedible parts of the animal. At Knepp, we have a high-temperature aerobic composting machine that converts all waste into compost within a few days. Even larger bones can be composted if they are crushed first. The resulting rich compost can be mixed with topsoil – the equivalent of the gardener's blood and bonemeal.

Bison

Licensing for release

Bison, prized for their meat, are farmed in several places in the UK, and can also be found in zoos and private collections. But they are listed under the UK Dangerous Wild Animals Act (1976) and require a licence for release. At present, only one licence has been granted by the UK government for importing and releasing bison for nature restoration. We sincerely hope there will be more in the near future – and in more ambitious numbers and over larger areas – as the benefits to the ecosystem, and heathland in particular, are recognised.

As with many large herbivores, releasing translocated or captive-bred bison as a family or social group is best practice, and the larger the group the better.[81] Introducing them into holding pens on site rather than directly into the wider landscape – known as a 'soft release' – gives them a chance to acclimatise.

Management

Bison have a reputation for being aggressive. But, as is the case with so many species with which we're no longer familiar, this is wildly exaggerated.[82] Every year 4,000 tourists walk the bison trail in the Kraansvlak in the Netherlands. There, the bison have even become acclimatised to walkers with dogs.

The fencing required for bison is not excessive. In the Kraansvlak, they use just four strands of electric wire, similar to the amount used for fencing in cattle. Bison could also be managed using virtual fencing to avoid conflicts with roads, crops and humans, with the older females – the leaders of the herd – wearing GPS collars.[83]

Water buffalo

Management

Unlike bison, water buffalo do not require a license in the UK. They are covered by the same regulations as domestic cattle, including requirements for ear-tagging and TB testing. Indeed, a number of farms in the UK manage water buffalo for meat, milk and mozzarella. In some countries they have a reputation for being aggressive, but often this is a result of bad handling. Water buffalo do not respond well to being driven like cattle, or being treated harshly. Experienced managers of water buffalo consider them to be exceptionally intelligent, sensitive, social creatures, with long memories and the ability to pass on knowledge to the rest of the herd. Michel Jacobi, the German ecologist with a reputation as a water buffalo whisperer, reared the animals that were introduced into the Danube Delta rewilding project. He says they behave more like elephants than cattle.[84]

Water buffalo eat twice as much as cattle, so have an even greater influence on the landscape. Their impact in wetlands, in particular, is dramatic, opening up vegetation and creating pools, which benefits birds, aquatic insects and flowering plants.

Water buffalo love rubbing against trees, sometimes pushing them over.

Their strong herd bond helps them resist even wolf attacks.

As well as eating dominant reeds, buffalo toss clods of reeds with their horns.

Mud-rolling creates ephemeral ponds – habitat for amphibians and drinking spots for birds.

Unlike cattle, their digestive system can cope with rush, sedges and alder.

A water buffalo's skin is six times thicker than cattle, with one-sixth of the sweat glands, so it needs to cool down in water.

Leading water buffalo from the front with a bucket of sugar beet pellets or similar treat is the best way to train and coax them into handling systems. They will learn to come to a particular call, and always respond best when treated gently and with patience. That said, some bloodlines can be more challenging and temperamental. As with cattle, these can be selected out of the herd over time. Their lack of flight distance and fear of humans may be a concern for projects with public access, especially dog walkers. It may be possible to manage water buffalo and keep them away from public footpaths and bridleways using No Fence GPS collars, perhaps even targeting reedbed and other areas in need of their disturbance. In conventional systems a single visible white electric strand is all that is needed by way of fencing. The animals learn very quickly to keep within it. As far as other animals are concerned, while not antagonistic towards them, water buffalo tend simply to avoid herds of cattle and ponies.

One final thing to be aware of is that water, of course, is no barrier to water buffalo, so if a site is on one side of an unbounded river or lake they will easily escape.

7

Becoming the Herbivore

Mimicking nature in smaller-scale rewilding

On smaller-scale rewilding projects, where dynamic natural processes are severely restricted or absent altogether, it is down to humans to mimic the kind of natural disturbances that play out in the wild and create habitat complexity. Many of these interventions are familiar to conventional conservation.

These volunteers hacking back spindle, hawthorn and wild clematis, or 'old man's beard', on Ditchling Beacon in the South Downs National Park are, essentially, performing the role of large herbivores. The rewilding mindset takes the process further. By randomising and varying the intensity of these interventions, as happens in the wild, the ecosystem of even the smallest site can become more complex and dynamic.

The idea of human intervention can be counter-intuitive to the would-be rewilder. A common view is that rewilding a small plot involves removing all impact and management by people, and just letting it go. However, rewilding is not the same as abandonment. Particularly at a small scale (anywhere from around 20 hectares to the size of a small field, say), leaving land alone is to leave it to an artificially limited range of natural processes – notably being overtaken by competitive perennial plants.

Rewilding is about restoring other natural processes, including the disturbance created by large free-roaming herbivores and dynamic natural water systems. Where this is not possible – which is generally the case at small scale – the rewilder mimics those natural processes by artificial means and the use of proxies, such as scythes, spades, diggers, bulldozers, hedge-cutters and chainsaws.

In some ways, this idea of human intervention may begin to feel more like conventional 'hands-on' conservation, and, indeed, some of the techniques involved may be the same. The table on page 410 shows how the rewilding mindset differs from conventional conservation, and here it is worth reiterating the difference in the two approaches with an emphasis on the management of small projects.

For example, the management of a hay meadow for conservation would involve cutting and carting in the summer, removing the nutrients to maintain the meadow's characteristic biodiversity. The rewilding mindset would envisage the meadow in the wild, as a preferential grazing area used by free-roaming herds of herbivores within a wider system. The summer cut would be seen as mimicking grazing impact and could therefore benefit from variation in timing and height from year to year. Only part of the hay crop might be removed, leaving patches with a higher cut or not cut at all; and rolling some areas with a corrugated roller to perform the same function as animals treading in the seed. Instead of being carted off for storage, the hay could be broadcast on nearby areas in patches, spread thinly and then rolled, ensuring the seeds make contact with the soil so they can germinate. This would mimic the longer-distance transportation of seeds in the hooves, coats and guts of large herbivores, and the trampling of those seeds into the ground. It would achieve a wider dispersal of seeds beyond the meadow site, while ensuring the nutrients remain within the system.

As well as the natural processes that it is possible to mimic or re-establish at small scale through human and/or mechanical intervention, there are disturbances that occur more dramatically at a larger scale, and episodically, such as windstorms and outbreaks of disease. These will happen of their own accord over time. But it is worth acknowledging their value and appreciating that, as distressing as they may seem at the time, these forces of nature can be hugely beneficial to ecosystems, and create widespread opportunities. The hurricane in southeastern England that felled 15 million trees in 1987 has proved to have been predominantly beneficial for ecology, opening up clearings in closed-canopy woodland and creating new habitat.[1] The tendency at the time, though, was to regard the storm as creating 'damage' that had to be mended as quickly as possible. Fallen trees, even those that might have regenerated, were chain-sawed to pieces and cleared away, often burned, and the clearings were replanted. The clearing operation itself compacted soil that could have provided fertile ground for wildflowers.

The rewilding mindset would react differently to similar storm damage today, recognising it as part of the natural cycling of forests between closed canopy and open habitat. Fallen, dead and dying trees would be left in situ wherever possible, and clearings allowed to remain open. Accepting storm damage and imitating its impact through cutting, felling and disturbing could be part of the mix of techniques by which to achieve the age structure of trees and the general diversity of species, which would normally happen through natural processes on a large scale, in the space available in a smaller rewilding project.

What kind of land is suitable for small-scale rewilding?

Not all land is suitable for rewilding, and this is particularly the case at small scale. Many small, isolated sites support sensitive, restricted wildlife that relies on a particular type of management. Removing that management may cause the loss of much of that wildlife value. One would not want to rewild chalk grassland with rare orchids and skylarks, for example, and in Chapter 4 we discuss the uniqueness and inherent vulnerability of isolated patches of ancient woodland (see page 183).

Highly productive agricultural land is also unlikely to be the first choice for rewilding, since food production is a legitimate – indeed necessary – objective for much land. This is not to say that nature has no place in agricultural landscapes. As we have seen, rewilding is hugely beneficial for farming and we should be planning to enmesh the two. Small-scale rewilding, therefore, might apply to smaller patches within a larger landholding under farming or other productive management. There may be obvious areas within a large site

A hedge at Knepp that was allowed to billow out is now home to nightingales and dormice, among others. The cut section, maintained for the trail hunt, shows how the hedge used to look.

that are inaccessible, problematic or unproductive that would lend themselves to small-scale restoration: corners of fields that might become rough grassland, boggy areas that could be rewetted, hedges that could grow out, and oases of ancient woodland that could be expanded or linked through natural regeneration.

In general, the best areas for rewilding are those with low value for agriculture and forestry, and that contain no special or sensitive wildlife already. One way in which small-scale rewilding can be particularly important is in connecting one area of nature with another.

Connectivity

When considering buying land to rewild, it is important to think strategically about what type of land, and where, makes the biggest difference. Buying a piece of land for restoration that is surrounded by industrial agriculture will be challenging because of the lack of opportunities for connection with other areas of nature, and the edge effects of a hostile and polluting environment, not to mention the impact on the rewilder's morale when surrounded by land managers with conflicting aims. That's not to say the situation can't change. If and when the UK government's Environmental Land Management scheme is rolled out there will be incentives for agricultural land everywhere to be managed more sensitively, and to create wildlife corridors and stepping stones. Some heroic souls consider small-scale nature restoration in these hard-farmed plains important for the very fact that it could, one day, seed regeneration in the area, in both a physical and an inspirational sense. It might also be saving a remnant of natural or semi-natural habitat from destruction. These projects become islands of hope.

For an easier ride and exciting results much sooner, however, it's worth looking at land that directly abuts or can connect with existing areas of nature. The Sussex Wildlife Trust's purchase in 2001 of 80 hectares of former arable and potato fields adjacent to Ebernoe Common, a 230-hectare National Nature Reserve and Site of Special Scientific Interest north of Petworth in West Sussex, is a perfect example of how rewilding even at relatively small scale can have a synergistic effect on the surrounding area if the site is connected to a biodiversity hotspot. The ancient wood pasture of Ebernoe Common is one of the UK's most important roosting sites for barbastelles, one of the rarest bats in Europe. Barbastelles forage over an extensive area, and can commute up to 30 kilometres a night on the hunt for insects. To forage successfully, they need a network of interconnecting hedgerows, wetland and belts of trees to use as flyways. The regeneration of woody shrubs and flora on nearby farmland at Butcherland has provided the Ebernoe bats with a rich source of insects on their doorstep and a connection with habitat further afield. It is almost certainly because of the boost of the Butcherland site that, between 1998 and 2008, the population of barbastelles roosting at Ebernoe doubled. They can now be found foraging at Knepp, not far to the east. The emerging bramble and thorny scrub of Butcherland have benefited a huge range of songbirds, too, including threatened species such as Dartford warblers, whitethroats and nightingales.

Rewilding 80 hectares of former arable and potato fields at Butcherland next to the nature reserve of Ebernoe Common (230 hectares of ancient wood pasture in Sussex) has created a flyway for rare barbastelle bats, enabling them to forage for insects much further afield. It is almost certainly why the barbastelle bat population at Ebernoe has doubled since Butcherland was rewilded.

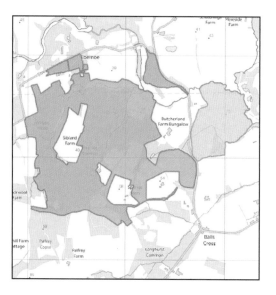

Patch rewilding and the passive-active-passive approach

We have mentioned the importance of habitat complexity and variation for biodiversity, particularly early successional stages of vegetation, such as near-bare ground with pioneer species, new grassland and thorny scrub.[2] At the landscape scale, where natural processes such as large herbivore interaction, natural colonisation and dynamic water systems are back in play, and boom-and-bust scenarios have freedom to perform, a kaleidoscope of shifting habitats emerges largely on its own. But at smaller scale this can be more difficult to

establish and is likely to need periodic intervention to keep the
system dynamic.

Patch rewilding from the start

It is worth thinking about how to trigger habitat complexity from the
very start. A powerful way to do this is to rewild in phases, rather than
committing the land to rewilding all at once. At Knepp – as described
in Chapter 4 – we realised this completely by accident. Withdrawing
our fields from arable over a period of six years produced far greater
complexity than if we had brought everything out of production at
the same time. Numerous factors influence what returns in a given
year, including weather, seed sources and the population dynamics
of wildlife, especially the field mice, songbirds and jays that disperse
the seeds. It's worth staggering the approach in this way even when
rewilding just a single field. Taking, say, one-sixth or one-quarter of
the field out of production per year, and thereby enabling the establishment of different communities of plants and varied vegetation
structure, propels biodiversity.

Most species benefit from the messy margins where one habitat
bleeds into another. Lack of habitat complexity is one of the main
reasons that biodiversity declines. Common lizards, slow-worms
and grass snakes, for example, need sunny, open patches where they
can bask, close to thorny scrub into which they can slither for protection. Turtle doves, which are critically endangered, need protective
thorny scrub for nesting that is no more than 120 metres away from
open areas of short, grazed turf (think of their short legs) where they
can feed on the protein-rich seeds of plants such as fumitory, vetches,
scarlet pimpernel and fluellen, while at the same time staying close
to their young. They also need clean standing water nearby for
drinking, and the males like dead branches of tall trees for territorial
posturing.

Different stages of vegetation succession favour different species.
The early rough-grassland stage of arable reversion is often colonised by skylarks, sometimes in large numbers. Once open, diverse
and scattered scrub exceeds 1 metre in height, yellowhammers,
linnets and whitethroats move in. As the scrub thickens, opportunities
present themselves for garden warblers and nightingales. Most of
these birds are rapidly declining in the UK.

In the absence of browsing and grazing animals, the question
arises: how to manage the growing scrub, preventing it from evolving

into species-poor closed-canopy trees? How to preserve the quality of the different habitats – particularly the early successional stages?

Episodic interventions

On many small nature reserves the conventional conservation practice is the regular rotational cutting of scrub in order to maintain early and mid-successional vegetation and prevent the end result of ubiquitous closed-canopy woodland. But this practice, akin to coppicing, does not disturb the soil, and can lead to uniformity of regeneration, losing vegetation complexity and structure across the site.[3] A more dynamic system can be generated through irregular, episodic interventions. These can be as dramatic as returning some areas to bare ground – with a digger or by hand with a work party – in order to trigger the natural colonisation process all over again. This is effectively what heavy-hitting browsers and wild boar would do in combination in the wild: trashing and uprooting woody shrubs, pawing out bull pits and sand-baths, scarifying the ground with hooves and antlers, and rootling up the soil with snouts. The patchy soil disturbance encourages different plants. On chalk and limestone, for instance, simply removing scrubby vegetation does not allow the area to revert to species-rich grassland. Disturbance opens the soil for early successional plants, such as kidney vetch and horseshoe vetch, that favour raw, disturbed chalk grassland. These are the food plants of the chalkhill, small and Adonis blue butterflies.

Episodic interventions mimic the random impact of herbivores in the wild, which is governed by population fluctuations and die-offs, long-distance migrations and the effect of apex predators. Allowing time between interventions gives the early and mid-successional stages of vegetation a chance to perform – the kind of respite the land would have after a natural crash in herbivore populations.

This mixed approach of neglect and episodic intervention has been described as the 'passive-active-passive' pattern of rewilding.[4] It is useful to bear in mind at larger scale, too. In terms of mental attitude, it's about being able to embrace broad limits of acceptable change and a style of management that is adaptive – being patient and allowing land to respond in perhaps unexpected ways, intervening only when biodiversity and dynamism are being compromised. The beauty of it is that there is plenty of room for experiment. In a sense, there is no such thing as a mistake, and no 'too late'. If something isn't working, the management can simply be changed a little and the 'mistake' becomes a springboard for something else. It might even have laid the foundations for an unexpected benefit.

Episodic intervention on the rewilding spectrum

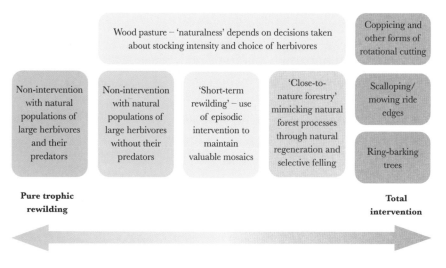

Another way of looking at the rewilding spectrum shows non-intervention at the wildest end on the left, with a sliding scale of human intervention down to the close, hands-on management of conventional conservation on the right.[5] In the middle, 'short-term rewilding' and 'close-to-nature forestry' offer a freer, more dynamic approach for the small-scale rewilder. This could include random and episodic inter-ventions that mimic animal disturbance and natural events such as storms, land-slides and floods.

The advantage to wildlife of rotating drastic management across different patches of land, especially within a loosely connected cluster of sites, is considerable. The approach has been implemented to great effect in two small-scale rewilding projects in East Anglia: a 17-hectare former arable field called Hullbacks, adjoining Arger Fen and Spouse's Grove ancient woodland reserve, and Black Bourn Valley reserve, which consists of 57 hectares of former arable plus flood meadows and planted woodland next to the river Black Bourn.[6] In both places, as at Knepp, there was no active preparation of the site following the abandonment of arable farming, and natural regeneration was favoured over tree planting. The remarkable diversity of habitat types that has developed at both sites in a short time is partly a consequence of the complex soils and topography. The creation and restoration of ponds has also played an important role, reintroducing wetlands that had been there before the land was drained for agriculture. But the way this has been done – restoring ponds in stages, combined with episodic small-scale scrub management aimed at maintaining different successional stages – has resulted in an astonishing rise in biodiversity, including colonisation by nightingales, turtle doves, yellowhammers, common lizards, grass snakes, great crested newts and numerous endangered aquatic plants and invertebrates.

Replicating natural processes

The kind of intervention that might be used in this episodic way will depend on the nature of the site and the stage of vegetation growth. But it is worth entering the mindset of the disturbing forces that would be active in a wider, wilder landscape. How often, when and to what intensity these interventions should happen will always be the subject of much discussion, but from a rewilding perspective variation over time, rather than regularity, is key.

In particular, studying the behaviour of large herbivores in the wild and understanding their impact on the ecosystem is helpful in imagining ways to mimic them. It is worth visiting rewilding projects with free-roaming animals, such as Wild Ken Hill in Norfolk, Wild Ennerdale in Cumbria, Wicken Fen in Cambridgeshire and Carrifran Wildwood in Scotland.[7] At Knepp we run various safaris with experts focusing on the behaviour of the free-roaming animals, as well as smaller-scale rewilding workshops covering all aspects of rewilding in smaller areas. Simply watching nature documentaries on wild horses, cattle, wild boar and deer can give a good feel for their behaviour.

On a small area of land, the interventions that can be carried out to mimic the actions of natural disturbance largely involve the rewilder becoming the herbivore (the keystone species):

- **Replicate wild grazing** by mechanical cutting of grassland. Patch-based rewilding does not have to involve scrub, especially on unproductive soils. Wensum Valley Nature Reserve in central Norfolk, an SSSI, consists mainly of fen and damp grassland. It had long been grazed by ponies belonging to a local scrap dealer. Now the ponies have gone, grazing is replicated by mechanical cutting. Grazing has also, sadly, stopped on many commons throughout England, among them Wood Green, Gissing Common, Brewer's Green and Boyland Common (all in Norfolk). These are now mechanically cut with the benefits for wildlife in mind. Using a patch approach (changing the height and frequency of the cuts as though mimicking the different mouths and grazing intensity of various herbivores) increases the diversity of the grassland.
- **Replicate herbivore browsing** by cutting and lopping trees. This could vary from cutting scrub and coppicing areas of woodland to sub-lethally damaging large trees (mimicking

deer in the rut using their antlers to break branches, and other browsers pulling down branches to feed on foliage) and perhaps even removing some bark (mimicking the winter bark gnawings of red deer, horses and bison).

- **Replicate the windblown uprooting of trees.** Standing dead trees are more valuable habitat than fallen trees for birds, bats and insects because these creatures take advantage of the height for nesting and feeding. In places where there is plenty of vertical deadwood, however, pulling the occasional dead tree over (perhaps where it might be dangerous to leave it standing) creates a deep, complex hole where the root ball was. On clay soils especially this will become an ephemeral pond, fantastic habitat in itself, and worth replicating artificially if there are no dead trees to spare.

- **Replicate storm damage in trees.** Strong winds can shatter the higher branches and crowns of trees. In the past, foresters replicated this process by pollarding (cutting the tops off) trees for livestock fodder and timber products, thus harvesting resources without harming the trees; indeed, pollarding can, like coppicing, significantly extend the life of a tree. The broken branches create cavities and niches for insects and nesting birds.

- **Replicate herbivore browsing on grown-out hedgerows** by using a hedge-cutter to randomly 'scallop' the edges of grown-out hedgerows, from the ground to about 2 metres in height. This mimics the undulating mantle created particularly by browsing cattle, creating a denser, thornier, more complex edge that is great nesting habitat for songbirds such as nightingales. It also provides wind protection for flying insects, which can always find still, calm places to feed within the scalloped margins.

- **Replicate herbivore browsing on open-grown thorny shrubs and small trees.** It's astonishing how woody shrubs, such as open-grown hawthorn and blackthorn, respond to 360-degree browsing. This is, after all, why gardeners prune rose bushes and garden shrubs: to increase their vigour, density, flowering and longevity. Where there are herbivores in a dynamic natural scrub landscape the effect can be startling. We call it 'goat topiary', after seeing the extraordinary, random, stunted, bulbous sculptures of holly (or kermes) oak heavily browsed by goats in ancient wood pasture in Crete. The trees grow up responding to browsing almost like bonsai, with tiny leaves and dense branches. This can continue for decades or even centuries, but the moment the tree is able to send up branches higher than the goats can reach, there is a sudden loosening of form, a rocketing of free, forking branches, which form a loose, feathered crown. We've noticed the same in wild crab apples, which reduce the size of their leaves, flood them with tannins, and turn their twigs into long, sharp thorns in response to browsing. Once out of the reach of herbivorous mouths, they don't bother any more and the leaves become large again. These lush leaves support a great variety of moths. For smaller areas, a hand-held hedge-trimmer is perfect for randomly sculpting shrubs from time to time, and can be great fun. For larger areas, a more brutal hedge-cutter, mounted on the back of a tractor, can be used. Local farmers often act as contractors for hedge-cutting, but would need to be directed as to how to behave as a herbivore proxy.

We think of topiary as human artistry, but herbivores have been doing it for millions of years. On the left, this oak in the mountains of Crete has been nibbled into an outlandish, almost unrecognisable form by goats. On the right, a hawthorn at Knepp 'feathers' out at the top, once it gets above the reach of deer, cattle and horses. Incredibly, intensive browsing extends the life of the plant, and creates an alternative form of dense, protective habitat for wildlife.

- **Replicate rootling disturbance from wild boar** by digging scrapes, stripping and rolling up turf, and physically churning the ground with a rotavator. The exposure of patches of bare soil is a key component of habitat creation, enabling new communities of plants and invertebrates to colonise. This is particularly valuable on sandy soils, where so many insects (from tiger beetle and bee wolf to paper wasp and velvet ant)

and annual plants (such as mossy stonecrop, yellow centaury, sand catchfly, allseed flax, chaffweed and the tiny grass called early sand grass) are restricted by monotonous turf. Exposing soil also replicates the excavation of territorial pits and wallows by bulls and bison, and the hoofing of bucks and stags. As well as exposing the soil, turning over turf – mimicking the clods of earth cast aside by the snouts of pigs – provides habitat for insects to colonise among the rolled roots and vegetation. Ants swiftly take advantage of turned-over clods to get a head start on their anthills.

- **Replicate pig nests.** Dried and rotting grass, twigs and leaves, nosed into mounds by sows creating warming nests for themselves and their piglets, provide wonderful habitat for snakes and reptiles, in much the same way as compost heaps in the garden. Leaving piles of debris from cutting shrubs and mowing or strimming paths can replicate this.

- **Replicate beaver activity** by coppicing bankside shrubs, creating leaky woody dams and increasing pond complexity. The watercourse can be the most artificial feature of a small site, straightened, deepened and often locked into an artificial water-scoured form (see Chapter 3). Disrupting streams with blockages of woody debris is relatively easy, but encouraging them to take a more dynamic, meandering course may, ironically, require more intensive up-front management, using mechanical diggers to restore a more natural landform and hydrology. If done well, however, this can be a one-time intervention that keeps on giving.

- **Replicate herbivore compaction** by digging shallow scrapes and ephemeral ponds. On wet land, horses, elk and red deer often dig and trample muddy puddles to roll in, and cattle, scratching on a favourite low-hanging branch, can compact the soil so that it begins to hold water. This effect can be mimicked by hand or using a digger to make a shallow scrape from a few square metres to 0.25 hectare. The bottom of the scrape can be left very rough, providing different water levels and niches for aquatic life.

- **Replicate the compaction of animal trails** with vehicles or footpaths. Compacted ground can be wonderful habitat for solitary bees and wasps. Red bartsia – the only food plant of the red bartsia bee – thrives in compacted soil on pathways. Horses create a very particular type of track – smoother and more uniform than those made by cloven-hoofed animals, such as cattle and deer – that is particularly favoured by solitary bees.

- **Replicate the transfer of seeds by herbivores.** Mammals both large and small transfer seeds across the landscape in their fur, and the larger herbivores in their hooves. Sweep nets can be used to move seeds from plants to other parts of the site, or scythed grass can be scattered over wider areas.
- **Replicate stallion latrines.** Stallions stake out their territory by dunging and urinating in particular places. This increases the nutrients in the soil, creating ideal conditions for stinging nettles. Nettles in turn are food for many insects, butterflies (or their caterpillars), moths and birds, and an egg-laying haven for ladybirds. This can be very easily replicated by wheelbarrowing in horse manure, leaving it in one spot and changing the spot next year.
- **Replicate bull pits.** Bulls and bison paw up the ground to make territorial pits, which they also use as dust-baths and wallows (see page 237). This is another way that bare soil is naturally exposed, creating opportunities for colonisation by insects, which nest in the walls of the pit, and, when the wallow is abandoned, by plants. Such pits can be replicated using a digger or by hand.
- **Replicate stone and gravel banks.** Stones have been systematically removed from agricultural fields over centuries, if not millennia. An abandoned stone farm building, made safe by pulling down unstable structures, provides wonderful 'brownfield' habitat. A farm building made of unsightly breeze blocks can be ground down using a concrete-crusher and left as a mound of rubble in a good spot where the sun hits it. Rubble is paradise for drought-tolerant plants, insects, snakes and lizards.

Large herbivore disturbance at small scale

Clearly, if the vegetation and size of the land can support even a few large herbivores, the time and effort spent mimicking their disturbance will be saved and the complexity of the habitat increased in ways that are almost impossible for humans to match. But in most cases, especially on land that has been farmed, it is important to allow time for vegetation to establish before adding large herbivores.

Wild herbivores may already have an influence on the land, in the form of deer, hares and rabbits. Geese and ducks are terrestrial herbivores too, of course, and they can have a significant influence in large numbers. Natural fluctuations in these populations will encourage patchiness of vegetation. But it may be necessary to ensure that the site is not a magnet for deer. Culling will keep the impact within acceptable levels, or the area can be deer-fenced for a period to allow thorny scrub and vegetation to develop (see page 174).

On land of 10 hectares or more, domesticated livestock could be brought in either seasonally, for just a short period, or every few years, perhaps rotating the species. For instance, a few cows might be allowed to graze one spring and summer, and ponies the next. It is better, if possible, to have a small herd of just one species – for herd dynamics and sociability – rather than a few individuals each of several species.

Much depends on whether there is a local farmer or horse owner willing to graze their animals on the land. It may be necessary to pay them, to cover the cost of transportation and managing stock on another site. Biosecurity will also be important, and, particularly, the concerns farming neighbours will have about importing livestock to the site. They will need to be reassured that the animals come from a reputable and disease-free source.

On a site of 10 hectares or more a pig or two might be brought in for a few weeks every now and then to churn things up. Autumn is the perfect time for this, when the soil is damp enough for rootling, and especially if there are oaks or beech and the pigs can fatten on the acorns and/or beech mast. Pig disturbance is particularly useful in the early stages of a project, opening up the soil to encourage the germination of wind- and bird-dispersed seeds of woody shrubs, such as hawthorn, blackthorn, dog rose and bramble, and trees such as ash and sycamore. Again, this depends on finding a farmer willing to let a couple of their pigs loose on the land, although as rewilding becomes more popular the availability of 'flying herds' and 'flying pigs' is likely to rise. Indeed, some wildlife trusts already run flying herds.

The pig is a useful unit to bear in mind when thinking about scale and disturbance. As we have seen, a mature sow can plough 20 hectares in one winter. But, of course, the minimum will always be two or three pigs free-roaming together, for social reasons and for their welfare. It may also be worth considering miniature breeds, such as Falabella horses, Dexter cattle and Kunekune pigs, for similar disturbances with less impact. These smaller animals could stay on the land for longer than the larger ones.

If the land does not contain standing ponds or other accessible water, the strategic siting of water troughs is a useful way of encouraging greater disturbance by herbivores in some areas over others.

One herbivore that can work wonders even at a small scale is the beaver. In just five years a pair of beavers has transformed 3 hectares of secondary woodland along 200 metres of canalised stream in Devon into a biodiverse wetland haven for wildlife, enabling the return of endangered Culm grassland. Hopefully, the granting of licences for releasing beavers will become common and the bureaucratic process easier as the benefits of these keystone species are more widely recognised. In even just a few hectares of wetland, a pair of beavers will be all that is needed to rewild.[8]

This leaky beaver dam of mud and sticks holds back water in heavy rains, storing it for slow release through drier periods. The watery kingdom the beavers have created is heaving with wildlife. Creating woody debris blockages – in other words, mimicking a beaver – is now a well-established method of improving the function and diversity of streams and wetlands.

Limits of acceptable change

In essence, the small-scale rewilder is working somewhere between the large-landscape philosophy at the broad, expansive end of rewilding, and the tight, controlled management of conventional conservation. The rewilder always aims to restore or copy natural processes, rather than to satisfy fixed conservation objectives. But at a small scale there will probably be a need for hands-on management, and the smaller the area, the more management. The emphasis for the small-scale rewilder must be on flexibility and broadmindedness: setting many objectives within a broad range of acceptability. This involves responding to the results with adaptive management rather than simply determining an outcome and sticking to it rigidly, as in conventional management.

The specific characteristics of a site, as well as the rewilder's own character and attitude, will determine the level of change they are prepared to accept. At Knepp, for example, in the larger-scale rewilding scenario, the abundance of bluebells in woodland has been reduced by the disturbance of pigs (proxies for wild boar). Many people would argue that common bluebells, being characteristic of the UK – a stronghold for the species – should be a conservation objective. But by not rushing in to protect them, and rather keeping a watchful eye on their long-term, overall survival, we have had unexpected results. Bluebells are now appearing in the newer areas of scrub. Restored natural processes have shifted their context in the landscape from being a limited feature of woodland only, to being part of overall ecosystem patchiness, which reduces and increases as part of overall diversity. The hands-off approach has allowed bluebells to demonstrate their ability to colonise unexpected areas.

By broadening the approach from the specific objectives and intensive management of conventional conservation, rewilders can embrace a more dynamic system of nature restoration, even at small scale. With fluid objectives and adaptive management, small rewilding sites can manifest some of the natural processes at work in the wider landscape, and this is likely to increase biodiversity and habitat complexity. Of course, in smaller areas, the dramatic interventions and periods of neglect characteristic of rewilding can be harder to live with. The 'limits of acceptable change' will be different for everyone and may shift with time as wildlife successes become apparent and rewilders grow in confidence.

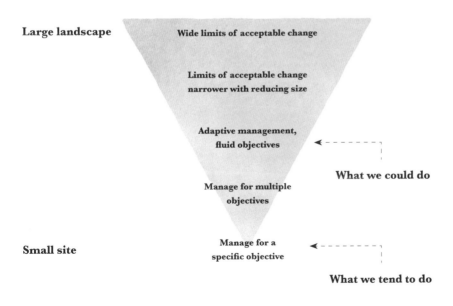

Large landscape

Wide limits of acceptable change

Limits of acceptable change
narrower with reducing size

Adaptive management,
fluid objectives

What we could do

Manage for multiple
objectives

Small site

Manage for a
specific objective

What we tend to do

At a small scale, though, this hands-off approach may not work. If there is not sufficient habitat around the woodland, pigs may rootle out the bluebells altogether. Traditionally, pigs would probably have been fenced out of the wood. This may still be the correct approach. But at an intermediate scale one could observe their effects using different criteria. It would be interesting to observe what else changes when the pigs are present. This may include other benefits, such as increased carbon storage in the soil.[9] There are likely to be limits of acceptable change at this scale – such as not wanting bluebells to disappear altogether. If the bluebell loss seems unacceptable, therefore, adaptive management comes into play and the pigs are removed.

Understanding the dynamic natural processes at play in the wider landscape is fundamental to rewilding at smaller scale. It is this ability of the small-scale rewilder to imagine and relate the disturbance of beavers, bison and wild boar, flash floods, landslips and episodic tree deaths to their site that will generate the energy and opportunities for life. The most unpromising field can, over time (and perhaps quite fast), become a complex ecosystem. Again, this is not about abandoning the land. And while considerably less work is involved in the rewilding approach than in conventional conservation, it still entails a fair amount of hands-on management. Being the proxy for a variety of keystone species can be challenging and time-consuming. Local low populations of deer and rabbits will be friends of the rewilder – as will volunteers. If grazing animals can be allowed in, even just every few years, they will do much of the job.

Perhaps the greatest challenge of all, for the small-scale rewilder, is to embrace randomness. As humans, we find routine comforting. It is tempting to fall into a predictable pattern, even with rewilding, and particularly at a smaller scale. Creating waves in the land-management routine with big, dramatic interventions and episodes of passivity – doing absolutely nothing for long periods – can be unnerving and feel counter-intuitive. It's the 'Don't just do something, stand there' approach again. But this is what nature thrives on. And no short-term disturbance will ever be too big for nature to contend with. An eminent ecologist friend of ours claims – provocatively – that there is no nature reserve in England, however small, that wouldn't benefit from having a bomb dropped on it.

But of course there will be limits of change that you, as a small-scale rewilder, will feel are unacceptable. This is your patch, and you have to live with it. If you feel you're not ready for some of the more dramatic interventions and/or extended periods of neglect, that is entirely up to you. You are still on the rewilding journey. Getting to know your land, how it responds to the changes you make, and the creatures that are stimulated by those changes over time, will inform and perhaps even embolden your future decisions. As always, looking at the land and the possibilities around you, the chances to connect with other areas of nature, will make the most of your site's potential.

At Abernethy Forest in Strathspey, Scotland, contractors ring-bark Scots pines with draw-knives to create standing dead wood. This experimental intervention to create diversity in even-aged Scots pine plantations and provide habitat for deadwood beetles and other creatures is part of Cairngorms Connect, the largest habitat-restoration project in Britain. Effectively, the contractors are mimicking various natural forms of damage to trees, such as bark-stripping by deer, bison, wild ponies and beavers, and lightning and storm damage.

8

Your Rewilding Project

First steps, planning, funding and income streams

A group arriving for a wildlife safari at Knepp. Rewilding projects provide enormous scope for nature-based enterprises. The demand for wildlife-watching and eco-tourism, in particular, has never been greater, especially post-pandemic, and excitement around nature restoration is burgeoning.

Whatever the size of your project – whether it's thousands of hectares with the potential for free-roaming herbivores, or just a few hectares where you'll be mimicking natural processes yourself – it can be daunting to make the leap into rewilding. But the process can be broken down into stages. Here we outline the initial steps you need to take to understand your options fully: the rewilding potential of your land, the funding that is available to you, and other potential income streams.

Getting to know your land

The way we look at the land is changing rapidly – for the better. There are now tools to help you delve into the history of your land, analyse the resources it currently provides and reveal all the nature designations in your area. These innovations and the evolving science associated with them can radically change the way you think about your land and reveal the options for rewilding. There are other obvious steps to take, too, which will help you make decisions. This is an inspiring time to be a landholder.

Topography and climate

If you own or are buying land for rewilding, it's important to understand the topography – the altitude and aspect of the land – since it
will affect the rate of change. Vegetation can grow astonishingly fast
in mild areas, such as the lowlands of southern England, but the rate
slows with decreasing mean temperatures, longer, harsher winters and
a shorter growing season further north and higher. In the uplands of
Wales and Scotland you can expect an even slower response for vegetation succession. Patience will be required in colder environments, but
there will also be economic considerations. For example, if very robust
fences are required because the slower growth of vegetation means
deer must be excluded from recovery areas for longer, this will have
a bearing on the cost of the fences. Those in upland areas may also
need to consider how reducing grazing pressure (having fewer animals
on the land) for longer, while vegetation recovers, will affect revenue.

Designations

Designations on the land, such as (in the UK) National Nature
Reserve (NNR), Local Nature Reserve (LNR), Site of Special
Scientific Interest (SSSI), Special Area of Conservation (SAC),
Special Protection Area (SPA), National Park and Area of
Outstanding Natural Beauty (AONB), will most likely constrain the
scope of rewilding. Equally, if the land abuts a designated site for
nature, this will help to attract funding and also facilitate the colonisa-
tion of the site by new species.

There may be a designated historic site and/or scheduled ancient
monument that must be protected from disturbance. Forests and
woods are subject to a felling licence if they are to be felled or otherwise modified, and the scope of action might be constrained by past
and current agreements with the Forestry Commission. Public footpaths and/or bridleways and open-access areas are designated under
the Countryside and Rights of Way Act, and this will have a bearing
on the choice and management of free-roaming animals.

If you've already entered a Woodland Grant Scheme agreement,
such as the Farm Woodland Premium Scheme, and have received
funding to create woodland plantations, you'll be under obligation
to maintain them. This means you must ring-fence any plantations
within the proposed rewilding area if you're intending to introduce

free-roaming animals – at least until the trees are robust enough to withstand browsing pressure and the agreement has completed its course. You can, of course, pay back the funding you've received, and you may also be able to negotiate a derogation for access to the plantations (depending on their maturity) of very low stocking densities of, say, just cattle and/or horses.

Other designations, such as Flood Zones and Green Belt, may help to attract funding to a rewilding project. Being in a Nitrate Vulnerable Zone (NVZ) may also be an advantage, at least for funding. About 55 per cent of land in England is designated as being at risk from agricultural nitrate pollution. These areas are governed by regulations to prevent the run-off of polluted water from farming. Rewilded land acts as a filter, purifying polluted water – an obvious benefit for NVZs.

The MAGIC Map app

Information on designations can be found easily on the Multi-agency Geographical Information Centre (MAGIC) Map app, an interactive mapping data website managed by Natural England.[1] It's basically a 'look and learn' site that enables the user to pull together all the designations from the various government departments and layer them with Ordnance Survey detail, such as field boundaries and footpaths. It's particularly good for showing your land in the context of other priority habitats and designations in the area – including the very latest directives – which will help with assessing the potential for wildlife corridors and connectivity. The latest update, for example, covers 'Open Mosaic Habitats on previously developed land', the newest priority habitat listed under the UK's Biodiversity Action Plan. These include railway sidings, quarries, former industrial works, slag heaps and brick pits: all brownfield sites that, unpromising as they may sound, are especially good for biodiversity.

The app also, crucially, shows all the Countryside Stewardship funding layers that may apply, such as Biodiversity, Water, Historic Environment, Landscape and Climate Change. The most pertinent layer for rewilding is Wood Pasture (WD6; see Funding section below). The app can also tell you if you are in an NVZ, for example, or in a priority area for lapwing, say, which might make it easier to win funding for wetlands. While these zones, designations, initiatives and pieces of history may lead to prescriptive measures that could constrain rewilding, they can provide a really good overview of the opportunities specific to the region and landholding.

The Land App

The Land App (www.thelandapp.com) was founded in 2015 by farmer and landowner Tim Hopkin after he found it difficult to appraise the different potential land uses of his family's Surrey farm so they could plan for the future. It provides some of the information given by the MAGIC Map app, including designations, but allows editing. Most land managers – controlling about 60 per cent of land in England – now use it. You can build your own project, mapping aspects such as hedges, wildlife corridors, water and vegetation cover, even individual trees, in as much detail as you like. While currently a little more challenging to get to grips with than the MAGIC Map app, it's an indispensable tool for building the concept of a rewilding project. The Land App help team responds swiftly to live enquiries online.

The app also enables you to download Rural Payments Agency (RPA) field-parcel data from the RPA website – an essential and time-saving tool for mapping and facilitating Countryside Stewardship applications or, as they are due to become,

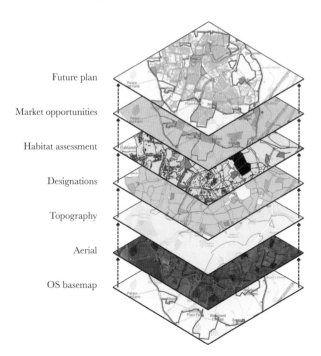

Future plan

Market opportunities

Habitat assessment

Designations

Topography

Aerial

OS basemap

The Land App: Building layers of information over your land. Here, the complex mapping and planning involved in the transformation of the land at Knepp is broken down. Detailed insights into the landscape can help identify areas suitable for nature restoration, as well as map the potential biodiversity 'uplift' and monitor the project as it progresses.

Environmental Land Management schemes (ELM). These are the schemes by which farmers and other land managers will be paid by the UK government to deliver sustainable farming, local nature recovery and landscape recovery (see Funding section below).

The Land App is being used on a variety of ELM trials and looks set to become the preferred mapping system for natural capital (natural assets such as soil, water, trees and biodiversity) and public goods (the services to society that these assets provide, such as carbon sequestration, flood mitigation, water purification and improved health and well-being) – both fundamental principles of ELM. It is also being developed to work in tandem with UK Habitat mapping, so that the delivery of natural capital and public goods can be layered on to individual landholdings. Using the Land App at this stage will stand you in good stead to make the most of ELM when it comes into being.

Connectivity and wildlife corridors

Becoming acquainted with local nature reserves through MAGIC and the Land App will reveal opportunities for becoming part of a wildlife corridor or a stepping stone connecting with one or more conservation sites. It's useful to think of nature as infrastructure with societal benefits, in the same way as roads and railways, and to look at how connectivity can facilitate the flow and movement of wildlife.

Visit local nature reserves. This will familiarise you with the kind of natural habitat you might expect to evolve on your land, and may influence your plans. Gaining as much knowledge as possible of the functioning ecosystems that already exist in your landscape will help you decide on your early interventions.

Become a member of your local Wildlife Trust. They hold extensive knowledge of nature in your area and are closely connected with local wildlife experts – it's astonishing how many there are, both amateur and professional – who may be able to advise you in either a voluntary or consultancy capacity. There may be local target species, such as nightingales or brown hairstreak butterflies, that you could help by expanding their habitat. Allowing hedges to grow out on your land, and perhaps even connecting with hedgerows belonging to your neighbours (especially if they are willing to do the same), creates a wildlife corridor in itself, which is helpful for myriad species.

Old railway lines, embankments, streams, canals, roadside or track verges and river catchments all have the potential for rewilding as wildlife corridors, so it's worth looking at these features in the landscape and identifying landowners and land managers along them with whom you might be able to connect.

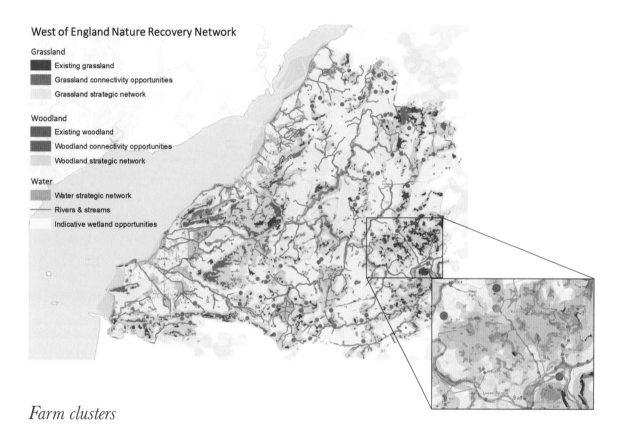

West of England Nature Recovery Network

Grassland
- Existing grassland
- Grassland connectivity opportunities
- Grassland strategic network

Woodland
- Existing woodland
- Woodland connectivity opportunities
- Woodland strategic network

Water
- Water strategic network
- Rivers & streams
- Indicative wetland opportunities

Farm clusters

If you've recently bought a farm to rewild or simply improve for nature, make friends with the local farming community at ploughing matches, open farm Sundays and country shows.[2] In particular, joining your local farm group or cluster, if there is one, is a good move.

The 'farmer clusters' concept was developed by the Game & Wildlife Conservation Trust and is supported by Natural England. Over 100 facilitated cluster groups, working with more than 1,000 land managers, have been set up across the UK since the scheme was piloted in 2013.[3] Some are privately funded, but the majority receive grants from the Natural England Countryside Stewardship Facilitation Fund. Despite in practice being the same thing, farm

This Nature Recovery Network map produced by the West of England partnership provides detail of existing natural habitats and suggests how they could be connected. This kind of intelligent, visionary mapping is hugely valuable, inspiring people to join the dots on the ground.

'groups' are funded by the facilitation fund, while farm 'clusters' tend to be funded privately or by the farmer(s) themselves.

The aim is to empower farmers to make changes on their farms specifically to improve the environment. Farm groups/clusters choose a conservation advisor, or 'facilitator', who helps members devise their own targets and make positive changes for wildlife on their farms, and advises on how to record progress. This landscape-scale approach is obviously far more effective than single farms working in isolation. It also enables farmers to pool their knowledge and encourage each other and, for the first time, devolves decision-making for the environment away from central government. Unsurprisingly, farmers have never been keen on being told what to do on their land by an environmental advisor with a clipboard, and having the reins back in their own hands has proved galvanising. The Farmer Cluster website shows the location of existing groups and provides information on how to set up a farm group/cluster and apply for funding.[4]

It may be that in your farm cluster few farmers, if any, are interested in rewilding. But they will be interested in improving their land for nature. They can provide the stepping stones and wildlife corridors that feed into your rewilding project, perhaps even connecting you with nearby nature reserves. They may be willing to allow a corner of a field that has never been productive to scrub up, for example, to grow out their hedgerows, to plant new ones, to allow a floodplain to rewet, and/or to allow some natural regeneration of trees. They may even be converting to organic or regenerative agriculture, making their land overall more permeable to wildlife. Coming together and thinking strategically about how to get 'more, bigger, better and joined up' land for nature, in any way a farmer is prepared to embrace, is the way forward for habitat restoration.

Our aim at Knepp, now, is to connect with the sea and the 200-square-kilometre Sussex Kelp Restoration Project off the coast. Our Weald to Waves project, initiated in collaboration with James Baird, who farms land abutting the sea at Climping, involves a number of farmers and landowners who have agreed to create habitat on their land that will, eventually, form a 30-kilometre green corridor between us. The corridor will also extend another 30 kilometres northeast from Knepp to the 5,600-hectare heathland of Ashdown Forest. The participation of farm clusters, including the Upper Adur Farm Group – of which Knepp is a member – is key to achieving this ambition.

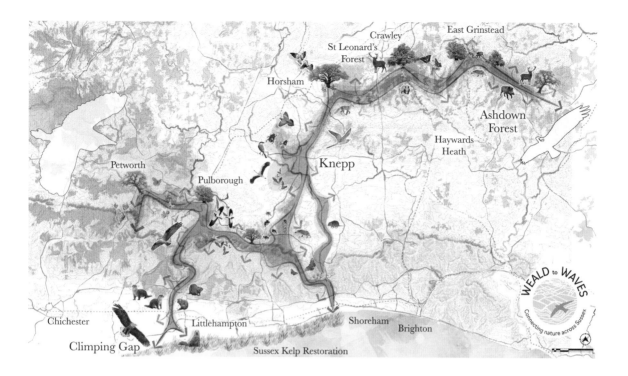

The labels visible on the map:
East Grinstead, Crawley, St Leonard's Forest, Horsham, Ashdown Forest, Haywards Heath, Petworth, Pulborough, Knepp, Chichester, Littlehampton, Shoreham, Brighton, Climping Gap, Sussex Kelp Restoration, WEALD to WAVES, Connecting nature across Sussex

Cropping history

The way land responds to rewilding may be affected by all kinds of historical management, not least intensive farming and chemical inputs, which change the characteristics of the soil. At Knepp the response of former arable land, in particular, has varied dramatically from field to field.

Part of this is undoubtedly to do with the year in which a field was taken out of production – a process that, at Knepp, happened over the course of six years. But even fields released from agriculture in the same year, affected by the same weather conditions and the same seed sources, can look dramatically different. This suggests that the cropping history of each field may have an influence on vegetation succession. Whether the last harvest was maize, wheat or barley – and, in particular, the chemical inputs associated with that crop – is likely to have a bearing on which plants are able to colonise in the early years of reversion. Some soils are more prone to compaction by heavy farm machinery than others. Using machinery in the wrong conditions – when the soil is wet, for example – can create compaction and 'panning' (layers of compaction beneath the topsoil), which the roots of plants cannot penetrate. Water, unable to filter

The Weald to Waves project aims to create wildlife corridors between Knepp and the coast, reaching the sea at Climping and Shoreham, and from Knepp northeast to the heathland of Ashdown Forest.[5]

317

through the compacted layers, then sits within the soil, creating waterlogged areas.

	Field A crop	Field B crop
2000	Winter bean	Spring rape
2001	Winter wheat	Spring barley
2002	Winter wheat	Winter wheat
2003	Winter rape	Winter wheat
2004	Winter wheat	Winter wheat

The suggestion here isn't to try to do anything about the historic effects of land management and agricultural input. Very little is known yet about whether past agricultural management affects vegetation succession, and how, and whether it is even possible or desirable to change the outcome. But it is useful to keep records of past management and monitor how individual fields respond to rewilding. Fixed-point photography (see page 373) is the easiest way to record the evolution of vegetation on the site.

Having seen the benefits of releasing fields from arable over an extended period, we recommend that rewilders on arable do the same, if possible. The mosaic of habitats that results is hugely beneficial for biodiversity – more so than if fields are all rewilded in the same year, even with differences in cropping history. Withdrawing from farming in stages may be a little complicated, but there are benefits in a gradual approach. It will have less impact on neighbours, for example, allowing them time to get used to the idea. For us, it meant a gentler transition. We took the problematic, low-yielding fields out of production first – an easy decision – and, having sold our farm machinery, contract-farmed the better fields for a few more years with a local farmer.

The inspiration of history

While it's important to familiarise yourself with existing areas of nature in your region, remember that these are likely to be just a shadow of what once existed. Looking at landscape change through local history can be hugely informative.

The National Library of Scotland has an amazing tool, the side-by-side viewer, which lays historic maps next to modern ones, including satellite data, so that you can slide history backwards and forwards.[6] You might see ancient hedgerows disappearing as fields become larger in the 1960s and 1970s, or uncover secrets such as old sandpits, chalk pits and ponds. It's an invaluable resource to help you understand the context of your land.

The screenshot on the next page shows part of Knepp in the late nineteenth century next to current satellite imagery, with the old deer park clearly visible around the castle and lake.

The local records centre will also have maps, publications and legal land documents that can offer insights into what your area and, particularly, your land was like in the days before intensive agriculture. If it is possible, going beyond the industrialised Victorian era to medieval times may provide a richer natural baseline to steer your thoughts, as well as flagging up cultural connections that might help with funding. Look for records of ancient woods, coppicing, hunting forests, parkland, commons, hay meadows, scrubland, permanent pasture or plantations on your land. Check whether systems of water functioned differently in the past. If you find evidence of boggy patches, ponds, marshes, floodplains and/or streams, you can try to restore them. Look for ditches and drains that are still functioning, and decide whether their purpose is still relevant; you might be able to consider getting rid of them (see page 134).

Investigate whether there are any unmarked archaeological sites – Iron Age settlements, long barrows or ridge-and-furrow tillage systems, for example – that might need to be protected, especially from the rootling of pigs. There may be other cultural landscape

A present-day satellite image of Knepp (left) aligns with a map of the park and mill pond (right) in the National Library of Scotland's side-by-side viewer. The parkland delineations in the old maps were crucial in providing evidence for Countryside Stewardship funding for the park restoration in 2001 (see designation on MAGIC map below).

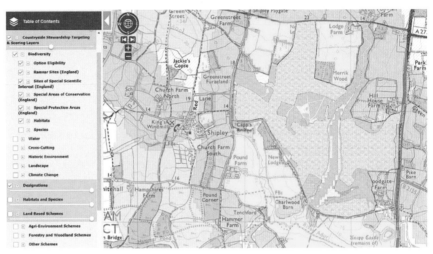

MAGIC map showing the English Heritage-listed park on either side of Knepp lake in mottled green, 'Habitat Restoration' areas in yellow, 'Priority Habitat Lowland Fens' in dusky pink to the north of the lake, and 'Priority Habitat – good-quality semi-improved grass-land' areas in bright pink.

features to preserve or restore, such as dry-stone walls, vistas, old banks and green lanes, which tell a story and, in themselves, provide important habitat.

Old field names can be instructive, bearing in mind that names often change over time with alterations in land use and field boundaries. At Knepp almost all our fields were, at one time or another, associated with scrub, gorse and reeds, although under modern farming no sign of it was left. We still have Benton's Gorse and Broomers Corner, but in the distant past we had High Reeds, Little Thornhill, Great Thornhill, Rushett's and lots of Furzefields ('furze' being the local name for gorse).

Many local museums have natural-history collections that can give an indication of past ecology. A stuffed beaver, a rare bird's egg or the bones of some long-lost species can provide an inspiring snapshot of the past – something you could aspire to revive.

The meticulous field notes of Victorian naturalists, such as John Gould, published in copious volumes, provide vital information about the behaviour and distribution of native species in a richer landscape. It's hard to understand why the observations of past naturalists are almost entirely ignored by modern science. They provide important evidence, particularly for birds, that is in many ways far more useful for rewilding purposes than the British Trust for Ornithology's 1970 baseline. Species that are assigned strict habitat designations today – for example as 'farmland', 'woodland', 'heathland', 'meadow' or 'wetland' species – often show themselves to have been far wider-ranging a hundred or more years ago, when more varied and extensive habitat was available.

Past publications by local naturalist authors are also worth seeking out. Thanks to John Arthington Walpole-Bond – a Sussex ornithologist and son of a vicar in Horsham, who produced *A History of Sussex Birds* in 1938 in three scrupulously detailed volumes – we discovered that the countryside around Knepp had been teeming with nightingales, turtle doves and nightjars less than a century ago, and we have the exact details of the habitats in which they were nesting and feeding.

While it's important in terms of the dynamics of rewilding and natural processes not to be led into a fixation on any particular species, background information of this kind helps shift our thinking away from the impoverished baselines we've grown up with. It can raise the bar for ambition across the board.[7]

Place names can provide links with lost species, and are a useful connection if you're considering reintroductions, and another way of telling a story. Beverley and Bewerley in Yorkshire, Beverston in Gloucestershire and Beverley Brook running through Richmond Park in west London all immortalise the presence of beavers in the past. Cranleigh, Cranfield and Cranmere were named for the common cranes that used to bugle from wetlands before these were drained in the seventeenth century. The village of Arncliffe in the Yorkshire Dales derives its name from *erne*, the Old English name for the sea eagle. Storrington, a local village to us at Knepp, comes from 'Estorcheton', 'homestead of the white storks' in Old English – something that has proved a springboard for local support for our project to reintroduce these birds.[8] Some place names may remain, for the moment at least, aspirational; 22 kilometres to the north of Knepp is Wolves Hill.

Knowing the history of your land, especially if you can take it back many centuries, can free the mind of the tyrannies of the present and of living memory. Understanding how very different the landscape has been – that today's countryside is not as enduring, secure or immutable as we often think it to be – can be a source of confidence for change. It can also offer pointers for planning and future management.

At the same time, it's important not to let these ideas root too deeply. Rewilding, as we've already discussed, isn't about recovering a blueprint of the past, but rather working with the material and tools that are available today. We can never re-create what has gone before, but we can continue the story and encourage nature itself to create the future.

This avian rewilding manifesto for the Peak District, designed by Dr Alexander C. Lees, went viral when it was first published on Twitter in 2018. Its ambitious projection of species reintroductions and revival is characteristic of the rewilding vision.

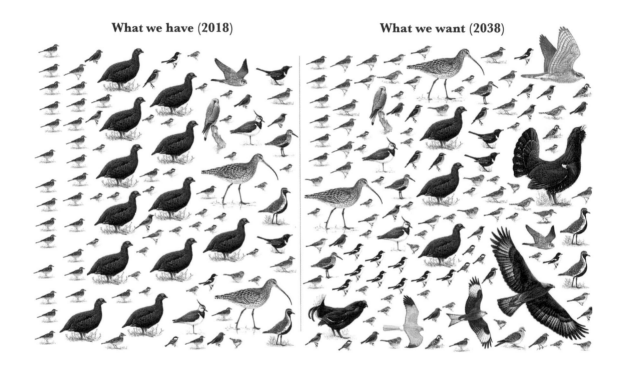

What we have (2018)

What we want (2038)

Conservation evidence

Rewilding is still in its infancy, and it will take a while for evidence to accumulate in the scientific literature. Since it is also a dynamic process, involving a vast number of variables – from the historical management of a site and the species that are present or able to colonise naturally, to details of microclimate, soil and topography – the results are unpredictable and fluctuate over time. This makes rewilding very different from the usual conservation mindset of specific targets and measurements, and, inevitably, more challenging to analyse.

There are, however, interventions that can be taken in rewilding, particularly at the outset, which are akin to conventional conservation measures, and about which there is already considerable evidence-based knowledge. We've touched on some of these actions in earlier chapters – such as restoring rivers to their floodplains, connecting areas of natural habitat and reintroducing species – and we've provided references to some of the relevant studies. But it is worth continuing to research scientific papers and reviews to find out more about topics that are of specific interest to you, and how they might apply to your project.

Unfortunately – and, we believe, unacceptably – the vast majority of nature-conservation literature is closed access.[9] Shadow libraries are an increasingly popular resource that provide free access to academic articles without regard for copyright. Being essentially piracy, they are controversial, but are the most efficient way around the paywall problem. Almost everyone, including the NHS, policymakers and professional researchers, now uses one of them.[10] Unpaywall has smaller coverage than most shadow libraries, but has the benefit of being legal, as well as free.[11] The third option, which has a lower success rate, is to write to the author of an article and ask for a copy. Authors are generally very happy to send a PDF, but the older the article, the harder it is to track down the writer. If you are not looking for a specific article, you can do a keyword-based search of research literature through Google Scholar or the Lens (select scholarly works, not patents).[12] Using websites that provide summaries of the evidence, such as Conservation Evidence and CEEDER (Collaboration for Environmental Evidence Database of Evidence Reviews), can be particularly helpful, providing a foundation on to which personal experience and knowledge of the site, its physical and biological characteristics, and advice from practising ecologists can be overlaid.[13] Evidence-based knowledge can provide the springboard from which a rewilding project can leap into the unknown.[14]

It is important to be aware of measures that have already been proved *not* to work. Bat gantries over roads, for example, were used in the UK for nine years without ever being tested, at a total cost of £1 million, before being found to be ineffective.[15] It is surprising how an idea can sometimes be so compelling that it overrides the evidence and continues to be funded. Likewise, many of the EU's agri-environment schemes for the Common Agricultural Policy had little evidence underpinning them and were used despite the existence of other suitable options with substantial evidence for their efficacy, simply because the evidence wasn't examined.[16]

However, given that much of rewilding is novel, and therefore has little or no relevant evidence yet to support it, and that so much of what evolves is both random and site-specific, monitoring the changes on your land will be crucial for generating new evidence to inform the decisions you and others make in the future. We discuss how to record these changes in Chapter 9.

Consultants

Depending on the size of the project, it can be worth employing a consultant to gather all the background detail in a report, including the funding options described below. Most land agencies have a specialist in this field, and the local Wildlife Trust and Farming and Wildlife Advisory Group may offer this service. There is also a growing number of rewilding consultants with varying levels of experience and cost.

Funding

The public and private sectors are – at last – beginning to respond to the environmental crisis, and the world of farming and land management is changing rapidly. In this fast-evolving situation it is difficult to keep up to date. Funding opportunities, particularly in the charitable sector, tend to come and go. Here, we explain the background, the basics and the direction of travel in the UK. You will inevitably need to research further or seek advice on the latest funding developments and opportunities when it comes to planning your project.

Central to the 25-Year Environment Plan unveiled by the UK government in 2018 is the concept of 'public money for public goods': paying landholders to provide 'ecosystem services' such as clean air, soil restoration, increased biodiversity, carbon capture, sustainable food production, flood mitigation, pollination encouragement and clean water.[17] This is a sea change from a subsidy system that has since the mid-twentieth century been geared almost entirely towards incentivising land managers and farmers to generate vast quantities of cheap food from their land, irrespective of the cost to the environment.

Alongside this shift in policy from government – which, it is hoped, will weather changes in leadership – the private sector is also rapidly changing. Wider and tighter legislation is being brought to bear on businesses to account for their environmental impact. Since 2022 more than 1,300 of the largest UK-registered banks, insurers and private companies with over 500 employees and £500 million turnover – from Tesco to Unilever – have had to disclose 'climate-related financial information', and it is likely this will begin to apply to a wider range of businesses in the near future.[18] From 2023, under the broader Corporate Sustainability Reporting Directive in the EU, all large and listed companies trading in Europe are required to report their impact on society and the environment.[19] As environmental, social and governance considerations become more pressing and mainstream, customers and shareholders are demanding that businesses take proactive responsibility, irrespective of whether they are legally required to do so. But companies themselves are also beginning to recognise how their own business is dependent on a healthy and resilient natural ecosystem.

In another paradigm shift, many companies are beginning to appreciate that being green can be good for business, and are keen to be seen as sustainability leaders in their industries. Not only can savings be made through energy efficiency, recycling, renewable technology and innovative thinking, but also businesses demonstrating robust action towards reducing their environmental impact will attract preferential investment. Those that do not innovate and/or account for their emissions and ecosystem degradation will fast fall behind and lose out.

Alongside the mapping of climate-related risks via the global Task Force on Climate-Related Financial Disclosures, since 2015 the Science Based Targets initiative (SBTi) has been assessing companies' carbon-reduction and net-zero targets against strict criteria aligned with the 2015 Paris Agreement, with the aim of achieving a net-zero society worldwide by 2050 and keeping the global temperature rise 'well below 1.5°C'.[20] This potent combination has led to a real change in corporate ambition; at the end of 2021 more than 2,200 companies covering over a third of the global economy's market capitalisation were working with the SBTi to set targets.

In the last few years, setting an SBTi-aligned net-zero target for a company's carbon footprint has moved from being innovative to being the basic starting point. The more ambitious and aware corporations are now looking to their risks as well as their own impact on the natural world. These risks include physical risks, such as extreme weather patterns and rising global temperatures; transitional risks, such as potential legal and policy changes; and reputational risk. All this is a much more complex task. When it comes to nature, we are now seeing initial frameworks from the Taskforce on Nature-related Financial Disclosures and the Science Based Targets for Nature, which together are expected to drive corporate ambition in that direction. In October 2022 the British department-store cooperative the John Lewis Partnership publicly announced that it was aiming to be 'the first British business to set verified, science-based targets for nature as soon as the framework becomes available'.[21] After typically nail-biting eleventh-hour negotiations, the COP15 summit on biodiversity in Montreal in December 2022 (the second part of a summit first held in October 2021 which, unlike the climate COPs, focused purely on biodiversity) created a 'Paris Agreement for Nature', galvanising global action in the way Paris did in 2015. A particular focus of COP15 was the '30 by 30' campaign: assigning at least 30 per cent of the planet's land and sea to nature recovery by 2030, and recognising the rights and contributions of indigenous peoples and local communities to the climate and environmental crisis.

Against this backdrop, both government and the private sector are promoting the idea that businesses should look to their own negative impact on the climate and on nature, reducing where possible (and where that is not feasible, offsetting) the remaining impact. Areas being looked at include carbon emissions, the destruction of natural habitat, pollution and the depletion of natural resources and biodiversity. While not a solution by itself, offsetting can be achieved by participating in schemes designed to compensate for negative impact, such as by buying carbon and biodiversity credits. Again, the most innovative companies are going beyond just offsetting the harm they cause (which is a zero-sum game); rather, they are looking for net-positive or nature-uplift opportunities, and giving back to the natural world more than they take. Get Nature Positive, Finance for Biodiversity and the Race to Zero are all initiatives for restoring nature being driven by financial institutions and the business community.

At the same time, companies are beginning to work with landowners to provide ecosystem services that save their businesses money in the long run. An example is water utility companies paying farmers to reduce nitrate pollution and soil run-off. Preventing the pollution of water at source, some water companies have calculated, is much cheaper than filtering those nitrates and soil particulates from the water at their end to make it fit for use. This is an example of Payments for Ecosystem Services (PES; see page 324).

Rewilding projects can play a role in both environmental 'offsetting' schemes and PES, and are therefore well placed to receive both private- and public-sector funding. The funding schemes are based on the concept of natural capital.

Natural capital

The Natural Capital Protocol is an independent, global collaboration of businesses, academics and governments that has drawn up a framework for companies seeking to measure and value their direct and indirect impacts on the natural world. It defines natural capital as 'the stock of renewable and non-renewable natural resources (for example plants, animals, air, water, soils, minerals) that combine to yield a flow of benefits to people.[22] Natural resources also include ecosystems, and natural processes and functions.'[23] The stocks of natural capital provide flows of ecosystem services, which translate into value for society. Ecosystem services, simply defined, are the benefits we as humans derive from the natural environment, among them the provision of food, water, timber and fibre; regulation of air quality, climate and flood risk; opportunities for recreation, tourism and cultural development; and such underlying functions as soil formation and nutrient cycling.

Those advocating a natural-capital approach believe that the best way to value nature is by relating it to human well-being, by translating impact and dependency into economically recognisable units: money. Giving a value to what the land can do for us as a society allows us to include it on the balance sheet. When it is not given a price in the economic system by which we live, that system destroys it.

This approach is not without controversy. For many people, putting a financial value on nature is not only immoral but also logistically impossible. Some conservationists insist that monetising nature leads what we most want to protect into the lion's den of capitalism, into the self-interest of the financial markets, to arbitrary pricing and trade-offs that replace nature with an impoverished ghost of itself.

How can one put a price on beauty or pure air, on a sense
of harmony and well-being? Should these things be tradeable?

Natural capital is the stock of
nature – soil, trees, air, water
and so on – found on a piece
of land. The natural capital
itself isn't sold (that would
involve a land sale) but the
ecosystem services it provides
can be sold. The challenge
is to assign a value to these
services by working out how
to measure the benefits they
provide to society.

Campaigners have made the moral case for protecting nature for
its own sake at least since the middle of the twentieth century. They
have argued that nature is beautiful and fundamentally important,
and that we have no right to destroy it. But their appeals have fallen
on deaf ears. However, at Knepp we believe that it is possible to set
a cost for destroying nature without reducing the overarching sense
that nature is, ultimately, priceless; without eroding its mystery and
enchantment. After all, we already do. Hospitals and health services
know how much improvements to air quality and access to green
space reduce the healthcare bill. Councils and insurance companies
can calculate what renaturalising rivers and restoring watersheds and
floodplains would save them in terms of flood-damage costs. Water
companies know how much renaturalising uplands saves them in
filtering silt, pesticides and nitrates out of the water supply. In 2015
the Natural Capital Committee, the independent body set up in 2012
to advise the UK government, suggested planting 250,000 hectares
of woodland close to urban centres in Britain, an undertaking that it
said would bring a net economic benefit of nearly £550 million
in recreation and carbon sequestration.[24] Thanks to economists such
as Partha Dasgupta, Dieter Helm and Jonny Hughes, natural capital
accounting is becoming part of mainstream policy, as something
that decision-makers, politicians and markets can quantify.[25] For
the rewilder, it means incorporating natural capital stocks into the
balance sheet so as to be able to identify a natural capital baseline at
the start of the project.[26] Knowing what ecosystem services your land
is currently providing (purifying or polluting; sequestering or emitting;
allowing biodiversity to recover or damaging it) and identifying the

potential for improvement will enable you to sell the provision
of these services and get funding for increases to natural capital.

Natural capital baselines

A natural capital baseline is a powerful tool that can tell the story
of the land, provide evidence of how nature responds to rewilding
and identify the wider benefits it brings to people – in other words,
the public goods or ecosystem services it generates. It is one of the
reasons Knepp has managed to change the way people think about
nature conservation.

One of the building blocks of the natural capital baseline is a
biodiversity assessment. In many ways, biodiversity is an indicator of
how well an ecosystem is functioning. It is a useful guide to broader
aspects of ecosystem services, such as water purity and soil function.
Many aquatic invertebrates and plants are indicators of clean water,
since they die in polluted systems. Fruiting fungi, orchids and earthworms are indicators of healthy soil, which in turn is key to carbon
sequestration, water storage and flood mitigation. Pollinating insects
and insect predators, being vital for agriculture, obviously benefit
society. It is important to consider not just the variety of species in a
biodiversity assessment, but the abundance of species' populations
(the biomass), and their lifecycles too. The presence of creatures with
complex lifecycles is a good indicator of habitat diversity and ecosystem resilience. Their rarity in relation to region – for example, a
nightingale in Lincolnshire, where the species is rapidly declining – is
more significant than a nightingale in Kent, where it is still doing well.

Building a baseline of natural-capital stocks also positions a project
to receive payment for services. By demonstrating the current state
of play, it makes clear the case for improving it and therefore shows
when payments are due. A major aspect of this is ecological monitoring, but it also tries to account for other ecosystem services. For
instance, six key pillars might be looked at in a new project: water
(quality/flow), biodiversity (birds/mammals/invertebrates), soil
(health/carbon), social (public access/education), habitats (types/
quantities/condition) and carbon (emissions/sequestration). The
way to do this can be as simple as mapping habitat and completing
a carbon calculator, or as involved as paying for a detailed, multi-
faceted report from an external auditing service.

The reasons for establishing a baseline – for funding or to provide scientific evidence for the project, or both – will inform how much data is needed and who can provide it. Since this is a new sector in the market, there is not yet a fixed accounting standard, but some front-runners are emerging and innovative methods are being developed. Depending on budget, available time and capacity, you can pay someone to do it or run it yourself.

External auditors and consultants

External auditors and consulting companies can provide a natural-capital appraisal. The cost ranges from £500 for a simple score to £10,000 and more for a detailed audit using on-the-ground data gathering, satellite and drone imagery, as well as collecting data available through the Environment Agency, Natural England, Ordnance Survey and other public sources. Such an appraisal can reveal, in varying scales of detail and reliability, how much carbon the land is sequestering and what ecosystem services it is providing. The closer to the ground you get, the better the detail. For example, testing organic matter from the site's soil using lab analysis will be far more reliable and interesting than simply using national datasets based on the soil types that are typical in the region. Ultimately, the more detail the better – but it can be difficult to know where to stop.

These reports can be very informative, but they don't necessarily translate into funding. The companies that provide these services, such as Environment Bank, Nature Capital, NatCap Research, Triage, Regenagri and the Land App, are now positioning themselves to broker agreements between landowners and businesses. There is a lot to be said for getting a helping hand, despite the cost – but any company you choose should be able to show what deals it has brokered, and that it has actually been able to raise money for projects.

The do-it-at-home baseline

For us at Knepp, biodiversity monitoring is the most important aspect of the natural capital baseline and so constitutes a large proportion of our time and budget. Others might prefer to focus on public access, or clean air and natural flood management. There is currently no universal way of measuring the natural capital baseline, and any information at all is an asset.

There are low-budget options. As a minimum, mapping habitats on the Land App using UK Habitat is a great place to start. It is a little technical, but you can go into as much or as little detail as you like. Even mapping the available data from MAGIC on to the Land App is valuable. Filling out a carbon calculator using the Cool Farm Tool, Agrecalc or the Farm Carbon Toolkit is another option. This is a good starting point from which to measure carbon emissions versus carbon capture, although the science is still evolving and many contributors to carbon capture, such as hedgerow trees and woody shrubs, are not yet accounted for. Over the twenty years that we have been rewilding, 3.7 million cubic metres of additional woody vegetation has evolved at Knepp.[27] We assume this will have resulted in net carbon capture, and are now working with Exeter, Oxford, Queen Mary London and Cranfield universities to see if we can measure it.

In the next chapter we discuss how to set up biodiversity surveys using local wildlife charities and volunteers, fixed-point photography, water-quality testing kits, the soil mentor app and so on. All this contributes to a natural capital baseline. For those with a bigger budget or access to funding already, comprehensive soil and biodiversity monitoring alongside in-depth habitat surveys will be highly advantageous. In particular, those intending to raise funding from natural capital are advised to design their baseline survey work with an accreditation company (see below for carbon and biodiversity credits).

Ultimately, whether you pay a consultant to run a comprehensive baseline, pay ecologists to conduct surveys or map habitats at home on your computer, all you need do at this stage is *something*. As you continue rewilding, you will learn what you are missing and add more data and aspects to your project. As the markets for ecosystem services and public goods develop, so will the accounting methods.

Finding the right funder or combination of funders

It's important to get to know your funder(s) and really understand their expectations. Public, private and philanthropic funding is often focused on targets and outcomes. You may need to unpick that mindset, to explain how rewilding is different and ensure there is no misunderstanding with regard to goals. The funder must appreciate the long-term nature of the investment, for example, and the need to keep the aims very broad and open-ended. The risk of getting the wrong funder on board is that they may want to direct the project in a certain way from the start, compromising its values. Communication is hugely important. This should all become easier as rewilding enters the mainstream and funders appreciate the different approach.

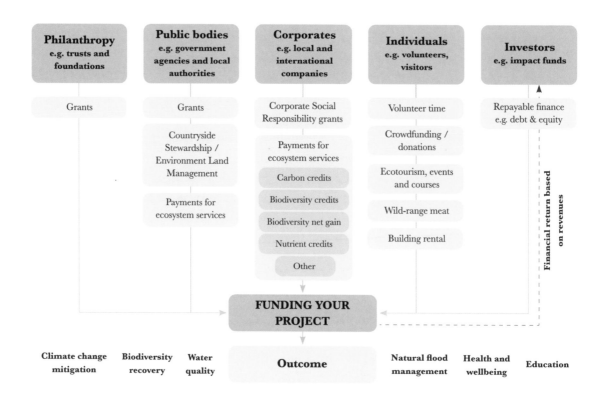

Philanthropy e.g. trusts and foundations	Public bodies e.g. government agencies and local authorities	Corporates e.g. local and international companies	Individuals e.g. volunteers, visitors	Investors e.g. impact funds

Grants — Grants — Corporate Social Responsibility grants — Volunteer time — Repayable finance e.g. debt & equity

Countryside Stewardship / Environment Land Management — Payments for ecosystem services — Crowdfunding / donations

Carbon credits — Ecotourism, events and courses

Payments for ecosystem services — Biodiversity credits — Wild-range meat

Biodiversity net gain — Building rental

Nutrient credits

Other

Financial return based on revenues

FUNDING YOUR PROJECT

Climate change mitigation — Biodiversity recovery — Water quality — **Outcome** — Natural flood management — Health and wellbeing — Education

A rewilding project can derive funding from a number of different sources ranging from grants and philanthropy to payments for ecosystem services and repayable finance. The challenge is to decide the most suitable blend of funding for your project.

There may be a lot of talk about ecosystem services, natural capital, biodiversity net gain, corporate natural-capital account funding and public money for public goods. But this is still a new and rapidly emerging 'industry'. Although new funding mechanisms are coming at us from all directions, the market is far from settled and the rules around how funding can be blended and stacked in a credible way are not yet clearly defined. As the understanding of natural-capital assets and how to monitor them evolves, opportunities will become clearer and greater accountability, better accreditation and higher standards will emerge. For now, different concepts and ways of accounting for natural capital are battling for recognition in a crowded space, and few projects are benefiting from the opportunities that are available.

In the UK, it is unlikely that grant funding from traditional sources, such as charitable foundations and government, will continue to be the primary driver of nature recovery in the long term. The small, fragmented efforts provided by grant funding that we have seen to date will simply not produce the required increase in biodiversity. The UK has now legally committed through the Environment Act to meet its ambitions for nature by 2050. But governmental financing falls woefully short of the required spending needed to meet its targets. According to the Green Finance Institute, if the UK government is to meet its 25-Year Environment Plan objectives, it needs to find an additional £44–97 billion over the next ten years. Private-sector finance must play the defining role if this funding gap is to be filled.[28]

Already this is picking up speed. The demand for funding-management interventions for nature recovery under private land-holding is set to increase dramatically. However, while billions of pounds in natural capital funding are being announced, there are currently not enough reputable projects to match the money available from the private sector. To meet this demand, landholders must work out how to make their projects attractive to corporate entities.

UK financing gap

£6.1 billion average annual funding gap for UK nature

Required spending £8.8 billion annually for the UK (average estimated)

£2.7 billion committed and planned spend

Approximately £8.8 billion of annual investment is required to safeguard the natural environment in the UK but current planned and committed spending from the public sector is only around £2.7 billion, leaving £6.1 billion left to find every year. The private sector has an important role to play in filling this funding gap.

Public funding

Hopefully, in the UK at least, some public money will continue to be available in the coming years, and land managers will be able to apply for it to help them achieve their rewilding objectives. Even as private-sector investment begins to take off, making use of public-sector money should be an important part of the funding business model.

Payments for Ecosystem Services

A significant funding mechanism that is opening up in both the public sector and private sectors is Payments for Ecosystem Services (PES). Maintaining and enhancing ecosystem services – and restoring them where they have been lost or degraded – is increasingly recognised as essential for sustainable economic growth, prosperous and stable communities, and human health and well-being.

Examples are water companies paying farmers to change to a regenerative agricultural system to improve the health of their soil and the quality of their water; and the Environment Agency or the originator of a new development in a flood-prone area working with a rewilder upstream to build natural flood management into the catchment. Some projects use online 'reverse' auctions, in which landowners can quote their price for delivering mitigations such as natural flood management and improvements to water quality. This opportunity may well develop for air quality, as well. However, one of the key problems with these auctions at present is the detail of the prescriptions that explain what a landowner has to deliver. Poor pre-scriptions generate a race to the bottom as landholders reduce their prices to win the bid but then cannot deliver an effective habitat type. This has been seen recently in relation to some great-crested newt pond creation schemes in the UK. The NatureSpace Partnership, a private-sector company that works with local planning authorities to deliver district licensing for newts, stresses the importance of proper funding and prescription for landowners to produce high-quality ponds that will stand the test of time. As their Chief Executive puts it, incentives should be for creating 'the best, not the cheapest'.[29]

While funding opportunities for PES are still limited – they often require entire catchments of landowners to work together – we expect this market to grow. There is currently a dearth of experience in the sector, so the relevant players find it difficult to come together. Some PES can be engaged by government, but a larger market is likely to stem from the private sector once the individual corporates identify their dependence on ecosystem services and resilient ecosystems. This under-standing is slowly developing, largely under the aegis of investment-fund managers placing far greater emphasis on environment, social and governance factors in assessing the investment worth of a corporate.

Such incentives would work well alongside the 'polluter pays' principle. Arguably this should be happening anyway, but sadly governments do not often have the will or integrity (or, perhaps, the funds) to police, charge and impose fines on polluters, which include water companies, industry, landowners and farmers.

Countryside Stewardship

Until now in the UK, rewilding initiatives have been funded through agri-environment schemes, originally through Environmental Stewardship and latterly through the Countryside Stewardship scheme. As part of the transition to the post-Brexit Environmental Land Management scheme (ELM), Countryside Stewardship will be phased out. It is, however, still worth mentioning as a funding resource, since Natural England will be accepting applications in 2023 for agreement start dates of 1 January 2024. The current option within Countryside Stewardship that fits best with rewilding is WD6 in the Higher Tier, for the 'creation of lowland wood pasture'.[30] At the time of writing, it is worth £499 per hectare per year.[31] However, although WD6 is suitable for rewilding projects aimed at restoring natural processes, it is restricted by parameters and criteria that only some projects are able to meet. Firstly, Higher Tier agreements are targeted at what is considered to be the most environmentally important land. They are competitively scored, with a threshold score set based on the nationally available budget, meaning that not all applicants may be offered an agreement. Being a larger site or a special area of any kind (a national park, SSSI, AONB or some other designation) is considered a plus point for entry of a Higher-Tier Stewardship agreement of any type. Knepp struggled to attract Higher Level Stewardship funding in the early days because ours was not a target area, not being deemed to have any real potential for nature.

Next are the strict parameters of WD6 itself. It can apply to land 'where wood pasture was once present, including ones under arable land' or 'where it extends, links or buffers existing wood pasture or priority woodland habitats'. This tends to be historic landscaped parks and ancient hunting 'forests' associated with large country houses and estates (such as Knepp), or historic commons that used to be wood pasture with grazing livestock and pollarded trees. Beyond these rather limited examples, Britain has few parcels of 'provable' wood pasture, so immediately things start to get tough.

Knepp was granted Countryside Stewardship funding in January 2021 largely for WD6, although it also included some WD4 ('maintenance of wood pasture') and a 'rare breed' supplement for our free-roaming Exmoors. We were lucky to be awarded the grant. The main reason we qualified was that the old deer park around the castle is, according to Natural England, historically designated parkland (identifiable as such on ancient maps). Therefore we could argue that the Southern and Northern blocks can be seen as 'extending existing wood pasture'. The reason for labouring this point is that currently Countryside Stewardship is not set up to help the majority of landholdings convert to a process-led system such as rewilding. It is hoped that the Local Nature Recovery and Landscape Recovery tiers of the new ELM will identify rewilding as a qualifiable system of land management and therefore broaden accessibility to funding.

It is possible to identify designations on your land that might qualify you for Countryside Stewardship funding by using the DEFRA MAGIC Map app described above. WD6 is currently the best option for rewilding. However, if your land does not qualify and you are desperate to get started (without waiting for the new ELM to come into operation), and particularly if you have a small site that may not be suitable for free-roaming herds of herbivores, Mid-Tier Countryside Stewardship offers a range of options that are designed to help farmers adopt wildlife-friendly practices.[32] This is not rewilding as such, but it does encourage a more diverse and nature-friendly system that can be a useful stopgap before ELM comes into operation. Mid-Tier Countryside Stewardship includes support for the creation and restoration of scrapes and wetlands, capital items such as fencing and water troughs, and some extensive grazing systems.

Environmental Land Management scheme
Since leaving the European Union the UK government has undertaken a radical reform of agricultural subsidies, which were formerly part of the EU's Common Agricultural Policy. The Basic Payment Scheme – a payment per hectare of land farmed – is due to be phased out by 2027. Both Countryside Stewardship and the Basic Payment System will be replaced by the Environmental Land Management scheme (ELM), which will have three levels, or 'components':[33]

1. **The Sustainable Farming Incentive** Open to all farmers, focusing on healthy soil, low pollution and lower input, encouraging movement towards more sustainable agriculture.

2. **Local Nature Recovery** Likely to be renamed Countryside Stewardship Plus, with payments for the nature-recovery aspects of sustainable and (we hope) smaller rewilding projects. This will probably be based on coordinated restoration efforts involving farm clusters.
3. **Landscape Recovery** Aimed at really big, ambitious nature restoration on large landholdings, large farms, estates and partnerships of several farms; we hope this will include wildlife corridors.

Early drafts, consultations and reports indicate that rewilding will be named explicitly in both the Local Nature Recovery and Landscape Scale Recovery components. This will provide broad incentives for smaller and larger projects in a way that Countryside Stewardship, with the selective focus of its wood-pasture grant, was not designed to do. However, we still don't know how ELM will look and what exactly it will support. Indeed, we don't know if ELM will even be delivered on time, or remain in its current form, or how the funding will be split. Originally, it was thought that the overall agriculture budget would be divided equally across the three schemes. But following pressure from the intensive agriculture lobby it now looks likely that most of the funding will go to the Sustainable Farming Incentive, with most of the remainder allocated to Local Nature Recovery/Countryside Stewardship Plus, and a much smaller pot for Landscape Recovery. As always, in times of economic and political uncertainty, and under pressure from big business interests, the more ambitious aims for nature restoration draw the short straw.

One further problem, inherent in the scheme itself, is to do with 'additionality', which means that a particular project has to produce something additional to what would have otherwise happened without it. If payments are made only for improvements, the effort of farmers and land managers who are already doing a good job for nature will not be recognised. It might even, perversely, incentivise land managers to reverse good management in order then to demonstrate the uplift in soil function, water quality, biodiversity and so on that is required for payments. This must be addressed in terms of when and how the baseline is set.

Private funding

There is a growing number of ways to attract private-sector finance into rewilding. This can range from investors interested in funding nature to receive a financial return, to companies seeking to buy nature-based credits or save on business costs, to private individuals who simply want to do something positive for the planet.

Carbon credits

Following the United Nations Framework Convention on Climate Change's Paris Agreement, in 2019 the UK became the first major economy to set a target of 'net zero by 2050'.[34] How serious the commitment is to this target, and whether it is achievable, is a matter of considerable debate. However, most large companies are now adopting decarbonisation targets, and a growing number of smaller ones are following suit. They are driven by pressure from shareholders and consumers, as well as voluntary initiatives such as the Task Force on Climate-Related Financial Disclosure and Science Based Targets initiative. Corporate decarbonisation roadmaps are laid out in stages, starting with reducing emissions within their own operations, then

Measuring carbon capture in the soil. Creating a quantifiable baseline for the existing carbon on your land is essential if you are to understand how you can benefit from carbon trading by proving a reduction in carbon emissions or an increase in carbon storage over time. The company Agricarbon has developed an affordable way to take soil samples and then measure the carbon and organic matter in its laboratory.

turning to their full value chain. For some industries (such as steel and some manufacturing industries), it will be impossible to cut emissions entirely. These companies can offset their residual carbon footprint by investing in carbon-reduction and carbon-storage projects elsewhere, via 'carbon credits'. The demand for these is already huge, and it is expected to grow. In 2021 a record $851 billion was spent on the global voluntary carbon market, a rise of 164 per cent on the previous year.[35] It is expected to grow exponentially in the coming decade.

Several markets are developing around the offsetting of carbon, and one is carbon sequestration brought about through the regeneration of woodland, soil, grassland, peat, wetland, mangrove, salt marsh and sea grass. This type of carbon credit is the most expensive and considered the best 'quality'. Other types of carbon credit include 'avoided loss', which is used to stop the destruction of existing habitats that might otherwise have been lost to agriculture or development; and 'carbon not produced', in which if X kilowatts of energy are produced from renewable energy, such as solar or wind, Y amount of carbon has not been used to produce that energy, and can therefore be sold as credits. The last type was originally intended to boost the generation of renewable energy, but it is a spurious way of accounting. Thankfully, now that renewable energy is well underway (largely because of new legal requirements and subsidy in most countries), the largest Voluntary Carbon Market credit organisations no longer issue these renewable-energy credits.

A carbon 'credit' refers to 1 tonne of carbon dioxide removed from the atmosphere or not emitted. There are two main types of credit: ex-post and ex-ante. Ex-post is for carbon that has already been sequestered – carbon locked up in existing peatland and woodlands, for example, and paid 'on delivery'. This is useful for protecting (among others) areas under threat from illegal tree clearance, such as the cloud forest in Cusuco National Park, Honduras, or those at risk of conversion to big arable agriculture, such as the traditionally farmed flower-rich meadows in the valleys in the Carpathian foothills of Romania.[36] Ex-ante is for carbon to be sequestered in the future: selling peatland restoration, tree planting or rewilding in advance, for the carbon that will be locked up as a result. The UK Woodland Carbon Code calls these Pending Issuance Certificates.

Creating a quantifiable baseline for the existing carbon on your land is essential if you are to understand how you can benefit from carbon trading by proving a reduction in carbon emissions or an increase in carbon storage over time. In the UK the established tools for measuring carbon for offsetting are the Woodland Carbon Code, the Peatland Code, the Woodland Carbon Guarantee and i-Tree Eco.[37] The Farm Soil Carbon Code, currently in development, will aim to provide best-practice guidelines for quantifying the carbon captured in farmland soils.

Unfortunately, the prevailing model for carbon offsetting in the UK is currently tree planting, which is unlikely to be helpful for bio-diversity (see page 153). Under the offsetting model, a business pays for planting trees that, it is calculated, will store an amount of carbon equivalent to the carbon the company emits. Businesses cannot offset emissions until the trees have actually sequestered carbon and this has been verified, so this approach only mitigates for future emissions. The really major carbon emitters, such as oil companies, steel industries and airlines, tend first to look for offsetting opportunities in the renewables market, and then buy a remainder of nature-based carbon credits in the tropics, where trees grow rapidly and carbon is sequestered faster and in larger quantities than in temperate-zone Europe. Other companies, especially lesser emitters, will look to buy carbon stored through tree planting local to their operations.

However, the tree-planting 'solution' can be simplistic, and it is by no means certain that this form of accounting has got it right. Calculating how carbon is sequestered in nature is complicated, and science is only beginning to understand it. It is not only trees that pull in carbon from the atmosphere, but also all other plants, soil, grassland, peatland and wetland – if the ecosystems are functioning properly. The current vogue for planting trees is driven by the private sector, governments keen to meet the carbon targets they've signed up to, the charitable sector and the vested interests of forestry and commercial propagation. It can in fact be damaging to the environment, and can lead to the planting of too many trees, trees in the wrong place and short-term commercial plantations (often of fast-growing softwoods), which will release carbon once they are harvested.[38] It has also been known to displace residents, and provides limited benefits to local communities, as much of the funding goes to intermediaries. Rewilding Britain, Knepp and other rewilding projects, the Royal Botanic Gardens, Kew, and others are lobbying for natural regeneration (or 'natural colonisation', as some describe it) to be the default mechanism for re-establishing trees worldwide.[39] But the carbon market in the UK still, disappointingly, favours tree planting.

Underpinning this plantation bias in the UK is the Woodland Carbon Code, the country's main certification system. Driven by the Forestry Commission and its particular forestry culture, the code does not adequately reward natural regeneration and dynamic wood-pasture systems for establishing trees. Natural regeneration is also too unpredictable for conventional, target-led accounting, since it's impossible to say how many trees will establish, let alone what species they will be. And carbon codes, with their narrow focus, do not account for the additional ecosystem services that natural regeneration provides.

At Knepp we are working with Oxford, Exeter, Cranfield and Queen Mary universities to try to influence this debate by calculating the net figure for greenhouse gases sequestered and released by our rewilding project since it began. It will take into account carbon sequestered through the natural regeneration of trees, woody species and other plants both above ground and below, in their root systems. It will measure carbon in the soil, and in wetlands. And it will examine the influence (both positive and negative) of free-roaming animals on vegetation growth/carbon storage above and below ground. We hope this study will provide evidence to show that rewilding provides considerable gains in carbon storage, especially when transitioning from intensive arable systems.

Many companies, however, are now concerned by the limitations and environmental problems of carbon offsetting, particularly through tree planting. Rewilding appeals to some companies as being more 'honest' and trustworthy than carbon-focused offsetting schemes because it provides ecosystem services, including biodiversity and wild spaces for increasing human well-being, on top of carbon capture. Crucially, in developing countries, it can also drive capital into underfunded communities.[40] This is an opportunity for rewilders to enter into partnerships and agreements with businesses to offset their behaviour and present a greener front to their customers.

Of course, there is still room for greenwashing and abuse of the system. Above all, it is important that pressure be brought to bear on companies to change their behaviour and reduce their carbon footprint and environmental impact – not simply to offset them. But there are, we believe, real opportunities for positive change. In due course, as soil restoration, judiciously grazed grasslands and natural regeneration become increasingly recognised for their carbon-sequestering potential, there will be significant opportunities for carbon payments in rewilding.[41]

Biodiversity credits

Some carbon-credit standards organisations, such as BeZero, the Global Biodiversity Standard (run by Botanic Gardens Conservation International) and Wilder Carbon in the UK, have begun grading carbon credits by taking into account their biodiversity co-benefits. Credit issuers such as Verra and Gold Standard offer different premiums on projects that include measurements for biodiversity. However, they still prioritise carbon over biodiversity. Climate change and biodiversity loss are intimately, indivisibly connected and addressing them must be of equal importance if we are to avert the crisis that threatens our very survival. Sadly, this relationship has so far failed to register among governments around the world. This is powerfully illustrated by the fact that the global summits known as COP (Conference of the Parties) for climate change and for biodiversity have always been staged separately. However, there are now signs that these agendas are being brought closer together.

Until recently, climate change has assumed centre stage. This is perhaps partly because the scenario of global warming and its threats to human existence are easier to envisage, and the fundamental importance of biodiversity to the planet and the systems that sustain human life are much harder to communicate. But also, measuring carbon is – despite its challenges – more straightforward than measuring biodiversity. Biodiversity is supremely complex and, of course, variable from one region to another. There is no single species, or group of species, that can identify biodiversity and abundance of life across the planet in the way that carbon dioxide can be used as the measure for the global warming of all other greenhouse gases.

One of the leading solutions to this problem has been created by the Wallacea Trust. It has come up with a methodology for biodiversity credits that can be independently verified and validated, that is applicable to all 1,300 eco-regions across the world, and that can be sold using the same architecture as the carbon credit market. It works on the same principle as the Consumer Price Index (CPI), whereby calculations are made from customised 'baskets' of goods according to what people tend to buy in different countries. Each country has different goods in the basket, but a final, comparable unit by which inflation rates across different countries can be compared. The biodiversity credit methodology works in much the same way: by comparing baskets containing at least five metrics specially selected to represent the intactness, complexity, rarity and abundance of biodiversity. Each site has different metrics, but they are comparable across the board, producing a tradeable credit.

For example, a lowland farm in England with a river running through it, which is planning to rewild, could use the following metrics:

- DEFRA Biodiversity Metric (habitat-based)
- Richness and abundance of breeding bird species, weighted towards Red and Amber List species
- Richness and abundance of butterfly and macro-moth species
- eDNA to quantify total species richness of aquatic invertebrates (a measure of water quality)
- Richness and abundance of pollinator bee and hoverfly species (a measure of pollinator activity)
- Abundance of soil macro-invertebrates (a measure of improvement in soil condition using eDNA)

This basket of metrics is very different from the one that would be used for a coral reef, for example, or a tropical forest.

Each of the metrics is measured at time zero and again after, say, five years, to determine the percentage change. The median percentage increase of the basket of metrics indicates the uplift, with one biodiversity credit defined as a 1 per cent increase in the median value of the basket of metrics per hectare. Credits can be issued ex-post (measuring the increases over time) or ex-ante, using a predicted increase based on a similar reference site. In the latter case, for a lowland farm on clay soil going into rewilding, the reference site could be the twenty-year experiment at Knepp, which shows how much biodiversity is likely to increase over a twenty-year period. If this study indicated a median 200 per cent increase in the basket of biodiversity metrics over twenty years (a conservative estimate), a 1,000-hectare farm would get 200,000 biodiversity credits.

It is expected that biodiversity credits could be stacked on top of carbon credits, as long as the rules governing 'additionality' are clearly maintained. And because both can be forward sold, they can provide crucial funding for rewilding projects during that start-up phase, before payments for other ecosystem services (such as nitrate pollution mitigation) can be received and while alternative income from such enterprises as ecotourism, meat production and barn conversions is still emerging.

Finally, a number of innovative companies are exploring the use of blockchain credits and/or non-fungible tokens (NFTs) to raise money for nature-restoration projects. Angry Teenagers, for example, sells NFTs that are used to fund reforestation in Ghana. The resulting project will produce verified carbon credits, and a percentage of the income will be used to pay dividends to the NFT owners.[42] We can see the potential of these models for novel marketing and fundraising for rewilding sites, such as the 'Butterflies of Knepp' NFT series. It will be interesting to see how this area evolves, or if it is just a fad.

Biodiversity net gain

The UK Environment Act 2021 stipulates that it is a legal requirement for developers to provide a minimum 10 per cent biodiversity net gain (BNG) to compensate for the displacement of habitats on land that is to be developed. 'Net gain' means there must be compensation over and above the damage to the environment. This is a question not just of offsetting the damage caused, but of providing a measurable gain in biodiversity as a result of the development. It is designed to be secured for at least thirty years through obligations or a conservation covenant. The initiative was mandated into law in the autumn of 2021 with a two-year transition to allow local planning authorities, which will be required to ensure BNG is delivered, to embed the mechanism into their local plans.

Under BNG regulation, the developer has the option of creating and managing land within the development or purchasing biodiversity credits from 'habitat banks' (such as the Environment Bank) established under management contracts with an external landholder.[43] Currently, developers and their ecological advisors assume they can deliver much of the BNG on site through, for example, turf roofs, the creation of ponds and meadows, and tree planting. This means sacrificing expensive, developable land on or next to the site for the provision of 'nature', and being locked into a thirty-year contract to manage it for biodiversity.

In reality, developers are likely to prefer to pay for habitat creation and management off site. Not only is this more cost-effective, but also it offers far greater potential for biodiversity. One might even argue that it is impossible to deliver genuine uplift for nature on any development site. Inevitably, existing habitat on such land is broken up by infrastructure. It is also disastrously affected by light, noise, air, water and soil pollution, human disturbance and predation by domestic pets, even if orders are issued to protect it. Those areas that qualify as 'biodiverse' within a development rarely survive for more than three or four years. The management company, left in charge after the developer has left, invariably – often under pressure from residents – introduces a 'tidiness' culture. Habitat is gradually homogenised into amenity grassland, trees and flower borders. Large commercial developments, of course, which are basically concrete slabs, have even less capacity to deliver any gain for biodiversity.

It is imperative that we move away from the assumption that BNG can be achieved by developers on site. This does not mean that developers should not continue to be held to best-practice principles for nature-friendly design. Bird and bat holes in buildings, tree and vegetation cover, and wildlife-friendly communal areas – all of which contribute to human health and well-being, carbon storage and energy efficiency – should be standard for any new development.

Off-site BNG should, in the coming years, bring opportunities for landholders who can offer land to developers for nature uplift. Rewilding is one of the easiest, most obvious ways to provide this uplift, and marginal farmland is likely to be in demand. Instead of the fragmented scraps of greenery on a development site, off-site provision of BNG can facilitate genuine nature restoration, creating functioning ecosystems. Developers could club together to create large-scale projects, creating even greater uplift. Sites can be chosen strategically, for example to connect existing areas of nature. To explore whether your rewilded land could be part of a BNG scheme, contact your local council, brokers such as the Environment Bank, and local developers. Local wildlife trusts may also be able to help on delivery.

A further serious concern about BNG is the time-limited covenant. Developers in the UK must currently secure the required biodiversity uplift for only thirty years. What will happen to the habitat when that time is up? Could even newly rewilded areas be opened up to development again? Clearly it would be far better for nature if the aim were to create and protect habitat in perpetuity. In the USA, for example, a system known as 'conservation easements' is used, whereby landholdings are conserved permanently for nature.[44] It is a powerful alternative to selling unproductive or unprofitable farmland for development, particularly in urban fringe areas. The landowner enters a voluntary legal agreement that permanently limits uses of the land in order to protect its conservation values. The tax concessions associated with easements often enable the landowner to keep the land for their descendants without selling off large tracts to pay estate taxes.[45] A conservation easement may also include the requirement to allow recreational or agri-tourism activities on the land. The easement is paid for by federal, state and local governments, non-profit organisations and/or private donors. Real-estate developers also sometimes pay for easements as a requirement of the planning process allowing them to develop land elsewhere.

Another tool used in the US is an endowment whereby revenues from invested capital provide funds for the maintenance of a site for nature in perpetuity.[46]

The DEFRA Biodiversity Metric

Biodiversity net gain is currently measured and accounted for in England and Wales using the DEFRA Biodiversity Metric 3.1, published in April 2022. This spreadsheet looks at the areas of different habitat types (according to the UK Habitat survey) on the land to be developed and assigns them habitat distinctiveness and condition scores, which provides an overall biodiversity unit score. Other multipliers may be necessary, depending on the location of the development.

For example, a developer destroying fifty units' worth of habitat to build houses, and able to reinstate five units' worth of biodiversity on the site by planting a small wildflower meadow and a copse of trees, will be left with fifty units to generate (the extra five represent the 10 per cent biodiversity net gain). That developer must then find a piece of land capable of generating these as BNG credits. The market is still developing, and David Hill, chair of the Environment Bank, has indicated that a conservation credit would be worth anything between £19,000 and £40,000 or more, subject to location and the habitat type created and managed. There may be up to two credits (for arable margins) or six or more credits (for richer wildflower meadows) per hectare, depending on the baseline of the site.

Let's imagine that wood pasture through rewilding generates a conservative three credits per hectare at £10,000 per credit (the markets will determine the price). This would mean that for the example above (where fifty credits are outstanding), a landowner would need to provide 17 hectares to generate £510,000. This agreement would be over a thirty-year covenant. The upfront restoration cost – which would include a full BNG evaluation identifying the uplift potential using a UK Habitat baseline survey of the site, registration of the site on the government's BNG Registry, the drawing up of a conservation covenant, legal fees, a habitat management plan demonstrating how the site will be restored, all ongoing monitoring and any initial interventions such as the introduction of free-roaming animals and the restoration of natural water bodies – might amount to, say, £12,000 per hectare, a total of £204,000 for the 17 hectares. That leaves £306,000 for the thirty-year agreement, equivalent to a 'profit' of £600 per hectare per year.

There are serious criticisms of the DEFRA Biodiversity Metric, however. The main one is that it uses habitat as a proxy for bio-diversity – in other words, it focuses on plants rather than birds, pollinators, mammals or other wildlife. While it considers rarity of habitat, it doesn't take into account the rarity of specific species or communities of interdependent species, nor does it take into account ecosystem services that generate wider societal and economic benefits. Its view of what constitutes good habitat is also flawed. It remains in many ways rooted in the agricultural mindset. Scrubland, for example, scores very low, although we know it is one of the most pro-ductive and biodiverse habitats there is. Ragwort and brambles are considered indicators of 'poor condition' grassland. On the DEFRA metric scoring system, Knepp's habitats rate relatively poorly for biodiversity, although this is improving with each new iteration of the metric.[47] Indeed, when challenged on these points on BBC's Countryfile programme, the author of the scheme said that the DEFRA system should be used alongside other metrics when assessing overall biodiversity.[48]

The accounting system of the metric itself is also open to broad interpretation and manipulation. For example, a developer might argue that the provision of a certain acreage of gardens within a housing development on a site that was previously conventional arable land is guaranteed to increase biodiversity because of the increase in garden habitat. In terms of species, this may be true. Gardens – especially if they are old and organic – can encourage and sustain large numbers of invertebrates (including pollinators) and birds. But most people still use pesticides and herbicides in their gardens. Many homeowners lay large patios or tarmac their lawns for extra car parking after they have bought their house. Some gardeners mow the grass incessantly, prefer non-pollinating exotic flowers and are obsessively tidy; some may even put down plastic grass. Developers surely do not factor predation by domestic cats into the equation, and it is likely that many of the small mammals, birds and butterflies that make it into these gardens will be caught by cats. For all these reasons the biodiversity measurement for one garden will be dramatically different from that of another. And what about all the other birds, including 'farmland' birds that would be lost to the site, and those that require entirely different habitats, or a complex mosaic of habitats, to thrive? How much disturbance will domestic dogs cause to wildlife in any scraps of woodland, grass-land, scrub or wetland habitat that have been preserved on the site? Is life in the soil accounted for – all the biota lost for ever under bricks and concrete?

One mitigating action that is often trumpeted by housing devel-opers in areas known for bats is the provision of bat boxes. But if the site is on a bat flyway between feeding and breeding grounds, the presence of houses – with noise and light pollution – will inev-itably prevent bats from being able to use it. Bat flyways must be broad, dark and undisturbed, and have continuous vegetation. Who will check that the bat boxes (and the birdboxes and bug hotels and hedgehog homes) in and around the new houses are actually being used? After how many years will such a mitigation measure be deemed to have failed, and will it even be possible to hold the devel-oper accountable years down the line? There must be an insurance scheme into which developers pay, which would provide money for further nature-restoration projects should the original schemes fail.

Because of all these pitfalls in the DEFRA Biodiversity Metric, several groups of academics, ecologists and private institutions are trying to develop different systems of credits, biodiversity scores and natural-capital calculators. Unfortunately, it is likely that the DEFRA metric will continue to be used in BNG calculations. But for the rest of the biodiversity offsetting market different calculation tools, such as 'voluntary biodiversity credits' (discussed above), could be used. This would create a huge injection of private finance into the nature-conservation movement. It would provide an alternative, or an addi-tion, to ELM funding for farmers, and a chance to sell the delivery of biodiversity gains across large areas through rewilding.

In particular, forward-selling carbon and biodiversity credits can play a significant role in funding the first few years of rewilding, bridging the gap before alternative income streams such as glamping, safari tours, wild meat, field courses and farm shops become available, and while ecosystem services eligible for ELM payments are still evolving.

Other potential income streams

Ecotourism

Tourism is one of the biggest potential winners of rewilding. With increasing urbanisation, more and more people are seeking out rural holidays. Nature-based tourism in Europe accounts for around 10 per cent of European GDP and 20 million jobs.[49] In the USA, recreation and tourism in national forests alone contribute $2.5–3 billion a year to national GDP.[50] In England, nature-based tourism is now believed to be worth around £12 billion a year.[51] In Scotland, walking and landscape enjoyment alone (around 1.8 million trips) nets £900 million, and nature-focused tourism – including bird-watching, whale watching, guided walks and practical conservation holidays – is worth £1.4 billion and generates 39,000 jobs.[52] In Wales, where 'wildlife-based activity' generates £1.9 billion, walking, on its own, contributes £500 million to the economy – £100 million more than subsidised farming revenues. In recent years COVID-19 has underlined the mental and physical benefits of access to nature, and ecotourism is set to continue on its steep upward trajectory.

At the same time, a growing interest in wildlife and visitors' increasing knowledge of the natural world means that many tourism venues that have been selling themselves as 'eco' are now found to be lacking. Visitors are becoming more discerning. And, in a world where the carbon cost and true worth of material goods are coming under scrutiny, the gift of an experience, a memory to be shared, is growing more popular. There is an educational aspect, too, and experiences are often considered more valuable when they increase knowledge or appreciation. The University of the Third Age (U3A) in the UK is just one organisation that recognises the attraction of providing its members with enjoyable activities that have an educa-tional aspect, such as identifying birdsong or wildflowers.

Larger rewilding projects provide the perfect context for ecotour-ism, such as glamping and camping, guided walks and vehicle-based wildlife tours. As recently as 2010 it would have been advisable for a rewilding project to wait for a few years, until some ecological gains had been made, before launching the tourism aspect of the business. Now, however, such is the interest in rewilding that many projects are launching tours in the very first years. It is clear that visitors are as interested in the early stages of rewilding as they are in seeing head-line species that may take time to arrive. Seeing nature springing back to life is a draw in itself.

Reintroductions

Species-focused tourism is enormously popular, and the reappearance of lost and rare creatures has special power to enrapture. Some 82 per cent of the British public support the reintroduction of extinct animals – a key feature of rewilding.[53] The osprey nest at Loch Garten in the Scottish Highlands – the first in Britain for forty years – has at the time of writing been visited by more than 2.75 million people, sometimes 90,000 in a single summer, with 300,000 tuning in to the centre's live webcam.[54] It is the most watched bird's nest in history anywhere on the planet.

The financial rewards for a rewilding project and the surrounding local economy can be immense. Across the UK, 290,000 people a year visit nine key osprey-watching sites, generating £3.5 million for local economies. White-tailed eagles on the Isle of Mull attract a tourist spend of between £5 and 8 million a year and support 160 full-time jobs.[55] On the Isle of Skye, white-tailed eagles bring in £2.4 million a year, and the introduction of these spectacular birds to the Isle of Wight in 2019 looks set to have the same effect.

As apex predators, birds of prey are keystone species, important for their ripple effects on the food chain, and beloved of birdwatchers and nature lovers generally. But charismatic species are important, too. They help to promote nature recovery and engage the public in ways that can also, incidentally, help rewilding projects financially. The success of the White Stork Project, which began in 2016 and of which Knepp is a partner, has been spectacular. In 2020 the first white storks to breed in Britain since 1416 nested successfully in oak trees at Knepp. They became a sensation, one of the few upbeat news items during the COVID-19 crisis.[56] We opened our mobile Stork Café largely to cater for crowds arriving to see the storks when lock-down was lifted. Many of the thousands of visitors every year who continue to flock to Knepp to see the storks and their fledglings are not archetypal birders at all, but non-binocular-owning newcomers to nature, often from cities. The storks have done an enormous amount to raise wider awareness of our rewilding project, and as a result we now feel that we are in a position to expand what we offer visitors, with a proper café/restaurant and farm shop.

The first white storks to nest successfully in the wild in the UK for over 600 years inspired this incredible mural by artist Sinna One in Brighton in 2020.

Public interest in species reintroductions is not limited to large-scale projects. On the river Otter in Devon, smallholdings offer accommodation and guided tours to visitors flocking to see free-living beavers. Such accommodation can be low-key. Bed and breakfasts or holiday rentals – even a single shepherd's hut – can bring in significant income all year round, especially if you manage it yourself.

It is worth considering whether your site is appropriate for a species reintroduction, and contacting NGOs, such as the Wildlife Trusts, to see if there are any projects nearby you could be involved with. The England Species Reintroduction Task Force, launched in 2023, brings together experts, landowners and NGOs to introduce species such as dormice, corncrakes, curlew and beavers. In time, it is hoped, it will establish projects for golden eagles, wild cats and lynx. The Vincent Wildlife Trust specialises in the reinforcement of rarer species of bat, as well as small carnivores such as pine martens, stoats and weasels. But for many rewilders it is a case of starting a project according to their own particular interest. At Knepp we have undertaken a scoping report on the potential to reintroduce red-backed shrike, a scrubland bird that is on the verge of extinction in Britain. We are also looking at regional-scale reintroductions of water voles and pine martens.

Even small creatures have reintroduction potential, such as wood ants – a keystone species.[57] In the 1940s Winston Churchill attempted to reintroduce the black-veined white, an extinct native butterfly, to his estate in Kent, but was thwarted by his gardener, who mistook his intentions.[58] In 2018 Citizen Zoo, a social enterprise promoting community-led rewilding projects, successfully reintroduced the endangered large marsh grasshopper to a wetland in Norfolk.

Reintroductions, especially of rare or endangered species, are of course not to be entered into lightly. They require enormous commitment, and must follow best practice, as laid out in the IUCN's complex but excellent guidelines.[59] Projects must work with experienced, knowledgeable reintroduction specialists, and in conjunction – or communication at least – with other reintroduction projects for that species, rather than working solo. The process of applying for a licence to release a species into the environment can be long and convoluted, and there are often considerable costs relating to holding facilities and introduction pens, feeding and ongoing monitoring – all of which may continue for years.

Fortunately, many charismatic creatures – including the rarest – are capable of arriving under their own steam as soon as their population is viable and there is habitat and food for them. In the 1990s red kites, which had been extinct in Britain since the nineteenth century, were re-established in the Chiltern hills in southern England and the Black Isle in northern Scotland. They are now spontaneously recolonising southern England, including Knepp, and other parts of the UK. Simply providing food for them can draw them to a site and create a wildlife spectacle. Some 150,000 people a year visit Gigrin Farm in Wales, where the farmer puts out offal to feed as many as 2,500 visiting red kites each week.

As we saw in Chapter 6, a vital natural process missing from today's landscape involves carcasses. Changing the regulations that apply to this would be a boon to rewilding. Allowing fallen stock to remain on the land as carrion would provide food for spectacular white-tailed eagles, golden eagles, red kites, buzzards and ravens, as well as white storks and other birds and, of course, many mammals, fungi and insects. Roy Dennis, an expert in bird reintroductions, suggests a role for dedicated carrion dumps – areas fenced to exclude domestic dogs and cats – along the Channel coast, from Kent to Devon, to attract vultures from Europe. In 2016 and again in 2020 a bearded vulture, or lammergeier, visited England for several weeks. Flocks of griffon vultures have been prospecting along the north French coast and in the Netherlands. In Spain and Portugal, 'vulture larders' have proved key to reinforcing populations of griffon, black, Egyptian and cinereous vultures, and are very popular with tourists.

Wildlife watching

Providing wildlife hides – sheltered, hidden places for wildlife watching – also attracts visitors. However, in a rewilding project with dynamic, shifting habitats, it is not easy to predict the favoured locations of species. Unless a hide overlooks a body of water, it may be that time and energy are invested into building a structure that is in the wrong place a few years later. Nature reserves often have this problem, and on RSPB sites habitat is often managed at huge expense to keep the wildlife within sight of permanent viewing structures.

Temporary hides in the form of tiny camouflage tents with netting and room for a small stool are more useful. Of course, if the visitor is to be guaranteed a sighting and that dream photo, wildlife – such as kingfishers, owls, buzzards, red kites, otters, foxes, hedgehogs and badgers – will have to be 'trained' to come near the hide through routine baiting at a given spot. While this may be counter-intuitive to the idea of rewilding, we see no problem if it is done sensitively. Supplementary feeding should remain minimal, for example, so that wildlife does not become dependent on it. We feel the rewards – a magical experience for the photographer, and revenue and high-quality wildlife images for a project – justify the 'unnatural' intervention.

Wild-range meat

We've already discussed the potential for producing wild-range meat from larger rewilding projects (page 353), but there is also huge potential for adding value to rewilded meat. As the anti-feedlot movement grows, interest is increasing in pasture-fed, organic meat produced sustainably and to high standards of animal welfare. Such meat has the added benefit of containing fats that contribute positively to human health, rather than the unhealthy fats put down by intensively farmed, grain-fed animals.[60] Rewilding projects produce meat with other dimensions, too, that appeal to the ethical and health-conscious consumer: contributing to the restoration of nature and the sequestration of carbon; and knowing that the animals led a semi-wild life.

It is relatively easy to sell organic meat on the open market in England, as it is in Europe and the United States, where the interest in organic produce continues to rise.[61] However, rewilded meat is 'beyond organic' and should, in theory, fetch a higher premium. But since there is as yet no certification for rewilded meat, the story, brand power and added value of the product tend to be lost on the open market. There can be a huge advantage in taking control of your meat and selling it direct to the consumer. The wholesale meat price of an old English longhorn, for example, is around £1,000 (the price at the time of writing). Selling through meat boxes, a farm shop, your own café or restaurant, or online, can add as much as three times the wholesale price (that is to say, £4,000). Selling direct to the consumer, cutting out the intermediaries and shrinking the supply chain not only increases the profit, but also cuts down on carbon and transport miles. It is also possible to add value by processing meat – either on the site or through a third party – into sausages, bacon, burgers, charcuterie and ready meals, and through catering and hospitality, such as pasture-to-plate events, pop-up restaurants and butchery courses.

Conversion of post-agricultural buildings

This has become a significant income stream at Knepp, where we have converted more than 12,000 square metres over twenty years for office space, light industrial use, storage, farm shops and events venues. Of course, most conversions involve significant capital outlay. There may be grants available for this, especially if the building is going to be used for public engagement or education. Ultimately, a great opportunity presents itself here to transform a building that was once part of the farm infrastructure, cost an arm and a leg to maintain and did not, itself, bring in any income, into a revenue-generating asset.

Events and courses

Rewilding projects provide great scope for nature-inspired activities, from art, poetry and sculpture courses to whittling, woodwork, coppicing, forestry, foraging and survival courses. It can be hard to decide which suit the ethos of a project without affecting it too much. Rewilders should also consider whether there is profit to be made from third-party arrangements, or whether it is better to do it themselves.

Sporting events – such as wild runs and cycling, and raft races and triathlons if there is a large body of water – can be scheduled to ensure the least disturbance to wildlife, generally towards the end of summer. Sports fishing can also be managed as a low-impact activity. Wild swimming is very popular, but can be much more disruptive to wildlife. We have dredged a pond specifically for this purpose at our campsite, in order to deter wild swimming in remoter ponds and lakes that are focal points for migrating birds and numerous other species. Deer stalking can be a lucrative way of achieving a task that you would need to do anyway. If you have fallow bucks and red stags with spectacular antlers, there may be considerable demand from trophy hunters.

Foraging is popular, too, and 'wild' produce, from juniper berries and sloes to fungi, samphire, seaweed, wild garlic and elderflower, is in great demand commercially, not only for food and drink but also for health and beauty products. Before embarking on an enterprise that could become a large operation, however, remember that harvesting what may seem plentiful at the outset will end up depriving wildlife of food. It is the story of our planet. Foraging could be worth considering as a source of income for a rewilding project, but it is crucial to set an acceptable threshold. For example, at Knepp, we have a maximum of twenty hives for producing honey, on advice from Dave Goulson that this is the number our project can support before the honeybees' requirement for nectar and pollen begins to affect native pollinators.

Corporate events can work well in a rewilding project, providing companies with a great backdrop for encouraging team building, thinking outside the box, and inspiring creative and transformative ideas. Wild weddings, green funerals and woodland burials are increasingly popular, too. Rewilding projects also have enormous potential for hosting adventure holidays and field courses for young people. Existing nature-based activity companies are often looking for new locations and joint ventures, although greater rewards may be gained by setting up a business yourself. From pond dipping and dung-beetle collecting to bivouacs and cooking on an open fire, programmes can be tailored to any age, and include anything from birthday parties to holiday clubs, bushcraft courses to summer camps.

While the recreation and leisure sector has been busy rewilding itself, the UK education system is, sadly, worlds away from the concept of *udeskole* – teaching and learning in the natural environment – that underpins the curriculum in Norway, Sweden and Denmark.[62] During the 1990s, however, a movement for 'alternative' educational methods introduced Forest Schools to the UK, based on the Scandinavian model.[63] It has been hugely successful, particularly for children with behavioural conditions. We very much enjoy hosting Forest Schools at Knepp.

There is clearly a need to get more children out of the classroom and connecting their academic subjects – particularly geography, biology and natural history – with experience in the field. But when funding for field trips and the teaching staff to organise them are at an all-time low, it is hard to imagine how this can happen unless the state starts funding state schools more generously.

University field trips are also at their lowest level for decades. This is of considerable concern on many counts, not least given the huge increase in jobs in wildlife conservation and climate change that is likely to come. Currently, universities are not turning out students with the relevant skills, particularly in how to measure biodiversity and carbon capture, to ground-truth the new credit and offsetting systems. At Knepp we have partnered with the Wallacea Trust to try to address this problem. Over six weeks every summer we host 150 school leavers and undergraduates at a campsite in the rewilding project. They spend a week learning from top university teachers and our own ecologists how to conduct experiments and surveys in the field. We were nervous for our first season, in 2021, that the students – unable to travel to Operation Wallacea's more exotic destinations because of COVID-19 – might be disappointed with a field trip in the UK. But the experience scored top marks on the feedback forms and we were tickled by the thought that the students, enamoured with rewilding, might go on to rewild their teachers and institutions.

When we began rewilding, we could never have imagined the income streams it would open up for us, or the employment opportunities. When we were farming, we had twenty-three full-time-equivalent employees. Under rewilding we have fifty, and we expect this to continue to rise when we open our café and farm shop in 2023. Knepp has gone from being a loss-making business to being profitable. It has also brought advantages for the local economy. Our ecotourism has boosted trade for local shops, services, tourist attractions and pubs. Our closest pub has just opened four rooms for bed and breakfast. The businesses that rent office and light industrial space in our converted agricultural buildings employ 200 people, bringing further trade into the local economy.

This turnaround is not unique to Knepp. Rewilding Britain's report *Rewilding and the Rural Economy* in 2021 analysed thirty-three rewilding projects in England, large and small, on land and sea. Many of them have been going for only a few years, yet already they have seen a 50 per cent rise in jobs, and a thirteenfold increase in opportunities for volunteers.[64] Rewilding is not about favouring nature to the exclusion of people, as its detractors sometimes claim. Quite the opposite. It brings nature and people together. It provides greater public access to nature, stimulates nature-based businesses that can revitalise rural communities, and drives forward a greener, more sustainable economy.

A business enterprise doesn't have to be hi-tech or cost the earth. The Stork Café, cobbled together from an old trailer by our son Ned and his girlfriend Lia during lockdown, is still doing good trade during our visitor season.

9

Recording and Monitoring Wildlife

How to use science in your rewilding project

Measuring the abundance and variety of life on a rewilding site is the most important way to document a recovering ecosystem. The evidence can be essential to funding a project, and it will certainly provide valuable information for others. It is also a wonderful way to engage with people, and immensely exciting to see proof of wildlife returning.

At Knepp we engage with numerous groups of volunteers, species specialists, university students and academics, who look at everything from bats and birds to the tiniest water snail.

Recording changes in the variety and abundance of wildlife is of fundamental importance to any rewilding project. Improvements in biodiversity are proof of a recovering ecosystem. There are plenty of other things that can be monitored (and we have touched on some of them in other chapters) such as air quality, water quality, water flow and hydrology, carbon storage and even social benefits and employment. But biodiversity is the litmus test. It is the most obvious indicator that everything else, both seen and unseen, is on track.

Structured, reliable records are the passport to funding. They will also be of great help to other rewilders and conservationists, adding to the knowledge base. To policy makers and the wider public, sound evidence speaks volumes, articulating the case for rewilding more convincingly than theory alone. Monitoring wildlife is a great way to engage with naturalists in the local community and to promote interest in the project. But it is also great fun and immensely gratifying. To be able to prove that your patch of the world is now a livelier, healthier, more abundant place is, arguably, the greatest reward of all.

The recording of wildlife can be done in many ways. At its most basic, you can use a simple spreadsheet to record the male stag beetle you saw when you were walking your dog, or the barn owl you spotted on your evening stroll. Such casual recording is a great way to build up a species list (see page 380) for a site, but is no substitute for detailed and considered surveying and monitoring – although the two approaches can work well together.

Recording days are a slightly more formal way of obtaining data from the land and adding to the species list. Surveys are still more structured and focused, based on considered methodology. A baseline survey (see page 368), undertaken at the beginning of the project, will provide a 'before' snapshot against which to measure any developments.

Ongoing surveys can then be undertaken as time goes on, chosen according to what is likely to change on the site, in both the short and the long term. If, for example, you're expecting the big change to come about from alterations to the management of grassland, you might hope to see an improvement in botanical species-richness, density and diversity of flowers, and the number of invertebrates that feed on those flowers and/or live for some or all of their life cycle in grassland, as well as small mammals taking advantage of seed sources and the protection of a denser, thicker sward. Think about the kind of analysis that would be needed to prove the point and 'reverse-engineer' a suite of surveys that could be done as part of the original baseline survey and repeated in, say, five years' time, for comparison.

Monitoring is essentially the same as surveying, but with regular follow-ups and with some level or criteria being assessed which, if breached, should raise a red flag. On grassland, for example, if moni-toring indicates a long-term decline in species diversity, grazing might be too heavy and require adjusting. Monitoring is, in some sense, associated with action.

It is important to remember, though, that with rewilding one tends to take a more distant and abstract view than with conventional con-servation, taking into consideration the complexity and dynamism of the entire habitat mosaic; encouraging, if possible, fluctuating densi-ties of herbivores over longer periods; and recognising that, however thorough, monitoring will inevitably present a limited snapshot of the whole picture (see page 366). Moreover, some species may benefit from what may seem, to the conventional eye, an excessively dynamic or even hostile habitat, and many species need access to different hab-itats over their life cycle. The eye should never be on a single habitat alone, but on the functioning, connected ecosystem. On balance, if there is dynamism in the system and biodiversity is improving across the board, it is best not to intervene, and certainly not to be tempted into knee-jerk interventions.

Monitoring strategies

Even though a rewilding project is, by its very nature, open-ended and not driven by targets, care should be taken at the start to incorporate a monitoring strategy into the overall management plan. A primary aim of rewilding is to increase biodiversity. Without a careful and rigorous system of ongoing monitoring, it will not be possible to say whether this has been achieved, or how.

It is clearly impossible to cover everything, so a monitoring strategy must be focused and streamlined. It should be designed to get the essential information with minimum effort. Extra surveys can be added according to the availability of resources and expertise. It is therefore worth prioritising a survey list, from 'essential' at the top, down to 'nice-to-do' at the bottom. Ideally, focus on species groups at the lower end of the food chain – invertebrates, plants and, to some extent, birds – because, being particularly rich and diverse, they cover many habitat niches that are sensitive to change. This allows greater complexity of analysis and interpretation. Invertebrates, however, are particularly difficult to survey and, as a result, more costly – something that should be taken into consideration at the outset.

For larger rewilding projects, consider applying a spatial element to the data collection – something that would not normally be done for a small nature reserve. A simple two-dimensional map can be used to reflect any number of features that you may previously have been unaware of: perhaps the distribution of anthills or of key trees or shrubs such as elm and wych elm (food plant of the white-letter hairstreak butterfly). These patterns can then be charted as they evolve under rewilding. Sampling in this way is often so useful that, in some cases, it is worth forgoing sampling randomly in order to generate grid maps of the data.

Future-proofing must be another consideration in an open-ended and changeable project such as rewilding. It is important to future-proof your methods, in terms of both successive surveyors and changing vegetation. Some are likely to take longer (and cost more) as the vegetation structure becomes more complex.

Observer bias

The greatest level of bias that enters monitoring is from using different observers. Everyone has a slightly different way of measuring and perceiving things. These effects, especially if many observers are involved over time, can be greater than the changes that have happened on the ground, resulting in misleading results and interpretation. Standardised surveys with rigorously defined methodologies do help to minimise bias, but they don't remove it entirely. It is therefore ideal to have the same person repeating a survey for as long as possible, so select your surveyors wisely. It is also the reason that, at Knepp, we've found university students are not ideal for monitoring, unless you can ensure the long-term commitment of the university. A good local volunteer can often provide greater continuity and a longer run on a survey than the staff who manage them.

Keep it simple and consistent

It is tempting to use new technology and trial techniques, especially if they're being advocated by a university or a consultant, but bear in mind that this institution or that individual is unlikely to be able to support you for ever, and you may need to introduce someone unfamiliar with the systems to take over their work. It is important to use conventional kit of the kind used by other surveyors, so that the sampling technique remains consistent and comparable to other surveys.

Crucially, using tried-and-tested methods, particularly those promoted by national schemes, means they can be compared to national datasets. One of our mistakes at Knepp in the early days was to use a hybrid method for our bird survey – combining the British Trust for Ornithology's (BTO) Breeding Bird Survey and the Common Bird Census – so we were unable to compare our results with either scheme. It also involved unnecessary cost and visits. We've now switched to a single method – the Breeding Bird Survey – which means we can easily correlate our data with that of the BTO across the country.

Working with consultants, volunteers and students

The level of detail in terms of monitoring and surveying will depend on the size and scope of the project, on your own interests and, perhaps, on conditions imposed by funding bodies. Some of the more basic work, such as fixed-point photography, simple soil sampling and water-quality testing, can be done with little or no training. But most other surveys must be carried out by experienced ecological surveyors and/or specialists in their field. Employing ecological surveyors is expensive, and there will be key surveys that are worth forking out for.

For larger rewilding projects, having a small four-wheel-drive vehicle is indispensable. As well, of course, as saving time (and time is money), it enables a surveyor to complete visits to a number of sites across the project on the same day during similar conditions. Without a vehicle a meaningful comparison between sites would be impossible.

Ecology students on a summer camp run by Operation Wallacea learn how to record data in the field at Knepp. Their work contributes to our ongoing monitoring and surveys.

For many surveys, it may be possible to recruit volunteers. Species recorders are an enthusiastic tribe, many of whom spend every spare minute pursuing their passion in the field. In the UK there is a dedicated following for almost every species group, from the Dipterists Forum (for the study of flies) and the British Dragonfly Society to the British Bryological Society (mosses and liverworts), the British Myriapod and Isopod Group (millipedes and woodlice) and the Conchological Society of Great Britain and Ireland (molluscs and their shells). They are generally willing volunteers, particularly if they can be unleashed on a piece of new, under-recorded land with limited or no public access. In the UK, your local Environmental Records Centre will be able to put you in contact with local biological record-ers and specialist groups.

Our volunteer groups at Knepp have become one of our most treasured assets, and are appreciatively nurtured by our resident ecologist, who coordinates all our survey work. As well as plugging into the local recorders' networks, it's worth building a following on social media, where you can advertise events and post updates, create a database detailing your volunteers' qualifications, interests and spe-cialities, and celebrate rare and unusual records. The wider following that is gained in this way can be of practical use, too. We use amateur volunteers for our deer population surveys (simply stomping around fields in phalanxes, trying to count deer), water-quality testing and manual work, such as cutting back hedges to reinforce beaver fences or maintain green lanes. These events are a great way of engaging the general public, with a sense of camaraderie and achievement cel-ebrated over tea and cake at the end of the day. Our turtle-dove and winter bird surveys, on the other hand, are conducted by a hardcore group of fifteen experienced birders – now a close-knit group – who are dedicated enough to get up at 4 a.m. on a spring/early summer day. Birding breakfasts are an enjoyable way of pooling experiences after a two-hour yomp around the scrub with a clipboard.

Another resource for conducting research is university students. In our experience, studies by Master's students tend to be more valuable than undergraduate work, while PhD students' results can be a pain-fully long time coming. The downside to using students can be the lack of continuity, unless you have a university supervisor keen to stay engaged and run on-going research projects. The quality of students' dissertations varies considerably, but the best of the Master's projects at Knepp have been extremely useful. If you have an advisory board it's an idea to include one or two members with university connec-tions who can help to set this up. The value of students' work lies in the scientific weight that can be thrown behind a single question in a relatively short time. If you have a very specific, burning question about something that requires a rigorous experiment to answer it, this is a good approach. Whenever you have an idea for a dissertation topic, make a note of it so that you can present a list of suggestions to universities and students when the opportunity arises.

Recording days

A recording day is a voluntary event and a free-for-all, with recorders – usually from special-interest groups, such as a local fungi group – going where they want, at any time of the day or night, with no transects or specific timings involved. Hold a recording day or even weekend at the beginning of your project. It shouldn't be seen as a replacement for structured surveying and monitoring, but it will give you a broad overview for a baseline, and help build the species list for your site, introduce you to local biological recorders and collect specialised data. One visit and they're usually hooked, and will most likely want to be involved in future recording days or, if their interest is really piqued, help with more structured ongoing surveys.

The local records centre may be willing to help you set up a recording day or, at least, advertise your event in a newsletter or on their website. At Knepp we had fifty recorders in total over a weekend in June, including specialists in beetles, birds, moths, planthoppers, dragonflies and flora, pond surveyors and pan-species listers. We provided a converted barn for their overnight stay, gave them an area in which to pitch tents, and a barbecue for them. If you don't have the facilities to put people up, a simple recording day with enticing refreshments works too. It is worth offering to cover travel expenses for particular experts from further afield whom you would like to involve.

A recording day takes a fair amount of advance planning in terms of advertising and recruitment, as well as commitment from volunteers. It may be weeks or months after the event before the recorders are able to submit their findings, especially if they've taken away samples, such as obscure insects for dissection and identification under a microscope, so it's best to do this kind of thing no more fre-quently than once a year. It is wonderful, though, if it can become an annual event and a celebration of the previous year's discoveries – with, of course, the challenge to find the rarities again, and in greater numbers.

Ideally all data is sent to the local records centre, which can then send you a spreadsheet of what was recorded. Bear in mind, though, that the system of recording is rarely very structured and therefore the amount of interpretation and analysis that can be gleaned from the records is limited. One particular day or weekend of recording will provide only a snapshot of life on the land. Habitat changes over the seasons; new foliage and blossom are important in the spring, and later in the year the focus of attention moves to the sward. Some species will be invisible on a particular day, either because – as with migrating birds and butterflies – it's not the season for them, or because they're in a reclusive stage of their life cycle. If a recording day becomes an annual event, consider varying the dates to embrace different seasons.

Sometimes a recording day is referred to as a 'BioBlitz'. A BioBlitz is an intense biological survey done by groups of scientists, naturalists and volunteers in an attempt to record all the living species within a designated area over a single period (usually twenty-four hours). We don't tend to use the term, because in England it's often more loosely applied to nature days organised as social or educational events. The National Trust, for example, holds family BioBlitz days as a way of connecting children with nature. While this is obviously a great thing to do (and you may want to consider holding a similar event), a family BioBlitz day is unlikely to provide the kind of reliable data that will come out of a professional BioBlitz or recording day.

Pan-species listers

Pan-species listers are omniscient creatures whose biological knowledge encompasses every conceivable taxa of wild life forms – excluding bacteria – starting with fungi and slime moulds. The idea of pan-species listing – the Olympic decathlon of natural history – was the brainchild of Mark Telfer and a number of other keen naturalists. It began in 2011 with eight people in the original rankings, and now has its own website and Facebook community, with 261 registered listers, although not all are active.[1] If you can attract one or more of these biological superheroes to your recording day, you'll have every avenue covered. But you may consider holding a pan-species lister day, when you can pit them against each other to see who can find the most species. If your land is relatively unrecorded and contains a variety of interesting habitats, it may well appeal to this phenomenal and irresistibly competitive genre. But, again, these specialist recording days, while fantastic for adding to the understanding of a site, should not be seen as a replacement for structured surveying.

Types of survey

Baseline surveys

The importance of establishing what species and habitats are on the land before you make any changes cannot be overstated. It is the keystone of any rewilding project. It provides the basis for ongoing monitoring and survey work, the evidence required for continued funding, the authority to tell your story and data that may be useful to science and other conservation efforts.

A baseline survey can be as simple or as complicated as you want, but the more surveying that is done before the project starts, the better. It is very important not to fall into the trap of thinking you don't want your land to look too good before rewilding starts. The unique assemblage of wildlife that is already there will start the story and provide the building blocks for whatever evolves. At the very least, it's a good idea to include a basic habitat map, soil health analysis, a breeding bird survey, butterfly survey and fixed-point photography – all of which can be relatively simple and inexpensive to do, and can be assisted by volunteers.

With hindsight, we made many mistakes at Knepp, partly because monitoring from scratch on depleted farmland had never been done before, but also because we had no idea how prolific and wide-ranging the resurgence of life on our land was going to be, or which sites and taxa would become particularly significant. We've learned some valuable lessons.

Breeding bird surveys

Birds are a key indicator of the quality and specifics of any habitat, and comparatively easy to see or hear and identify. The Breeding Bird Survey (BBS) methodology has been developed by the BTO, the Joint Nature Conservation Committee and the RSPB. It is straightforward and quick (observers usually spend no more than ninety minutes counting birds on each visit), and you may be able to find a volunteer from your local bird club or ornithological society to do it for you.

The BBS uses line transects or routes, which sample randomly selected units of the 1×1 kilometre squares of the Ordnance Survey national grid. Each is composed of two roughly parallel lines, 1 kilometre long and about 500 metres apart. Each of these lines is divided into five sections, making a total of ten 200-metre sections, and birds and habitats are recorded within these ten units. Mammal sightings can also be recorded.

Fieldwork involves three visits to each survey square each year. The first is to record habitat details, so as to allocate each square a BBS habitat code and establish or check the route. The second and third visits are to count birds, with the 'early' visit ideally in early May and the 'late' visit about four weeks later, in early June, to cover late-arriving migrants. You can then compare the national data with the trends on your site.[2]

It is possible to set up more targeted transects (rather than the squares randomly generated by the BBS) using the same survey method, and still compare the data you gather with the national dataset. However, choosing the location of a transect can be tricky in an early-stage rewilding project, when it is unclear how the habitat will evolve. Bearing in mind that vegetation and terrain may become increasingly complex, and in some places impenetrable, it's worth using existing footpaths, farm or access tracks, which will always be kept clear.

Having said that, we made a significant error of judgement when we established one of our transects in the Southern Block along a 'green lane', an ancient droving road that is now a public footpath. Compared to the surrounding arable fields, it had remained comparatively biodiverse. Bordered by steep banks, mature trees and ancient hedges and featuring the occasional pond – once watering holes for driven livestock – it was in no way indicative of the impoverished landscape of the wider farm, nor did it reflect increasing biodiversity as the rewilding project evolved, largely protected as it was from the influences of the habitat changes around it.

We grew concerned, as biodiversity rose dramatically elsewhere in the project, that species counts on this transect were not reflecting this trend. It was only when we finally decided to establish new transects, in 2018, that the extent of the anomaly was revealed. Our original green-lane transect of about 2 kilometres recorded 122 individuals of 20 species over ten visits in 2018, while the new transects of 6 kilometres in an area much more representative of the rewilding project recorded 886 individuals of 48 species in just two visits the same year. For example, the survey on the original transect recorded three singing male whitethroats for the ten visits in 2018, while the new transects recorded 44 on the first visit and 30 on the second. This mistake has essentially cost us twelve years of valuable data relating to the rising numbers of birds in that area of the project. The moral of the tale is to balance pragmatism with relevance to the rewilding project when choosing transects.

Invertebrate surveys

Invertebrates may not be as charismatic as birds, but they are a great indicator of the quality of habitat. There are so many more species of invertebrate than of other groups that it's typical to record between 500 and 1,000 species during a single year. This might seem daunting, but the analytical power it provides is great. Running the dataset through a resource-analysis database provides useful insights into the changes happening on the site. Is the number of nectar-feeding species increasing or decreasing, for example? Has there been a change in species that need bare ground, or those that eat only heathers?

This is another important aspect of the baseline survey and ongoing monitoring. It's not cheap, however, because it requires the work of experienced entomologists and arachnologists, and invertebrate identification is time-consuming, often involving microscope work. Repeat surveys are recommended every five years.

The UK's Wider Countryside Butterfly Survey (WCBS), run by Butterfly Conservation and the Centre for Ecology and Hydrology, is a 'reduced effort' method of surveying that uses the same transects as your BBS. Only two to four visits are required for the WCBS squares, with a minimum of two visits in high summer, and other visits encouraged, particularly earlier in the year for spring species such as orange-tip and the first broods of species that have two or more a year.

The UK Butterfly Monitoring Scheme is more labour-intensive, covering fixed-route transects using randomly selected squares through sites of high-quality habitat. The surveyor must commit to twenty-six surveys a year, which can be onerous, especially if there are many transects to cover. It is not the most suitable methodology for remote habitats, where weekly visits are not feasible, or for monitoring butterflies that may be difficult to detect, such as purple emperors or purple hairstreaks in the tops of tall trees, or white-letter hairstreaks, which are confined to localised areas near groves of elm.

Alternatively, if you have a keen lepidopterist volunteer you could still gain a valuable amount of data from simply conducting monthly surveys over the summer.

Butterfly monitoring may not be appropriate for every project. In the UK, away from the warm southeast, some sites may be fairly poor in butterfly species and the time commitment a survey requires may not be justifiable. A baseline survey will provide information about this.

Vegetation mapping

Vegetation is one of the building blocks of survey work. Untold species feed on vegetation, from leaves to flowers to living and decaying wood, and of course the vegetation has its own inherent value. It is also the part of the landscape that most clearly reflects the management approach, whether that be with free-roaming animals or by using human intervention to mimic their impact. British National Vegetation Classification (NVC) mapping provides a comprehensive picture of baseline vegetation and can show the intricacies of habitats and how plant communities evolve under certain soil and climatic conditions. We wish we had done this from the outset at Knepp.

There are 286 different plant communities under the NVC, grouped within twelve major categories, among them 'mire', 'heath', 'salt marsh' and 'woodland and scrub'. This is a very specialist survey, however, and may be too detailed and expensive for many projects. It requires employing a botanist with an eye for the structure of plant communities, and is ideally carried out before the project begins and subsequently every five to ten years. Some changes, in woods for example, are likely to be very slow.

Other levels of mapping, such as Phase 1 or the recently developed UK Habitat Classification, are a little less taxing on the observer and quicker to carry out. UK habitat mapping has an added advantage in that it underpins the DEFRA Biodiversity Metric 3.1, so will help projects large and small understand their net gain in biodiversity. The Land App helps to generate a quick and free version of this. It requires a lot of subsequent ground truthing (in other words, observation and fact-checking on the ground) and refining, but is a good place to start.

All these methods share the propensity to compel the observer to go to every corner of the site, even to areas that they would never normally visit. This is an excellent exercise in understanding what makes a site tick, and is likely to unearth a few surprises. For example, land can change considerably under different owners and types of land management over decades. But, generally, land along boundaries tends not to change so much, so interesting plants and features are often found at the very edge of a site.

If you're interested in creating your own maps or reports on vegetation it's worth doing a day course on how to use QGIS, the mapping software for the free, open-source geographic information system.[3] This is an extremely useful tool for creating maps that correlate species with habitat. For example, you might want to map where your scrub and hedgerows are and overlay this with data for nightingale territories or brown hairstreak butterfly colonies. (The Land App provides an easier but less technically detailed way of doing this.)

A Google satellite photograph of the Southern Block at Knepp marked up with the territories of whitethroat surveyed by the BTO Common Bird Census in 2019.

Original image © Google Earth

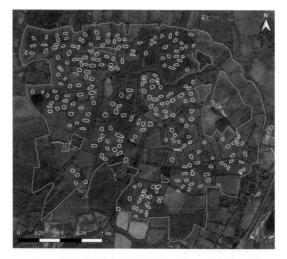

Overlaying whitethroat territories onto a vegetation map, rather than just a Google satellite photograph, provides more useful information. It shows the birds in highest densities in thorny scrub. (Scrubland vegetation map generated by Queen Mary University of London.)

Vegetation type:
☐ Bramble scrub
☐ Sallows
☐ Thorny scrub

Fixed-point photography

We kick ourselves for not establishing this sooner in the Southern Block at Knepp. In particular, we're missing those basic 'before' shots that make the 'after' images so compelling. Our sights were so much on the excitement of the future and we were so familiar with our conventionally farmed landscape that it didn't occur to us to take many boring photographs before we had something to show for our rewilding work. We've had to make do with satellite mapping images available on Google, which we can overlay for time-lapse sequences, showing the emergence of vegetation. Although this looks impressive as an overview, aerial imagery, especially that taken from directly overhead, gives little indication of the complexity, structure and height of vegetation. For one's own interest, but also for media use and funding bodies, the detail and atmosphere of contrasting ground-level images are worth their weight in gold.

Learning from this, we have been meticulous about fixed-point photography in advance of the release of beavers into a pen at Knepp in 2022 – anticipating that they will have a big impact on habitat – and in our Victorian walled garden, which we began to 'rewild' in 2020.

It is almost impossible, however, to anticipate the true value of fixed-point photography. One might set up a camera to focus on open ground in front of a hedge, for example, with the aim of recording blackthorn and other woody species expanding from it. But over that period catastrophic ash dieback happens, so in the event you have also recorded ash trees pre- and post-disease. This might be useful to help identify the percentage of ash trees of different forms and ages that are surviving the disease. It is therefore important to design fixed-point photography with as broad a range and outlook as possible, not just with specific expected changes in mind. Aim to capture the unpredictable, too, in keeping with rewilding's open-ended philosophy.

Plan to take photographs from fixed points at least once a year, but preferably in both summer and winter, or even all four seasons if you have time. If you can do it only once, do it at the height of the growth season, roughly in the middle of summer. Avoid times of the day when the sun is low on the horizon. A flat, grey, overcast day is perfect.

Setting up your camera on a permanent structure in the landscape will avoid the complication of carrying around a tripod and remembering the exact position each time. Using existing fence posts on a boundary saves the cost and effort of erecting new posts, and avoids problems with damage and maintenance. Cattle, in particular, are adept at pushing over free-standing posts, and you must never use wooden posts in a beaver area. Straining posts work especially well because they stand out on a fence line and are usually placed at a corner or angle in a fence.

A helpful system is to cut a slot in a fence post to fit a smartphone and/or the back of a conventional camera in landscape format. This way, you get exactly the same angle every time, and it's so easy to use that you can take a photo whenever you're passing. You can even use these slots to engage with visitors, requesting them to send in their photos from the footpaths.

Elsewhere, keep grid references and provide a direction. Using a compass and not going to a level of detail beyond 'northwest', for example, will make this easier. You can also take a printout of the original template photo to ensure you're taking exactly the same shot each time. Perhaps the most important thing to remember is rigorous labelling on your computer. Storing each sequence of photos in its own folder will allow the observer to scroll through the photos easily, showing the change over time. Again, keep it simple, and file them as you go to save having to sort them out later. If you take 200 photos, this becomes a huge job to catalogue in one go, making it less likely to happen and defeating the object. The beauty of fixed-point photography is ease and simplicity. However, many fixed-point photography projects fail simply because no one has thought of long-term storage. Files of photographs are easily lost with office moves, computer upgrades and staff changes. Think about this in advance, if you can, and set up a foolproof system that will stand the test of time.

Soil surveys

The soil is probably the most important aspect of any nature-restoration project as an indicator of change, ecosystem function and biodiversity. At Knepp, alas, we didn't undertake any soil surveys as part of our baseline. In 2002 we still didn't know how important soil was. Science, agriculture and conservation have neglected its relevance for a century or more, and as a result soil degradation is now one of the most pressing crises we face. Globally, over 80 per cent of farmland is described as 'moderately or severely eroded', and 75 billion tonnes of topsoil are lost every year through run-off or blown away by the wind. Land degradation costs up to $10.6 trillion a year, equivalent to 17 per cent of global gross domestic product.[4] In the UK, that cost is between £900 million and £1.4 billion a year – half of which is caused by the loss of organic matter, more than a third by compaction and about 13 per cent through erosion.

Thankfully, we're now beginning to wake up to this catastrophic neglect, and rewilding is proving to be one of the most significant and swiftest methods of soil restoration. At Knepp we are lucky enough to have friendly neighbours, on the same soils and farming in the same way as we were. They have allowed us to take samples from their land for comparative studies on soil changes on ours in recent years. Without this as a comparison we would not have been able to draw any meaningful conclusions. Using neighbouring land as a control, we've been able to show that the carbon content, organic matter and biomass of microbes in our soil have more than doubled, and soil fungi have more than tripled in the twenty years since we stopped farming.[5]

Ideally, however, comparisons should be made on site and in the same specific locations. That means including soil sampling and key surveys, if you can, in your initial baseline. Thankfully, there are now plenty of options for soil testing. We list some of these in the Further Resources section (page 531).

If you want to dig deeper into the changes in your soil using laboratory tests, you'll need to employ an agronomist. Talk to your local branch of the National Farmers' Union (in the UK) or simply search online for a local agronomist. The suite of tests recommended as a good baseline are: phosphorus (P), potassium (K), magnesium (Mg), pH, organic matter, carbon, soil texture, and microbiological health and mycorrhizae/fungal indicators – now possible using eDNA. eDNA uses a technique called 'metabarcoding', which identifies species within the soil. With soil tests it is particularly important to ensure consistency, using the same method, laboratory and, if possible, consultant.

Soil analysis is still an emerging science, and techniques are improving all the time. Also, soil is alive. Take it out of context, put it into test tubes and into a laboratory, and you can't be sure you're looking at the same substance that was once functioning underground in the field.

One aspect of study that is still in its infancy is that of mycorrhizal fungi: the networks of fungal cells that emanate from plant roots into the surrounding soil. These are a fundamental component of what is sometimes known as the Wood Wide Web. More than 90 per cent of vascular plants depend on symbiotic partnerships with mycorrhizae to provide them with nutrients, and, as well as storing carbon, mycorrhizae are a key indicator of the health and functioning of soil. Techniques for identifying and understanding mycorrhizae are likely to progress in leaps and bounds over the coming years, but in the meantime, simple surveys of fruiting bodies of fungi and orchids (which rely entirely on mycorrhizae for germination and to supply them with food) above ground can provide indications of mycorrhizal activity below the surface.

Soil life

Fungi, microbes and invertebrates break down dead plants and animals, returning nutrients to the food chain

Mycorrhizal fungi give water and minerals to plants in return for sugars rich in carbon

Sexton beetles bury a dead shrew

Burrowing animals bring oxygen and water into the soil, encouraging the growth of roots

Ants process dead animals and plants on a large scale

Dung beetle nest chamber

Grazing stimulates the release of sugars from roots. This attracts fungi, which stimulate the roots to grow

Plants communicate with each other through the fungal network

Earthworm and dung-beetle surveys

For an excellent indication of soil health and, hence, above-ground biodiversity, earthworm and dung-beetle surveys are both relatively simple and fun to do. You can carry out a simple count yourself (the app Soilmentor describes a useful method) and/or find a local specialist to help you identify the various species.[6]

As Charles Darwin recognised in the mid-nineteenth century, earthworms are a keystone species and their burrowing, consumption of organic matter and discharge of undigested material – or 'worm casts' – play a crucial role in the formation, structure and nutrient recycling of the soil.[7] There are three broad ecological groups: non-burrowing 'epigeic' earthworms, found on the soil surface within the layer of leaf litter; 'endogeic' earthworms occurring in the upper 50 centimetres of the soil, where they make branched, horizontal burrows and feed on decomposed plant material; and 'anecic' earthworms, such as the common earthworm *Lumbricus terrestris*, the true 'soil engineers', which drag organic material from the surface down through vertical burrows as far as 2 metres deep into the soil.

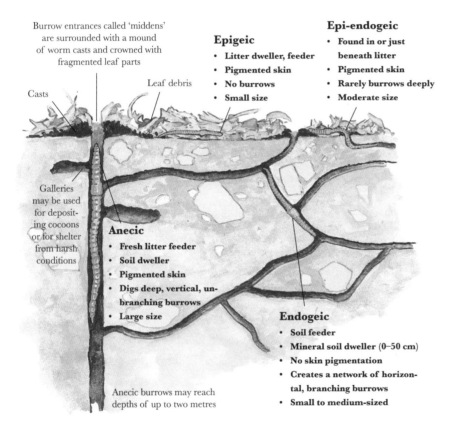

Burrow entrances called 'middens' are surrounded with a mound of worm casts and crowned with fragmented leaf parts

Leaf debris

Casts

Galleries may be used for depositing cocoons or for shelter from harsh conditions

Epigeic
- **Litter dweller, feeder**
- **Pigmented skin**
- **No burrows**
- **Small size**

Epi-endogeic
- **Found in or just beneath litter**
- **Pigmented skin**
- **Rarely burrows deeply**
- **Moderate size**

Anecic
- **Fresh litter feeder**
- **Soil dweller**
- **Pigmented skin**
- **Digs deep, vertical, un-branching burrows**
- **Large size**

Endogeic
- **Soil feeder**
- **Mineral soil dweller (0–50 cm)**
- **No skin pigmentation**
- **Creates a network of horizontal, branching burrows**
- **Small to medium-sized**

Anecic burrows may reach depths of up to two metres

The main ecological groups of earthworms. Their collective effects – burrowing, digestion of organic matter and contribution of beneficial bacteria – underpin healthy, functioning soil. As Darwin recognised in the nineeenth century, earthworms are keystone species.

Populations of earthworms on agricultural land are likely to have been severely depleted by the use of artificial fertiliser and chemicals, the clearing of leaf litter and surface residue, and ploughing, which cuts up the worms and exposes them to predation.[8] The diversity and speed of recolonisation can be dramatic, however. At the outset of our project we were advised by scientists that, while we might see a swift increase in epigeic and endogeic earthworms, the vertical-burrowing anecic earthworms would take longer to move out from the hedgerows and colonise the former arable fields. We were told they might colonise at the rate of about 1 metre per year – which meant it would take more than a century to reach the middle of many of the fields. Happily, this proved not to be the case, and after only a decade or so we were seeing the characteristic worm casts of *Lumbricus terrestris* in the middle of most of our fields. In 2013 Master's students from Imperial College London, using neighbouring farmland as a baseline, found a significant rise in the abundance and variety of all three categories of earthworm, and a total of nineteen species – an extraordinary diversity.[9]

Dung beetles are another keystone species, specialising in breaking down animal dung and recycling and drawing down nutrients into the soil. They have declined dramatically in the UK in recent decades because of the loss of permanent pasture and the use of antiparasitic drugs in livestock. A survey by Natural England of dung beetles across the UK in 2016 showed that 25 per cent of species were nationally rare, four had become extinct and sixteen were endangered, vulnerable or threatened.

One of the most numerous and conspicuous of the twenty-one species of dung beetle now active at Knepp is the beautiful, tunnelling violet dor beetle (*Geotrupes mutator*), an example of which is shown here. Before it appeared in our rewilding project in 2017 this beetle hadn't been seen in Sussex for fifty years.

But it's astonishing how fast dung beetles can recolonise and thrive in areas where organic dung is available all year round. In a recent study, the number of dung beetles caught in pitfall traps at two sites at Knepp over a three-day period (11,666 individuals) far outweighed the number trapped at two sites on nearby organic farms

(512 individuals).[10] We believe this is explained, at least in part, by the variety of dung-producing animals present – pigs, ponies, cattle, deer, badgers and foxes – and the varying habitats, which benefit different dung-beetle specialists. But it is also significant that we have animals in the landscape throughout the year. On farms, even organic ones, livestock is generally brought indoors in the winter; since dung beetles feed all year round, this starves them of food for many months.

Additional surveys

There's a litany of further surveys that could be undertaken. Specialist groups can record anything from amphibians and reptiles to dragonflies, moths, winter birds, mosses, lichens and beetles, and their knowledge will add valuable species records to your datasets. Some will even be able to set up regular monitoring to capture the changes. For example, in addition to the general invertebrate survey mentioned above, you might want to carry out surveys on aquatic invertebrates if you have water, and/or deadwood invertebrates if you have old trees. Specific methods and equipment are required to survey these specialised creatures.

Your choice of additional surveys will depend, to a large degree, on the nature of your land and what you might expect to evolve over time; on your own interests; and on the appearance of 'rising stars' – unexpected species/taxa of note. Consider surveying some species, such as bats, which may not be breeding on site but may be using the land as a flyway or feeding ground.

Consider also including specific surveys for breeding birds that are on the 'red list'. In the UK, this is officially the Birds of Conservation Concern list, and includes birds such as nightingales and turtle doves. Communally nesting birds, such as herons, jackdaws and rooks, might also be of interest. It's been fascinating to observe changes in the Knepp heronry as the rewilding project has matured. Nests that have historically been established at the very tops of the trees are now being built much lower down, under cover of the branches, presumably to avoid predation by rising numbers of buzzards, red kites, ravens and so on. We've used drones very effectively not just to survey the number of nests in the heronry, but also to count individual eggs and even chicks, and to observe splinter colonies starting up nearby.[11] However, you do need a licence to fly a drone, and it's crucial to be very cautious about keeping a respectful height and distance from the nest to avoid disturbing sitting adults, chicks and fledglings.

While it's important to undertake the basic surveys at the outset, there can be a danger in trying to do too much at the beginning. You can always add more specialist and detailed surveys as the project matures, as your knowledge increases, and as you meet and engage more volunteers.

A baseline survey is the fundamental first step for the new rewilding project on 600 hectares of marginal land at Boothby in Lincolnshire, which is managed by Nattergal, a company co-founded by Charlie to create large-scale, privately financed rewilding projects on the Knepp model. Here, entomologist Graeme Lyons conducts an insect survey using a sweep-net.

Site lists

Above all, it is important to maintain a complete list of every species recorded on your site. This exciting record embodies the scope and progress of the project and helps to sustain the interest of record- ers and the general public. Being able to say how many rare species you have and when they were last recorded is invaluable, and the pan-species approach to a site list is a way of presenting evidence of biodiversity that is immediately understandable to both professionals and laypeople, including in the context of natural capital.

There are numerous apps available for helping you identify species and they grow more accurate and sophisticated the more they are used (see Further Resources, page 531).

A record should include the following:
- The scientific name of the species (with some apps you can

enter the vernacular name into the database, which then populates it with the scientific name)

- The date the species was recorded
- Where the species was recorded (normally a six-figure grid reference, giving an area indication of 100×100 metres, for a flying bird of prey, down to a ten-figure grid reference for a closer area of 1×1 metre for a static, singing nightingale, say)
- The name of the person who recorded it or what survey it was recorded on
- The abundance of the species (in some circumstances this can be just 'present' or 'absent')

Additional information could include:
- Location name (such as the name of the field)
- Recording method (such as field record, pond-dipping, pitfall trap, net, 125W MV Robinson moth trap)
- Vernacular name, if there is one
- Sex, if known
- Life-cycle stage, if relevant (such as egg, larva, pupa, imago/adult)
- In cases of species that are extremely rare or notoriously difficult to distinguish from others, the name of the expert, other than the recorder, who has confirmed the identification

Citizen science

Using apps is a way of plugging into the growing phenomenon of citizen science. Engaging the public in the collection of data is revolutionising the way scientific information is gathered. It taps into the enormous observation potential of 'eyes on the ground', and it's free. Numerous breakthroughs have been made through citizen science, not just in natural history but also in such disciplines as astronomy and archaeology, where scientific organisations have been stretched too thinly or strapped for cash, or were simply looking in the wrong place. Not only does engaging with the public have huge potential to add to the data for your project, but also it helps raise environmental awareness more broadly and inspire a new generation of amateur naturalists.

10

Where do
We Fit in?
*How rewilding
affects people,
and how people
affect rewilding*

*Connecting people with nature is one of the greatest
joys of rewilding, but there is a balance to be struck
between human access and wildlife recovery. Even
passive activities such as birdwatching and walking
can affect the ability of creatures to feed, breed
and socially interact. Creating clear and accessible
walking routes, and communicating the need to keep
some areas free of human – and dog – disturbance,
is key to the success of a rewilding project.*

Access to nature is fundamental to human physical and psychological health, and the positive influences increase with habitat complexity and biodiversity.[1] Green space in the form of urban parks and farmland provides some health benefits, of course. But it is love of life that touches the human spirit – something the great American biologist E. O. Wilson calls 'biophilia'.[2] Wild creatures and varied, biodiverse habitats naturally attract us. People travel longer distances to visit forests with trees of different species and ages. It has been shown that in Europe people travel further to hike, ski, cycle and ride in wild areas where there are wolves, lynx and wolverines.[3]

At no time was this desire for living landscapes more apparent than during the outbreak of the COVID-19 pandemic in early 2020. Under strict lockdown, with little vehicle and plane traffic, and much of normal human activity suspended, the birdsong during those first eerie, disembodied months seemed almost supernaturally loud, and gave solace and joy to many. Although it's unlikely that the decibels actually increased, birds may have taken advantage of the lack of background noise to sing more complex, lower-amplitude, 'sexier' songs.[4]

Across Europe, when restrictions were relaxed, people flocked to the countryside like birds let out of a cage, seeking out places where they could restore morale and a sense of well-being by connecting with nature. As travel bans were lifted, wildlife suddenly found itself on the receiving end of the 'lockdown surge'. In the UK, Epping Forest on the outskirts of London – usually visited by about 4.3 million people a year – had received an estimated 12 million visitors by the end of 2020. The Wildlife Trusts, observing widespread disturbance to ground-nesting lapwings and skylarks, and fires from barbecues igniting heathland, appealed to visitors to 'love and look after' the countryside.[5] On the remote mountainsides of Ben Lomond and Ben Lawers, the National Trust for Scotland begged hikers – double the usual numbers – to stay on existing footpaths and in single file, to prevent galloping soil erosion and damage to rare alpine plants and carbon-sequestering peatland.[6] At Knepp, within easy striking distance of London and Brighton, we received 30,000 visitors in three months, ten times more than in the same period the previous year. The increase in footfall, combined with social-distancing regulations, encouraged people to wander off the footpaths. Many visitors, particularly those from urban areas, were unaware of the existence of, let alone the obligation to follow, the Countryside Code – a sort of best-practice protocol for visitors to rural areas.[7] Dogs nosing down the hedgerows and through the thorny scrub flushed nightingales and other songbirds from their nests. Quiet areas where the Exmoors, reds and fallow go to have their young were constantly disturbed.

And herein lies a fundamental dichotomy for rewilding and nature recovery. If people are to love nature, they need to experience it, feel part of it and be protective of it. Access to wild spaces and clean air, one might argue, is an inalienable right. Who wouldn't want the ability to swim, walk, kayak, forage or climb freely wherever nature presents itself, and to give their children that freedom?

But even 'passive', 'non-extractive' activities, such as birdwatching, cycling, running, climbing or even just walking and talking, can massively affect wildlife.[8] In woodland close to urban areas, recreational activities destroy the soil, insects and flora through trampling and compression.[9] Human presence affects the way birds, mammals, reptiles, amphibians, fish and arthropods behave and move. Elk in Sweden move thirty-three times faster in the hour after being disturbed by cross-country skiers.[10] The home ranges of endangered Bonelli's eagles in eastern Spain expand dramatically at weekends and on public holidays. The birds spend longer away from their nests, depriving their chicks of food.[11] Ecologists call this phenomenon the 'weekend effect'. At Mar Lodge in Aberdeenshire, capercaillie – once a common Scottish bird, now on the verge of extinction for the second time – steer 125 metres clear of busy footpaths and tracks.[12] Human disturbance can push wild creatures away from their pre-ferred feeding and breeding habitat – often beautiful places that we love to visit – reducing breeding rates and resulting in population decline. Our love of nature is itself becoming one of its most serious threats.

As the global population has risen, recreational activities and access to wild and natural areas, particularly in affluent countries, has increased. Yet areas of nature are still diminishing rapidly, particularly in densely populated countries, such as the UK. There is simply not enough nature to go round. Green belts, green fields and floodplains continue to be developed at an alarming rate. Only when rewilding has done its job and we have green corridors connecting urban areas with wild belts and regenerated farmland, and wildlife spilling out of reserves, rebounding in naturally regenerated woodlands and re-naturalised rivers, will nature be in a position to cope with increased human disturbance. Until then, a balance must be struck between public access and providing the tranquillity to enable the recovery of wildlife.

How to achieve tranquillity within a rewilding project must be an important consideration from the outset. Given the rising interest in nature (and rewilding in particular), an enthusiastic response must be anticipated. It is important to work out how to manage visitor numbers, particularly if there are public footpaths and bridleways on the land and/or if an ecotourism or other business will be set up in conjunction with the project. Even projects that intend to remain low-key must consider contingencies for public engagement.

At Knepp, with 26 kilometres of public rights of way and a high-profile project and ecotourism business, we have found our-selves increasingly involved in visitor management. Mostly, this is a hugely positive experience for all involved. The diversity of visitors is enormous, ranging from amateur and professional naturalists used to hiking out on their own and reading maps, to first-time visitors to the countryside who need more information and help to make the most of their trip. For us, people enjoying nature – especially if they're new to it – and the wonderful feedback we get are among the greatest rewards of rewilding.

Information and clear signage are key, and most people are considerate and responsive to information. But some aren't, and it's often surprising who they are. Competitive birders, for example, will go to any length to photograph a critically rare species, including scaring it off its nest. During the deer rut keen photographers endanger themselves by approaching roaring, testosterone-fuelled red stags and fallow bucks. Some dog-walkers allow their pets to run riot off the public footpaths, scaring up ground-nesting lapwings, chasing deer and livestock, bounding into water after ducks and coursing the thickets where nightingales nest. Such incidents are vexing, even distressing, but they form only a tiny percentage of our interaction with the public. Most people want to do the right thing, but may need direction. Raising awareness of the human impact on wildlife and the land, and how to behave around wild and semi-wild animals, has become a fundamental part of what we do.

Local communities

Winning over local support at the start can help spread the message of wildlife restoration and provide a source of volunteer rangers and wardens, should they become necessary. But the response to the project in the early days may be mixed. Although rewilding is becoming increasingly popular, there is bound to be some alarm about any change in land use, particularly among those with neighbouring properties. Emerging thorny scrub, free-roaming animals, swathes of 'weeds', dead trees, dynamic rivers – the general messiness of rewilding – challenge the aesthetic of Britain's manicured countryside. This is the antithesis of how some people think the British landscape should look.

Rewilding is all about shifting a mindset, and often this comes about only once people have had a chance to experience it at first hand, to get used to the changes and, particularly, to appreciate seeing wildlife return. A baseline survey and ongoing monitoring and species records (see Chapter 9) are crucial here because they can provide evidence of ecological improvements and wildlife arrivals, providing something to discuss with local interest groups and the media.

We put a lot of effort into reaching out to the local community in the beginning, offering free tractor-and-trailer rides, guided walks,

Open Farm Sundays and village-hall presentations. We also targeted groups such as the Young Farmers' Club, the Country Land and Business Association and our local wildlife trust, natural history society, and district and parish councils, as well as local people more generally. But the results were only partially successful. Often the take-up was from those who were already well disposed towards the project. This was in 2003, it must be said. But our most entrenched and vociferous critics did not want to engage in this way.

We were able to bring some of them onside by meeting them personally and simply listening to their concerns and addressing them where possible. Occasionally, there have even been ways of involving them in the project, and that has been enormously worthwhile. But equally, some critics have never wanted to engage. Instead, as the project has become more widely appreciated and rewilding itself has become more popular, their influence has simply fallen away.

Sometimes inviting public engagement too soon can generate more problems than it resolves, especially when dealing with concerns about the principles and consequences of rewilding. Rewilding is a leap into the unknown. Since there is no way of accurately predicting the outcome, one finds oneself in a position of defending the unknowable. Ultimately, the results speak for themselves. In particular, the fear element tends to subside when anticipated problems – such as protective behaviour by free-roaming cows with calves or aggressive red deer in the rut – fail to materialise, or are much less concerning than people expected.

Animal welfare is often a major concern for people who are unused to seeing free-roaming livestock living outside all year round. This can easily be addressed by explaining the management system and regulatory checks (see Chapter 6) on the project's website. Other concerns can be allayed by good communication and accessible information at the outset. There are a number of ways to achieve this, and a multi-pronged approach is likely to be the most effective.

Website and social media

The most important vehicle for first-base communication is a website. It's well worth the time and cost of setting up a well-designed, easy-to-navigate website – preferably one that you can update yourself – where people can find information about your project, especially on topics that may be stirring up concern. This, in itself, projects a reassuring spirit of openness, and you can, of course, add links to supporting information on other sites, scientific papers and so on.

Blogs, podcasts, Facebook, Instagram and Tiktok – especially if you post wonderful wildlife photos and video clips – can be a big draw for supporters, and a useful resource for recruiting volunteers. Twitter, on the other hand, is a double-edged sword. It reaches more people than most other platforms, but is more volatile. We've learned to stand back from frenzied outbursts of posturing or axe-grinding. More often, we've found, concerns expressed with a genuine desire to engage arrive by email or letter.

We also use SMS to warn local walkers who are known to us about activities such as deer culling or cattle round-ups. Our WhatsApp group for locals is now a vital tool to rally help from neighbours and volunteers, too, and get information out quickly on the ground.

Injurious weeds

There is great concern among farmers and gardeners about the spread of 'weeds', particularly the five listed as 'injurious' under the outdated and unscientific Weeds Act 1959: curled dock, broad-leaved dock, spear thistle, field or creeping thistle and common ragwort. Contrary to popular interpretation, the Act does *not* make it an offence for any of these five species to be present on land. It is primarily concerned with preventing their spread into agricultural crops. Ragwort is particularly vilified, although its reputation as a killer of livestock, especially horses, has been wildly exaggerated.[13] Its seed is more likely to come from the seed bank in the soil than to be blown any distance from the plant by the wind. Studies have shown that 60 per cent of ragwort seeds fall around the base of the plant. Seedfall decreases with distance, so on average, at a distance of 36 metres from a single plant producing 30,000 healthy seeds, only 1.5 seeds land. Moreover, seed that is blown on the wind is lighter and likely to be infertile.

Still, owners of livestock and horses will not want to see swathes of ragwort anywhere near their fields or paddocks. And, while most people might be happy with weeds in a nature reserve, they won't want the seeds blowing into their gardens (although it is to be hoped that, as people start rewilding their gardens, we'll see greater tolerance for native plants and wildflowers). We keep a 50-metre strip of land (the width recommended in DEFRA's Code of Practice[14]) regularly mown, within our boundary fence to prevent the seeds of ragwort and other weed species from drifting on to adjacent land. Although it is time-consuming and costly – and not a legal requirement – the conspicuous mown strip inside our boundary fence greatly reassures our tidier-minded neighbours. We also post our policies for the control of the spread of ragwort and other 'injurious' weeds on our website.[15] In areas of particular sensitivity, it may even be worth cutting a 100-metre strip, as we did for a number of years where our land abutted a llama farm.

Footpaths and rights of way

It is essential to encourage people to stay on public footpaths and bridleways, if nature is to have the space to recover. This needn't compromise the visitor's experience; indeed, it will probably improve it. The best footpaths, we feel, add to the wildness of the land, rather than taking away from it.

Higher footfall, though, is likely to increase damage and erosion to footpaths, and the landowner, in conjunction with the local authority, is responsible for repairing them. Free-roaming animals, especially if they are in the project all year, can also cause considerable trampling or 'poaching'. On our heavy clay soil, animal traffic and pig rootling over the winter can turn footpaths into quagmires. Therefore, we feel it is largely down to us to maintain and improve rights of way on the estate, particularly in areas frequented by our animals, by laying hardcore to make the paths weather-resistant and suitable for all seasons. Our local county council is also very helpful, within its own budgetary constraints. It frequently replaces wooden public-footpath posts that are broken or bent by our animals, and it occasionally supports improvements to the path surfaces through small grants.

Nevertheless, complaints about mud, uneven ground and the potential for twisted ankles seem to have risen in recent years – an indication, perhaps, of an increasingly urban attitude to the countryside. Sometimes, too, visitors, attracted by the wildlife, expect the quality of paths and infrastructure of a National Trust or RSPB public-access reserve. Managing these expectations is tricky. But the website is a good place to post information about accessibility. Posts on social media can also alert visitors to conditions in poor weather, and perhaps even dissuade them from venturing out and being disappointed if the going is particularly bad.

In some areas, where thorny scrub has taken off, it may be necessary to prune back vegetation from footpaths during the growing season. Landowners are responsible, too, for removing dead standing trees and branches that could pose a danger to people on public rights of way. With the increased frequency of diseases, such as ash dieback, affecting trees, this can be a huge expense. Where we can, we leave dead wood on the ground for wildlife, and the trunks and bigger branches provide seating spots for visitors.

Permissive paths

One of the reasons walkers stray off paths may be the path system itself. Footpaths are often ancient, dating back to different eras of land management and ownership. Sometimes there is room for improvement, such as diversions that avoid boggy patches or dangerous corners in roads. Paths generally came into existence as a means of getting from A to B for work or social reasons. But when an area of land becomes the destination, people are looking for circular routes, especially if they're arriving by car. Making sense of the footpath system on the land and, where necessary, linking the existing network with some 'permissive paths' can make the experience for walkers more enjoyable and avoid bottlenecks and confusion. Permissive paths are so called because a landowner grants permission for them to be used by the public. They can be created by formal agreement with the local council or, less formally, as a unilateral decision by the landowner – in which case the landowner is liable for injury to users caused by any defect in the surface of the path. It is up to the landowner to decide whether to open a permissive path to just walkers, or to cyclists and horse riders too. At Knepp, under our Countryside and Higher Level Stewardship agreements, we chose to add 6.5 kilometres of permissive footpaths to the existing 25.5 kilometres of public rights of way, and have decided to keep these as informal paths beyond the agreements. As landowners, we are liable for path maintenance and signage on our permissive paths. Maintenance of surfaces and signage on public rights of way are the local authority's responsibility.

Our old English Longhorns at Knepp are exceptionally docile, but it's important for visitors to be aware of the difference between free-roaming animals in a rewilding project, and conventional farmyard livestock or pets. And, in particular, not to try to feed or touch them.

Permissive bridleways

Wild or semi-feral free-roaming herds of horses can lead to conflicts with bridleway users. Riders can be pestered by ponies galloping around them or accosted by territorial or amorous stallions, particularly if riding a mare on heat. The size of the herd and its cohesion are therefore important. An 'inward-looking', cohesive herd of horses, absorbed in their own social interaction, is less likely to be interested in domesticated horses on the bridleways. Allowing the formation of family harems within a larger herd – usually including one or more adult males, between one and ten adult females and recent offspring – will forge stability in the herd.[16] The harems will tend to interact with each other, discouraging interaction with visiting horse riders.[17]

In the early stages of the project we added 7.5 kilometres of permissive bridleways to the existing bridleways at Knepp, and these were managed by the Toll Rides Off-road Trust (TROT).[18] However, we continued to have problems with riders straying off the permissive routes. In an effort to improve communication and understanding with local horse riders, we created the Knepp Exmoor Project (KEP). KEP currently has about thirty-five local members, who pay £25 per year for exclusive use of our permissive bridleways. Members agree to abide by certain regulations, which include sticking to designated routes, respecting wildlife and not riding their horses on the estate after using ivermectin-based wormers. We also advise them not to ride mares on heat in the rewilding project. We're able to notify members via WhatsApp if we need to close routes for bird-ringing or other reasons, and members can report to us any interaction – positive or negative – with the Exmoors. This contributes really useful observations to our own, alerting us to any problematic individuals – particularly stallions – in the herd that we might need to exclude from the project.

Engaging openly in this way with our local equestrian community has dispelled a considerable amount of anxiety and misunderstanding and encouraged their interest in the use of endangered equines in rewilding. Most riders enjoy riding in a wild landscape where there's the chance of encountering feral horses. But they also appreciate us acknowledging and mitigating the risks, while trying to maintain natural herd dynamics (see Chapter 6).

We're also in discussion with the British Horse Society, which produces excellent leaflets about riding with livestock and free-roaming horses. The Society is currently revising its guidance to include riding within rewilding and conservation-grazing projects.

Cyclists

Access laws in the UK permit cyclists to use bridleways, but not public footpaths. While it's great to see an expanding network of cycle routes linking towns and cities with the countryside, especially if it reduces the use of cars, it does add a burden of traffic to even remote areas of nature. Erosion and the degradation of paths are much higher under the traffic of bicycles, making paths far more costly to maintain. Unlike with walkers and riders, the pace of travel when cycling isn't conducive to a pensive, 'slow-take' appreciation of nature. Our particular concern is that cyclists tend to push riders and walkers off the bridleways, which can have a further impact on habitat and wildlife.

The impact of dogs and cats

The rules on dogs in the Countryside Code are, unfortunately, a little ambiguous. While dogs are categorically not permitted to chase wildlife, they do not have to be kept on a lead on public paths as long as the dog is 'under close control'. This provides ample room for dispute. While a dog should be on a lead in the presence of livestock, the Countryside Code recommends that it be let off the lead if there is reason to believe livestock are threatening it (since the dog has a better chance of escape if it is not attached to a person); again, this provides a potential excuse for transgression in areas with free-roaming animals.

Under lockdown during COVID-19 it was evident from the increase in birds nesting close to footpaths how big the impact of walkers, particularly those with roving dogs, generally is. There are 9.9 million dogs in the UK and 1.5 billion dog walks undertaken in a year. A study in 2021 showed that 64 per cent of dog owners let their pets roam free in the countryside, despite the fact that half the owners surveyed admitted their dog doesn't always come back when called.[19] The research revealed that 42 per cent of dog owners had been walking their pets more often during the pandemic. The cost of dog attacks on farm animals across the UK rose by over 10 per cent, to an estimated £1.3 million during 2020. There is no data for attacks on wild animals, such as deer, rabbits, hares, stoats and birds.

While most visitors to Knepp keep their dogs on leads, some – particularly those who own 'sporting' breeds, such as spaniels, lurchers, Labradors and retrievers, which love to scent and chase – are serial offenders. Some of our neighbours allow their dogs to run free and hunt through the rewilding project. In response, we try to catch the dogs (which can be time-consuming) and call the council's dog warden to collect them. There is a fine levied on the owner to collect the dog, which has had some limited success, at least, in reduc-ing repeat offences.

Nature has no defence against irresponsible dog owners. We would like to see regulations stipulating that all dogs be kept on leads in nature areas at all times of year. It's not just the spring nesting season when birds are vulnerable. In the summer, when waterfowl such as geese, ducks and swans moult, they are flightless for a period, and in the winter, too, any unnecessary evasive action uses up precious fat reserves. Even dogs on a lead have an impact. On small nature reserves, just the presence of dogs can cause a 40 per cent reduction in bird species across the whole site.[20]

Until we have a much more robust landscape in which nature is widely recovering and the cataclysmic downward trend of biodiversity has been reversed, wildlife needs all the help it can get. Curtailing disturbance by dogs seems a simple and obvious measure.

Dogs are also the definitive hosts of the parasite *Neospora*, and can transmit the disease it causes to livestock through their faeces. In 1999 some 12.5 per cent of abortions in dairy cattle in England and Wales were thought to have been attributable to the disease.[21] This is, naturally, a concern for rewilders with free-roaming animals. The routine use of wormers in dogs, as well as in cats and horses, is also of concern since, like ivermectins in livestock, it passes into the faeces and can kill insects and soil biota. Ordinarily, faeces from dogs not treated with drugs such as wormers or antibiotics should not affect the soil. On organic land, with functioning soils, it should biodegrade. But with large numbers of dogs on footpaths, faeces accumulate quickly. This can create nutrient enrichment, disrupting the delicately balanced ecology of nutrient-poor soils, such as acid grassland. Richmond Park in west London, designated a National Nature Reserve for its rare and fragile acid lowland grassland, is particularly vulnerable to nutrient loading from car fumes (containing nitrogen oxides) and dog fouling.[22] Almost worse, to our minds, than the owners who don't pick up their dogs' waste are those who go to the trouble of putting the faeces in a bag that they then leave beside the path or hang on a branch.

The impact of domestic cats is discussed in more detail in Chapter 11. There are over 12 million domestic cats in the UK, the majority of which are free to come and go from their owners' homes as they please. They catch up to 100 million prey items a year, 69 per cent of which are small mammals and 24 per cent birds. If there are domestic households in or near a rewilding project the local cats may well have a field day. The RSPB's advice is to put a bell on the cat's collar, but cats can still manage to kill when they are wearing one, and it can be annoying to live with. Although we do suggest a bell collar to cat-owning tenants at Knepp, there is rarely any take-up, and predation by domestic cats continues to be a drain on wildlife.

Signage and information boards

At the start of our project we were adamant that we didn't want to have bossy, civic-style signage everywhere. But as the number of visitors has risen we've had to revise our ethos. We've learned that minor public offences, such as wandering off footpaths, approaching animals or speeding in cars, are often a result of the lack of clear signs. There's a reason why road signs look the way they do; an enormous amount of research has gone into developing them and much of it is to do with subliminal awareness. They may look garish and out of place in the natural landscape, but that's why people notice them. Symbols are more immediate than words. Primary colours and shapes speak a universal language. A wordy text, however friendly and welcoming, has far less chance of hitting home.

We do, however, try to limit the boldest 'do' and 'don't' signs to entrance areas, such as car parks, drives and gates, so as not to intrude on nature and the aesthetics of the rewilding project itself. As always, there's a balance to find between enforcing the message and going over the top.

Many of our vehicle access tracks and well-used animal trails now look temptingly like footpaths, and were often leading people astray. To turn people back, we designed more low-key 'Wildlife Only' signs on posts set a short distance down the trail from the footpath.

It's also surprising how few people can read maps. Perhaps this was always so, but it feels as though satnav, online maps and smartphone apps have eclipsed the arts of orienteering and thinking on your feet. Wooden public-footpath posts on their own are no longer enough to keep walkers on course. Added to this, a rapidly rewilding landscape – with fields and ancient woods that no longer have clear boundaries, thorny scrub shielding long-distance views of landmarks, and a maze of animal trails – increases the chances of disorientation. On several occasions we've encountered people who have 'dropped a pin' on their car's location on their phone so they can find it again, and simply headed back to it in a straight line, irrespective of the fact that they're nowhere near a footpath.

The best system of path signage, we've found, is the European model whereby walking routes are emblazoned with splashes of coloured paint on natural features, such as trees, paths and rocks. While the paint requires regular touch-ups, this is easy to do and there can be as many trails (in different colours) as you like. Somehow it looks less intrusive than lots of signs or numbers, although in areas where there are no natural features it may be necessary to bash in posts. (Bear in mind that, as with public footpath signs, if you have cattle the posts must be deep enough to withstand the rubbing of a 650-kilogram cow.)

Emblazoning walking routes with splashes of colour on trees, rocks, gates and posts is the easiest system of path signage to maintain, and for visitors to follow.

Producing a map helps to reinforce the painted walking trails on the ground, as well as showing interconnecting public footpaths and bridleways. We provide these for free at our campsite shop and car parks, and they are downloadable from our website. We've found the map that works best is a colour aerial image from Google Earth, rather than the old-fashioned Ordnance Survey-style map, because it gives a sense of vegetation type and density. Perhaps 3D mapping is now more familiar to us, anyway.

Volunteers

For rewilding projects with public footpaths, volunteers can be invaluable, reaching places that maps and signs cannot. Since COVID, volunteers have been helping us with visitor management: directing people who are lost, providing information and pointing out things of interest, such as where one might see purple emperor butterflies, a stork nest or a family of pigs. They pick up litter and dog-poo bags, and help to mend wonky gates and signs.

We now consider our volunteers one of our most important and valued assets. They are great fun to work with, and an important bridge with the local community. We have 200 on rotation, many of them experienced birders who simply enjoy being out and about in nature. Our ranger provides training, manages the roster and fields their queries. On duty in pairs and wearing purple tabards or Knepp T-shirts to identify themselves, volunteers are our interface with the public – a much kinder, more sympathetic way of communicating than bossy signage, which, in any case, is often ignored. They can explain in a way that is more nuanced and engaging than a bullet point why, for example, it is important not to approach the animals, where nightingales nest which makes them so vulnerable to free-running dogs, and what to do if a herd of cows is blocking the foot-path. Many visitors have told us how chatting to a pair of volunteers made their walk more rewarding, and how much more they now understand about what they have come to see. And for us, volunteers help to pre-empt problems and reduce disturbance to wildlife.

Free-roaming animals

Apart from keeping to the footpaths and not letting dogs run free, the message we're keenest to impart at Knepp is to keep a respectful distance from the free-roaming animals. Increasingly, nature reserves use de-domesticated grazers as a conservation tool. But, even so, most people are unused to thinking of ponies and pigs as, essentially, wild. There is a huge temptation to get close enough to take a photo with a phone, and to pet or feed them.

One of our biggest problems is the risk of over-familiarity with humans and an expectation of food. This has caused some of the sows to charge up and prod walkers on the footpaths, sometimes quite forcefully, with their snouts. Some walkers and visitors feed them, so we have increased our signage requesting people not to do so.

Looking at free-roaming animals through a domesticated lens is what leads to the most dangerous situations, especially when mothers are protecting their young. Cute calves, foals and piglets are magnets for human interest. So, too, are red-deer stags and fallow bucks when they're battling it out in the rut and their testosterone levels are at their highest. Maintaining a natural flight distance protects both the humans and the animals, and helps to ensure the normal social behaviour of the herds. That said, the livestock manager must be able to catch the animals up. Cattle and ponies can grow accustomed to being walked up, but our pigs will come only to a rattling bucket of pig-nuts (see Chapter 6).

No matter how many times nature programmes such as BBC's *Springwatch* warn the public not to pick up young fawns they find, many people feel compelled to, thinking the young have been abandoned. However, a mother often leaves her young tucked up in a place she thinks is safe while she goes off to feed, often for long periods. As soon as a fawn or red-deer calf is touched by human hand, though, it is abandoned by its mother, who can smell the scent of a human on it. This can be a real problem in dry summers, when there is no long grass in which the deer can hide their newborns, leaving them more conspicuous to the human eye.

Parking and access

Our ecotourism business arose organically from modest beginnings, and we didn't anticipate how successful it would become. We have kept our glamping and camping capacity small in order to preserve the low-key, tranquil atmosphere of the site and reduce the impact on the project itself. Still, we underestimated the traffic involved at full capacity. We have woeful public transport service in our area, and anyway, most people transporting a large amount of kit for camping will come by car. For any business model, it's worth considering parking contingencies and overestimating numbers. Creating a car park can be hugely expensive, but the cost can be significantly reduced if you already have hardstanding. It is important to think about the visual, noise and other impact of a car park at an early stage in the project, even if there are no immediate plans to create one.

We didn't anticipate the number of people who would want to drive to us in order to walk the footpaths. As we became more popular, and especially when wildlife excitements like our white storks break in the news, cars thoughtlessly parked on verges and in private driveways have become a real irritant to villagers. We have found ourselves having to install a car park specifically for walkers. We have an honesty box suggesting a small fee for the upkeep of the car park and wear and tear on the track, and most people are kind enough to pay this. But still, some persist with parking illegally and often dangerously on surrounding lanes.

To encourage visitors not to come by car, there may be ways you can connect with existing cycle routes, and it is worth providing a cycle lock-up rack in the car park. But wildlife corridors and better footpath and cycle routes are desperately needed to integrate areas of nature with towns and cities – as well, of course, as much, much more space for nature.

Public liability and insurance

The prescriptive rules governing health and safety and the modern culture of litigation are a worry for rewilding. Some nature reserves have opted not to embark on conservation grazing with free-roaming animals, much as they would like to, because of a perceived risk to the public. Others try to mitigate the risk to humans by restricting the presence of stallions and bulls in their free-roaming herds, or even separating mothers from calves. Similarly, leaving dead standing trees, even off public rights of way, and other natural hazards such as dynamic rivers and floodplains, cliffs and landslides, quarries and bogs make conservation bodies and private landowners fearful of litigation. This risk-averse approach, which seeks to rein in and control natural processes, underpins much of the problem with nature conservation in general. Along with other rewilders, we are campaigning for wild areas to be recognised as such. While the 'duty of care' to warn the public of potential dangers is important, the public should also be encouraged to take responsibility for their own safety.

The feasibility of free-roaming animals in rewilding projects and nature reserves depends, to a great degree, on the considerate behaviour of the public. In Richmond Park, visitors are encouraged to give red stags a wide berth, especially during the rut.

11

Rewilding Your Garden

Applying rewilding principles in a small space

Creating humps and hollows in a garden, imitating the uneven surface of natural ground, radically increases opportunities for wildlife. In Knepp's Walled Garden, the previous monoculture croquet lawn is now a kaleidoscope of tiny habitats. Native wildflowers mingle with species from arid parts of the world, to reduce the need for watering.

As gardeners, we consider ourselves the keystone species – our interventions mimicking the snout of a pig or the pruning teeth of a pony. If a plant begins to dominate, we selectively rootle it out to retain the maximum chances for biodiversity. But as much as possible, nature leads the way.

Much of our mailbag at Knepp is from gardeners. Inspired by a visit, or by rewilding in general, they want to know if it's possible to apply the principles of rewilding, so often associated with large herbivores and the wider landscape, to a much smaller space – their own garden. We think it is. Gardens can play a crucial role in restoring biodiversity and bringing dynamic nature back into our lives. And if a garden can link with other gardens and green space, it can become part of that connective webbing that will allow species to interact and respond to climate change and other pressures.

On the rewilding scale, a garden is even further along the axis towards intensive human management than the smallholdings discussed in Chapter 7 because, naturally, the area is even smaller and more constrained. The average garden in the UK is 190 square metres, although the size varies widely, from 3.6 square metres to over 2,200 square metres. As far as natural processes are concerned, these little pockets of land tend to be almost completely isolated from the wider landscape. Some gardens may be affected by rabbits and, occasionally, deer. But many are not disturbed by herbivorous mammals at all, and dynamic natural water systems such as springs, streams and rivers are rare. From the rewilding perspective, therefore, a garden is where humans are the primary drivers of the ecosystem. As a rewilding gardener, you can mimic many of the natural processes that occur in the wild in your own backyard. You become the keystone species.

Collectively, the potential for gardens to contribute space for nature is colossal. On average across the regions, 87 per cent of households in the UK have gardens.[1] In large cities, where a higher proportion of people live in flats, it is lower. In London, about 61 per cent of households have gardens. But still, one-quarter of the area of a typical city in the UK – and half its green space – is private gardens.[2] Nationally, gardens total 23 million, covering an area of around 433,000 hectares or 4,330 square kilometres. This is one-fifth of the area of Wales, and bigger than the Exmoor, Dartmoor and Lake District National Parks and the Norfolk Broads put together. In England, the area of land under gardens is four and a half times larger than the area designated to National Nature Reserves.

This aerial view of suburban London taken in 2007 shows the huge potential of gardens to provide for nature in cities. Increasing connectivity between back gardens can create wildlife corridors. The challenge, then, is to connect these corridors with other green space such as avenues, parks, roadside verges, embankments, tow paths and inner-city nature reserves.

Most gardens are designed primarily for human enjoyment: for beauty (however we conceive it), for eating and sitting outside, playing ball games with children, and growing fruit and vegetables. A growing concern for and interest in nature, however, has brought about a rise in gardening to encourage wildlife. A nature-friendly approach to gardening includes not using pesticides, herbicides or peat; deliberately planting nectar-rich plants and fruiting trees; using hedges rather than fences; creating a pond; and turning lawns into wildflower meadows. Installing birdboxes, birdbaths and feeders, and hedgehog and bat boxes increases nesting and feeding opportunities for garden wildlife. This approach is clearly within the conventional conservation mindset and can be hugely successful. Numerous books have been published in recent years offering excellent advice on nature-friendly gardening, and we've included some of our favourites in the Further Resources section (pages 532–3).

However, rewilding can take a garden to another level of dynamism and species richness. This is not, as is often believed, about 'letting your garden go' or abandoning it. Certainly, relaxing the normal garden obsession with tidiness will almost always increase the potential for wildlife, and using traditional tools instead of labour-saving devices – swapping the leaf-blower for a rake, for example – can increase a garden's hospitality to wildlife in much the same way that traditional farming implements are more wildlife-friendly than industrialised agricultural machines. But relaxing management completely, allowing a garden to scrub up and become overwhelmed with trees and shade, will almost certainly decrease the diversity of plants and negatively affect insects, birds and other life, just as happens anywhere on land that is allowed to evolve into closed-canopy woodland free of any significant animal- or human-caused disturbance.

Rewilding a garden is more about focusing on ecological results. A main aim will be to establish a mosaic of different habitats and manage the garden to sustain that complexity. It also involves trying to bring natural processes and dynamism into a confined space in order to increase the variety of habitats, perhaps allowing some boom-and-bust dynamics to play out. This is where a person becomes a proxy for the other creative influences at play in nature (such as large herbivores) that are unlikely to be able to act within the limited area and location of a garden. Learning to think like a beaver, wild boar or browsing pony frees the mind from cultural constraints. It will almost certainly change the way you make decisions in the garden.

Naturally, there will be 'limits of acceptable change', and these will be different for every gardener. In the cultural world of the garden, with its many different priorities and activities, you may not want fluidity and looseness of management throughout the entire space. Editing is required, and this can be thought of in ecological terms as forms of guided or selective disturbance, such as allowing some corners to become 'self-willed' with nettles, brambles, dead branches and the like. It is a mistake to think of these areas as being 'messy' in the sense of being neglected or disorderly. Even here, there is method in apparent madness. Within a chaotic-looking tangle of weeds and scrub is vegetation structure – varying heights and depths, thickness and thorniness, light and shade, subtle layers and interconnections – that provides an array of niches and opportunities for life. You can even go one step further to foster a habitat mosaic by importing different soil types and substrates into a garden (such as chalk, limestone, sand and gravel, or even crushed brick and concrete), and varying topography, wetness and aridity.

Some gardeners may prefer to stay close to a conventional nature-friendly garden approach, with a fairly static, traditional layout of managed lawns, paths and beds. Others will feel able to explore a different aesthetic and a more radical array of interventions. Just as with nature restoration in the wider landscape, it's useful to remember the rewilding spectrum. Why not combine approaches working in tandem in different parts of the garden? Every step towards a wilder system is an important driver for change, starting with what we feel able to do. Rewilding one corner of a garden, for example, may inspire a desire for incremental changes elsewhere. It takes time to become comfortable with a different way of doing things.

1.

2.

3.

1. Conventional garden

A manicured, monoculture lawn, kept pristine with herbicides and pesticides, and a sterile concrete patio provide little chance for life in this garden. Flower beds are rigorously weeded, leaving bare earth to dry out in the sun. A petrol-powered leaf-blower rids the garden of natural compost and artificial light floods the area at night, disturbing bats and magnetising moths.

2. Nature-friendly garden

Going pesticide-free has brought insects back to this garden. The organic lawn, mown less often, is rich in wildflowers, and longer patches of grass give cover to small mammals and insects. Replacing tanalised wooden fencing with species-rich hedges provides nesting habitat for birds. A dead tree, trimmed for safety, has been left in situ for deadwood-loving beetles and woodpeckers. Its branches, stacked beside a leafy compost pile, give shelter to hedgehogs and insects. LED fairy lights minimise the disruption to bats and moths. The small pond is a honeypot for dragonflies, frogs and newts. Mown paths through the long grass create a circular walk of interest and pleasure.

3. Rewilded garden

Hedges, trimmed less frequently, have grown shaggy and complex, providing protective niches and berries for birds, small mammals and insects. Fallen fruit and seedheads are left. Greater areas of long grass and wildflowers create cover for voles, field mice and hedgehogs. These will be mown or scythed at the end of summer. Ivy and scramblers climb the walls and roof of the garden shed. A Bee Kind Hive (see page 449) provides natural habitat for honey bees. Sandy patches are havens for burrowing bees and beetles, and ants and amphibians find refuge under the wet patch stepping-stones. The end of the garden, fronting a field, is now porous for wildlife. This rich, complex treasure-trove of a garden is perfect for hide-and-seek and building dens or, for adults, sitting in a quiet corner, listening to life. With no light disturbance, the garden at night is filled with bats and moths.

Rewilding a garden is likely to include all aspects of nature-friendly gardening, but, with a different mindset, takes the aim further along the rewilding spectrum towards a more dynamic and biodiverse ecosystem. This table, devised by Charlie Harpur, head gardener at Knepp, shows how it could work.

Variable	Conventional Gardening	Nature-Friendly /Wildlife Gardening	Garden Rewilding
Mindset	Static	Nature-focused but static	Natural process-led/dynamic
Chemicals	Selective pesticide use	No pesticides	No pesticides, no additional pollution such as outside lights or pet wormers
Compost	Occasional peat use	Peat-free compost	Plant selection focusing on those requiring minimal fertiliser. Any compost used principally for productive growing
Plant Selection	Includes double-flowered cultivars	Selecting plants for pollinators/food for birds, etc.	Selecting plants for less publicised pollinators, too, such as night-flying moths, wasps and hoverflies, as well as insect predators including lacewings and ladybirds
Habitat	No additional habitat	Providing nesting boxes for birds, bats and hedgehogs	Providing thorny shrubs, as well as dense wall cover and climbers for natural nesting habitat
Water Sources	No additional water sources	Ponds	Shallow, ephemeral ponds, or bog gardens with interventions to mimic beaver activity and the puddling of herbivores
Water Sustainability	Mains irrigation	Water butts and rainwater collection for irrigation	Site-appropriate plants, prioritising those that require minimum or no irrigation
Tidiness	Leaf-blowers	Leaving heaps of deadwood and leaf litter	Where space allows, leaving a fallen tree to rot down over time for habitat
Lawns	Fine, grass-only lawn	Wildflower lawns; reduced mowing; allowing rougher areas of long grass as cover for wildlife	Increasing the surface area and floral diversity of a lawn by introducing lumps and bumps, allowing anthills and molehills to form
Boundaries	Fence or wall	Planting hedges instead of garden fences as biodiverse boundary markers	*Internal:* Preserving old infrastructure as habitat (decaying gates, sheds, old brick walls, etc.) *External:* Connectivity and corridors

The very intensity of human activity in gardens has the potential to encourage biodiversity. One study identifies more than twice as many plant species in urban gardens as in semi-natural areas.[3] Cultivating a huge variety of plants, and doing everything to maximise their survival, can stimulate a correspondingly high variety of insects, birds and other life.[4] Gardens are often much more biodiverse than the surrounding landscape, where wildlife may be struggling.

It must be emphasised, though, that diversity of plants alone – literally, the number of floral species – is not the best way to measure true biodiversity. Indeed, this has become a fundamental problem with the Biodiversity Net Gain calculations made by developers. It all depends on what kind of plant they are, what ecological roles they play and what other species they support. You could plant fifty extra species in your garden and claim that you have improved 'species richness'. But if those plants have little ecological contribution to make, and just take up space, you have not made significant gains for biodiversity. Much more important is to select plants specifically for their ecological usefulness: for fixing nitrogen in the soil (if you're growing produce); for supporting a large number of different insects (prey species for birds, mammals and other insects); or for hosting natural predators (such as dill and angelica to encourage green lacewings to control aphids and whiteflies), for example. Later, we discuss plants that benefit insects and other species, and those that will have almost no positive effect for wildlife.

Too much management in a garden – excessive weeding and pest control, zero tolerance for fallen leaves, dead wood, brambles, nettles and thorny shrubs, and excessive mowing and strimming – will also reduce opportunities for wildlife. As the environmental writer George Monbiot puts it, 'Between killing nothing and killing almost everything lies the pragmatic aim of maximising diversity and abundance on Earth.'[5] As always, it is about striking a balance between intensive management and letting go. Or, expressed another way, rewilding a garden is about using human agency to facilitate change and natural processes rather than trying to dominate in order to keep things static and unchanging.

Connectivity, another fundamental principle of rewilding, has synergistic effects, especially at the smaller end of the scale. Making garden boundaries permeable to non-flying wildlife turns a row of gardens into a nature corridor. Replacing fences with hedges or cutting holes in fences provides passage for hedgehogs and other

The pesticide-free garden of
Great Dixter in East Sussex
contains more insects and
other species than even the
surrounding wildflower
meadows. Dixter's Sunk
Garden, with old stone walls
and planted with minimal
disturbance to the soil, is a
hotspot for biodiversity.

small mammals. If a street of connected, wildlife-friendly gardens can connect with a nearby railway embankment, towpath or river, it becomes a bridge into the wider landscape; it can act as a conduit, funnelling life into urban areas, linking the greater countryside and green belts with inner-city parks. Even window boxes and plant pots on a doorstep can be stepping stones for insects and birds if there is good vegetation cover and functioning green space nearby. Context adds another dimension to the rewilding of a garden.

Going chemical-free

One of the most important steps you can take for biodiversity in a garden, as with all other places of nature, is going chemical-free. This is an obvious choice for wildlife-friendly gardening. But it is not always easy to change the habits of a gardening lifetime, and sometimes it can even be difficult to identify what substances are toxic. A rewilding approach – thinking 'outside the box' – helps re-evaluate the dangers and come up with holistic solutions.

Pesticides

Pesticides are an obvious offender. Around three million tons of pesticides enter the environment globally every year. Estimates are difficult to come by, but it is likely the world has lost 75 per cent of its insects since 1970, through pesticide use largely in agriculture but also in parks and gardens, and even areas designated for nature.[6] It is virtually impossible to target one pest with a lethal poison without hurting other life forms and, quite possibly, humans too. Pesticides also stimulate 'super-bugs', driving resistance to pesticides. Pesticide manufacturers are manifestly aware of the short-lived efficacy of their products and are continuously engineering the next generation of products, each more toxic than the last; it is a gravy train that works in the manufacturers' favour but loads increasingly harmful toxins on the environment. Proponents argue that modern pesticides are much safer than older, banned chemicals. But the reality is that most are thousands of times more lethal to insects than when Rachel Carson blew the whistle on DDT and other pesticides in her book *Silent Spring* of 1962. Recent research shows that the dose of DDT needed to kill a honeybee is 7,000 times higher than for neonicotinoids, the family of chemicals brought out in the 1990s (and now banned in the EU, but permitted again in Britain in spring 2022). There is every likelihood that new insecticide compounds coming on to the market to replace neonicotinoids are just as lethal, if not more so. So far, lobbying of politicians by the chemical industry has prevented the rigorous assessment of new pesticides.[7]

In the rewilding mind one might ask why insect 'pests' are prevalent in the first place. As the entomologist Dave Goulson points out in his book *The Garden Jungle* (2019), plants generally suffer pest infestations only if they are stressed, most probably because they are unsuited to the local climate or soil.[8] Accepting that your favourite plant does not like your garden may be tough, but it's ecologically astute. Planting to suit your conditions, and being led by what thrives there, will save a lot of distress, time and energy. Likewise, being fixated on a particular species, such as roses, and on perfection of appearance – being intolerant of blemishes of any description – is asking for trouble. Remarkably few of the 'great' gardens in the UK, especially those with showy seasonal displays and specimen collections, are pesticide-free, with honourable exceptions, such as Great Dixter in East Sussex and Sissinghurst in Kent.

Gardeners growing fruit and vegetables for their own consumption tend – rightly – to be cautious about using chemicals. In any case, just as in the wider landscape, if you exercise a little patience nature will often sort out the problem itself. This is very much part of the rewilding mindset: allowing natural predators to respond to an outbreak of a 'pest', so that a natural boom-and-bust scenario can play out. Goulson describes the plethora of predators waiting to prey on the dense clumps of blackfly or bean aphids clustering on the growing tips of his broad beans in mid-spring. Not reaching for a pesticide gives ladybird, hoverfly and pearly green lacewing larvae the chance to hatch and attack. Soldier beetles, tiny black parasitic wasps and earwigs labouring invisibly at night are all attracted by the aphid feast. Allowing this battle to arise without intervention may reduce the abundance of broad beans a little, but the predators that have bred among this early crop will be primed for efficiency as they fan out to attack aphids throughout the summer crops of vegetables.

Often, nature-friendly gardeners provide flowers for pollinating insects. But it is just as important to choose plants that encourage predators, providing a garden with natural pest control: carnivorous insects such as lacewings, ladybirds, ground beetles and braconid wasps.[9]

Sometimes it can take a season or two for predators to rise to the occasion, particularly if your garden is just going organic or the predators are slower reproducers, such as birds. Patience, though, will be rewarded; a family of blue tits, for example, can eat 100,000 aphids a year. Much of our haste in tackling pests is rooted in our lack of trust in natural processes, particularly in gardens where we are used to exerting absolute control. We dislike living with a problem even for the shortest time, and the chemical age has provided us with quick, but not necessarily long-term, fixes. The astonishing range of pesticides and lethal control that we are constantly being encouraged to buy generates in us a cultural intolerance for living things across the board, and has led to a profound misunderstanding of insects in particular. Public campaigns may remind us of the value of bees as pollinators, and we may be fond of butterflies because they're lovely (unless their caterpillars attack our cabbages) and ladybirds for their penchant for aphids.[10] But we overlook almost all other pollinators, including ants, wasps, hoverflies, mosquitoes and night-flying moths – arguably the best pollinators of all.[11] We find some insects repulsive, such as earwigs, one of the most effective garden pest predators there

Beneficial animals for pest control in the garden

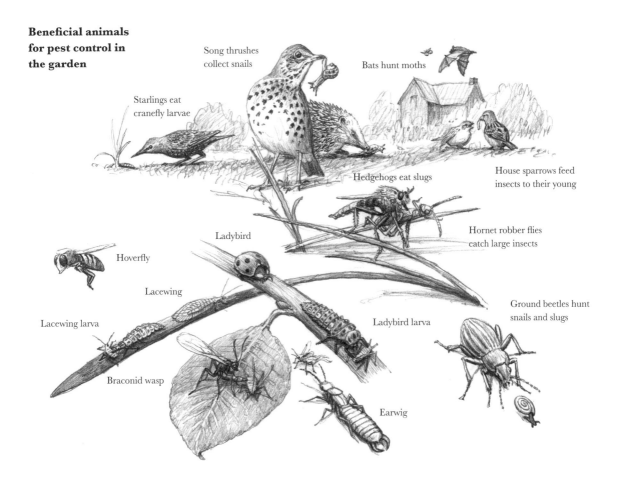

Song thrushes collect snails

Bats hunt moths

Starlings eat cranefly larvae

Hedgehogs eat slugs

House sparrows feed insects to their young

Ladybird

Hornet robber flies catch large insects

Hoverfly

Lacewing

Lacewing larva

Ladybird larva

Ground beetles hunt snails and slugs

Braconid wasp

Earwig

is, and others, such as hornets, terrifying, although they are passive in nature, effective pollinators and predate on spiders, aphids and other insects.[12]

Such is the pervasiveness of pesticides in modern life, however, that even the most conscientiously nature-friendly gardener may find themselves introducing lethal chemicals into their patch without knowing it. An investigation published in 2017 revealed that of twenty-nine plants, such as campanula, catmint, lavender and scabious, bought from garden centres – many of them, indeed, bearing a 'Perfect for Pollinators' logo – twenty-two contained at least one insecticide.[13] One flowering heather contained five different insecticides and five different fungicides. Ironically, the very plants that well-intentioned gardeners were being encouraged to buy for pollinators had been propagated using pesticides that would harm and possibly kill the insects they attracted. Following the report, which

Just as important as pollinators in a garden are natural predators. Providing food plants and habitat for carnivorous insects and other creatures that feed on common garden pests triggers a microcosm of trophic cascades – similar to those in the larger landscape – and helps create a healthier garden ecosystem. Hoverflies, earwigs, Braconid wasps, ladybirds and lacewings and their larvae, all eat aphids.

sparked a campaign by Friends of the Earth, big garden-centre chains in the UK, such as Hilliers and Wyevale, and general stores such as B&Q, Homebase and Aldi, agreed to stop selling products containing neonicotinoids, the main insecticide found in the plants. But commercial propagation continues to use other powerful chemicals, such as pyrethroid and organophosphate insecticides, and fungicides, which can also harm insects.

The only certain way to avoid accidental poisoning of this kind is to buy plants from an organic nursery (look online), grow them from organic seed or swap plants with friends, neighbours or local seed-swap groups.[14]

Domestic pet treatments

Pesticides can also arrive in our gardens through our pets. There are currently ten million dogs in the UK and eleven million cats.[15] Eighty per cent of these animals are treated regularly with flea treatments, whether they need it or not.[16] The most common flea treatment is a 'spot on' chemical anointed on the back of the animal's neck. Another is a collar impregnated with insecticide. In both cases the active ingredient is generally imidacloprid, one of the neonicotinoids that are highly toxic to bees and now banned by the EU from use on flowering crops.[17] A medium-sized dog will receive a monthly dose of 250 micrograms dripped on to its neck: enough to deliver a lethal dose to about 60 million honeybees or sixty partridges. Of course, neither the honeybees nor the partridges are likely to come into direct physical contact with the dog, although their owners, including small children, obviously do. But if the pesticide is present in the animal's urine – which is likely, considering the nature of the treatment, although, surprisingly, no research seems to have been done on this[18] – the neurotoxin will be absorbed by the roots and leaves of any plant on which the dog urinates, and from there taken up into pollen and nectar. Goulson goes so far as to recommend intensive mowing to prevent the flowering of clover, dandelions and other flowers in the lawn if you have a small garden where your dog or cat, regularly treated for fleas, tends to urinate.[19]

Flea and tick treatments are water-soluble, so it is particularly dangerous to allow treated pets to wander your garden or the wider countryside in the rain, and particularly hazardous to let them swim.

A study of dogs treated with another insecticide, fipronil, found that up to 86 per cent of the chemical washed off when a dog was bathed soon after treatment.[20] Research by Goulson and his colleagues using samples gathered by the Environment Agency in twenty English rivers, from the river Test in Hampshire to the river Eden in Cumbria, suggests that significant quantities of pesticides from veterinary flea products are entering waterways via household drains.[21]

Wormers are also routinely given to pets, whether they need them or not, and – just as in agricultural livestock – this will be having a lethal effect through faeces and urine on dung beetles and other waste-eating insects, as well as bacteria and fungi, and ultimately infiltrating the soil and watercourses. Disposing of dog and cat faeces in poo bags may only transfer the impact (and does not, anyway, remove contamination through urine). Even if the contaminated faeces end up in municipal landfill, pesticides may eventually leach out into the soil and watercourses in the wider environment. Thinking holistically about nutrient cycling, something that comes naturally with rewilding, drives home the reality of the persistence of chemicals once they are introduced into the system.

Clearly these highly toxic chemicals, whose effects on human health are poorly researched, should not be prescribed prophylactically for our pets. But it's obvious why they are. The veterinary industry makes a great deal of money from promoting them. There are plenty of nature-friendly alternatives, including herbal wormers, that can be used when and if a pet is afflicted, and herb supplements for pet food, such as Billy No Mates, which ward off fleas, ticks and mites.

Herbicides

Around 80,000 tonnes of the herbicide glyphosate, the most commonly used agricultural chemical in the world, are applied each year in the UK for 'non-agricultural use': that is to say, cosmetically, in parks and gardens. More commonly known by its brand name, Roundup, it is the favourite herbicide of gardeners. But long-term use of glyphosate is known to have adverse effects on earthworms, insects and bees, and upsets the balance of microbial communities in the soil, increasing the numbers of some micro-organisms and decreasing others, affecting soil fertility.[22] Despite Monsanto (the manufacturer of glyphosate) claiming that it is unlikely to pollute water systems, glyphosate has been found in wells, groundwater and reservoirs across Europe and the UK, and traces of it are almost universally present in human urine. Its effects on human health are, surprisingly (or perhaps unsurprisingly, given the vested interests of Monsanto and the pressure it can bring to bear on academic institutions and governments), poorly researched. But in the USA, successful claims have been brought against Monsanto for glyphosate causing cancer.[23]

While some countries in Europe have banned or begun to phase out the use of glyphosate, in the UK the chemical's role in agriculture is defended assiduously by the National Farmers' Union. It is also still on sale in garden centres, although alternative products are now more widely available. A glyphosate-free version of Roundup, based on acetic acid but – confusingly – with the same name, is now often sold side-by-side with the original. Some lawn fertilisers sold in the UK still include herbicides to target non-grass species.

Given that herbicides have been available to gardeners only since the 1970s, there are plenty of alternative ways of getting rid of weeds in a garden. They include hand-weeding; hoeing; topical applications of boiling water or horticultural vinegar (which can have a positive effect on soil biota); burning weeds in paths and driveways with a 'flame gun', propane torch (high carbon) or electric weed burner; or – in areas of high productivity – mulching to smother weeds. Most of these techniques are no more labour-intensive than applying Roundup itself.

Establishing the plants you want so they cover every inch of garden beds is another way of reducing colonisation by the unwanted ones. Not minding so much about the presence of the more benign and non-invasive weeds is, perhaps, an even happier approach, lifting the gardener's nagging burden of perfection. But it can be surprisingly hard to resist the prejudices one has grown up with, such as a pristine grass lawn. Remember, most 'weeds' are native wildflowers, beautiful and precious in their own right – or, as the nineteenth-century American philosopher Ralph Waldo Emerson aptly described them, 'plants whose virtues have not yet been discovered'.[24] What one doesn't want, in the interests of biodiversity, is a plant – be it a 'weed' or not – that dominates all the others.

The Pesticide Action Network offers excellent advice on alternatives to garden chemicals.[25] Encouragingly, as communities become more aware of the insidious nature of chemicals and their persistence in our ecosystems, garden societies, villages and even entire cities are choosing to go pesticide-free (a term that includes herbicides and fungicides).

Fertilisers

High productivity

Just as traditional farmers used to, experienced gardeners – particularly vegetable growers – understand about healthy soil, how it crumbles in the fingers, the way it smells. They know the importance of feeding it organically, rather than synthetically. Synthetic fertilisers cannot replenish trace elements such as magnesium, calcium, zinc, sulphur and selenium. Crops and garden vegetables grown in soil that is dependent on applications of artificial nitrogen, phosphorus and potassium (N, P and K) alone are probably deficient in essential nutrients that are also crucial for human health. Good compost is the best provider of nutrients, sustaining a healthy, functioning soil replete with bacteria, mycorrhizal fungi and all the other microbes and animals such as earthworms, protozoa, nematodes, mites and springtails. A healthy soil enables plants to take up all these micronutrients, and so provide the nutrients in our food.

Many gardeners create compost themselves in time-honoured fashion using various combinations of prunings, vegetable peelings, wood ash, straw, manure (from horses, cows or chickens), coffee grounds, the contents of dustpans, shredded paper and cardboard, lawn mowings, fallen leaves and woodchip, all mixed in a compost heap or bin. (The contents of vacuum cleaners and tumble-drier fluff must now be avoided, since they contain a high percentage of plastic fibres from clothes and carpets.) In the rewilding mindset, garden compost mimics the active decomposition of organic materials in nature, but in a 'tidy', efficient way. The process of decomposition is speeded up through the heat naturally generated by larger heaps, and by turning the pile regularly to increase aeration (think of a pig grubbing around in the leaf litter of a wood). Adding manure to compost heaps increases nutrient levels as well as the bacteria and other micro-organisms that speed up decomposition, in the same way that animal dung does in the wild. The segregation of the decomposition process into a tucked-away corner or separate bins enables a gardener to apply compost selectively, where it's most needed and once it has fully broken down, without messing up the orderliness of the garden.

The fact that compost can be bought very cheaply from garden centres – along with an array of additional fertilisers, some more sustainable than others – is a disincentive for gardeners to make their own. By far the worst offender is peat compost, which is valued by gardeners for its capacity to retain nutrients and moisture. Of the 3.9 million cubic metres of compost bought in the UK each year for garden and horticultural use, more than half – 2.9 million cubic metres – is peat-based.[26] Campaigners have been lobbying for a ban on peat compost for more than twenty years. It was only in 2021 that the UK government announced its intention to ban the sale of peat to gardeners in England and Wales by 2024, and to the professional horticulture industry by 2028. This slow response from government has been as frustrating as it is inexcusable. Some UK outlets, such as Dobbies and the Co-op, have initiated their own bans, and B&Q will cease peat sales by 2023.[27]

Another fertiliser beloved of gardeners is fish, blood and bone bought in a meal or powder, an application that, in effect, mimics the natural process of decomposing carcasses in the wild. Effective though this is, the origins of the beef bones and blood is rarely stated, and fish meal, in particular, is unsustainable. Nature-friendly gardeners are turning instead to plant-based liquid fertilisers or 'teas' made from comfrey (rich in potash and particularly useful for fruiting plants and vegetables) and nettles (high in nitrogen, especially in the spring). These can be easily made at home, and the comfrey and nettle plants benefit wildlife as they grow. Comfrey is a great source of nectar and food plant of the scarlet tiger moth. The brewing liquid, or tea, supports the wonderful rat-tailed maggots of the *Eristalini* and *Sericomyia* tribes of hoverflies. Nettles are the food plant of the caterpillars of small tortoiseshell and peacock butterflies.

Unlike industrial farmers, whose approach has been driven by agrochemicals for the past seventy years, most gardeners still have a strong connection with the earth and natural nutrient cycles, continuing to act as a benevolent keystone species and fostering good soils. It is from horticulture that the new techniques of vermiculture (cultivating worms to speed the composting process), bokashi fermentation (using fermentation to turn cooked food waste into a nutrient-rich tea for plants) and biochar (carbonised organic material used as a soil additive) have developed. All these wildlife-friendly technologies are now used in agriculture.

Low-nutrient systems for biodiversity and sustainability

The elephant in the room, so rarely discussed in relation to gardens, is our cultural bias in favour of unnaturally high levels of soil fertility. Of course, we need high fertility to grow fruit and vegetables. But generally, such 'eutrophication' is pursued everywhere in the garden, even on soils that are inherently poor. Gardeners tend to believe that rich soils and faster growth are fundamentally good, and our aesthetic has been groomed to favour plants that are large, leafy and 'productive'. It's a culture borrowed from agriculture and promoted by gardening shows, the media and gardening books.

From the point of view of biodiversity, however, this can channel a garden in the wrong direction. The more fertile the soil, the more difficult it becomes to maintain the complexity and diversity of plant species. Faster-growing species easily outcompete the smaller and slower ones. Essentially, gardeners end up creating much more work for themselves trying to maintain complex plant communities. If nutrients aren't applied, on the other hand, the growth of plants is less rapid and maintaining the complexity of the vegetation becomes much easier. As so often in rewilding, it's a question of 'less is more' – a difficult principle to take on board when the prevailing culture emphatically advocates 'more is more'. If your soil is naturally low in NPK, why not choose plants that require few nutrients? The flowers that benefit insects and other wildlife the most tend to be those that favour poor soils, so do not think of high levels of NPK as the holy grail.

Also to be considered is the carbon cost of buying in fertiliser, particularly if it is synthetic. Most garden soils are relatively fertile, with high levels of organic matter owing to historic mulching and root turnover. They do not need extra fertiliser for most things to grow. Our garden advisory board member James Hitchmough notes that when his garden is open to the public, the most common questions from visitors concern how much NPK, compost and other fertilisers he uses. His answer, 'none', is almost impossible for most gardeners to comprehend.

However, in one half of the Walled Garden at Knepp – the non-produce-growing part – we've moved away completely from the idea of productive soil and deliberately established conditions of low fertility. The concept has been driven by our advisory board, which comprises the garden designer Tom Stuart-Smith; James Hitchmough, professor of horticultural ecology at the University of Sheffield; Mick Crawley, emeritus professor of plant ecology at Imperial College London; and Jekka McVicar, vice-president of the Royal Horticultural Society (RHS) and organic herb expert. Inspired by Tom's work in Mediterranean gardens and James's creation of spectacular gardens on brownfield sites, we've created a patchwork of crushed-brick-and-concrete mounds and hollows with a shallow sand topping. All the topsoil has been removed. Into this matrix of poor, porous substrates we've planted a huge range of plants that love poor conditions and are wonderful for insects.

Our primary aim is to maximise biodiversity (discussed more fully on pages 428–9), but we also hope this garden will rise to the challenges of climate change and, particularly, minimise the need for watering. In the UK, a total of 205 billion litres of treated, drinkable mains water are used by homeowners outside each year. The RHS and Cranfield University have produced guidelines and an app providing advice on how to save water in the garden, as well as a 'Mains to Rains' pledge, which can be signed online.[28] One of their recommendations is to create a gravel garden using drought-tolerant Mediterranean plants. Using water to the extent that we do in our gardens is not only unsustainable, but also costly, and likely to become increasingly so as water reserves dwindle across the UK.

Becoming the herbivore

Looking at a garden with the rewilding mindset – thinking like a herbivore – will help to free your imagination from cultural constraints and embolden your decisions so that you can think creatively about its management.

What are we doing, for example, when we turn the earth over with a trowel, other than mimicking the snout of a wild boar or the pawing hoof of a territorial bull? Exposing the soil is a natural process that creates opportunities for plants and insects to colonise new ground, but, apart from on very sandy soils, nature never tolerates bare earth for long. A wild boar, once it has excavated the tubers, dock roots and grubs it is after, will not return to that patch until it is fully restored and has food to offer again. A bison rolling in the earth and creating a wallow will move on to the next site, leaving its parasites behind. Open soil is swiftly healed by a dressing of pioneer plants or even, on poorer soils, lichens and fungi, protecting soil biota from desiccation. Keeping continuously bare soil between selected plants in a border or vegetable patch may look like assiduous gardening, but it is essentially as unnatural and deleterious as a ploughed field. Like nature, we should be looking to disturb the earth judiciously and infrequently and, for the most part, provide continuous cover, conserving moisture, carbon and life in the soil.

Pruning is another technique that we have stolen directly from herbivores. Cutting with secateurs or pruning shears stimulates regrowth and structural density in a shrub because plants have evolved over millions of years to protect themselves from browsing animals. Secateurs can't quite replace the nipping teeth of a herbivore, though. The enzymes in an animal's saliva triggers even more dramatic bifurcation of branches and the production of protective chemicals and thorns, as well as 'emergency' flowering.[29]

Rethinking the lawn

Similarly, when we mow the lawn, we're mimicking the effect of large herbivores, geese or rabbits on a patch of sward. That is why grass responds so vigorously with new growth, and wildflowers, such as buttercups, daisies, self-heal, clover and even thistles, have evolved to respond by being able to flower even at miniature height.

Closely cropped grass benefits a range of insects, including crane flies (daddy-long-legs). They are considered a pest by some lawn-lovers because the larvae, known as leatherjackets, feed on the roots of grass, but they are also fantastic food for birds such as crows, magpies and starlings, which will peck them out of the lawn. Ants, too, are a keystone species. Anthills provide micro-habitats with a different soil composition and different directional aspects to catch the sun, benefiting a multitude of fungi, lichens, mosses, grasses and flowering plants, such as wild thyme, which help to bind the surface. If you can tolerate anthills in your lawn, you will be providing a basking spot for creatures such as small copper butterflies and common lizards, a place for grasshoppers to lay their eggs, and food for numerous birds, including green woodpeckers.

Grazing lawns, as discussed in Chapter 6, are a natural feature created by herbivores in the wild. In places such as the Serengeti, in northern Tanzania, under the influence of vast herds of animals, they can extend for hundreds of miles.[30] They are characteristic of the wood pasture/savannah ecosystem, something that is so deeply ingrained in us erstwhile hunter-gatherers that we subconsciously replicate it in our parks and gardens. Psychologists and evolutionary biologists have identified this natural parkland aesthetic – with its spatial openness, lone trees, leafy plants and greenery, and still or slow-moving water – as the landscape that makes us feel most at ease.[31]

The grazing lawns of the African savannah, typical of natural pasture systems where humans evolved, give us our love of cropped grassy areas in our gardens and urban parks.

But over the centuries we have taken this pleasing aesthetic and reduced, corseted and manicured it, stripping it of opportunities for life. The first lawns were cultivated by the wealthiest estate-owners, who could afford to replace herbivores with men wielding scythes. In the 1830s the first cylinder mowers were pulled by ponies wearing soft leather boots to protect the grass. But it was the invention of the petrol-driven mower in 1902, followed by the availability of synthetic fertilisers and herbicides to favour grass above other species, that brought about the cult of the immaculate, grass-only lawn.[32]

There are an estimated 200,000 hectares of lawns in the UK, and the lawn often occupies the largest footprint in the garden. Most are regarded, in the words of Professor Stefan Buczacki (former chairman of Radio 4's *Gardener's Question Time*), as the 'carpet of the room outside'. They are mown weekly in the growing season to create a striped or even chequered pattern. Their sterile uniformity – their very lack of diversity – is the pride and joy of the mower. A lucrative industry of lawn-care contractors exhorts us to 'bring your lawn to life' by killing everything in it but the grass. The website 'Grass Clippings' lists ways to control moss, moles, ants, chafer-beetle larvae, red thread fungi and weeds in a lawn.[33] In the same breath as promoting chemicals to control worm casts by suppressing the aerating activity of earthworms, it advises us how to aerate the lawn using mechanical tow plug or coring aerators, manually spiking holes with a garden fork or even walking up and down wearing spiked lawn-aerating shoes.

Such intensive management is not only high-carbon, time-consuming and costly, but also disastrous for wildlife. Instead of considering a garden lawn as our domestic domain, an extension of indoors where we exert complete control and have zero tolerance for life, we should see it for what it is: part of the natural environment. Why not imagine a lawn as a meadow, a prairie or even a wood pasture in miniature, and consider how those ecosystems are sustained in the wild?

In nature, lawns are dynamic, multi-species systems grazed at varying intensities over the course of the year. The herbivores that create them have different influences on the sward. They have preferences for different plants, and nutritional requirements that change throughout the year. Their mouthparts take plants in different ways, affecting the way they regrow. Herbivore populations fluctuate throughout the year, sometimes under pressure from predators. The animals may even migrate, leaving the lawn after weeks or months to regrow. When we mow a garden lawn repeatedly throughout the growing season, year on year, we behave like a herd of sheep restricted to a single pasture on a farm. There's no patchiness of grass length. Wildflowers have less chance to bloom (if they are tolerated at all); there is no time for ants to build their mounds; and plants that cannot survive as individuals less than 1 centimetre tall are eliminated.

Unnaturally high soil fertility itself – of the kind needed to grow an immaculate lawn – has a dramatic impact on biodiversity. The world's longest-running ecological experiment, the Park Grass Experiment at Rothamsted Research in Hertfordshire (begun in 1856), studies the rates and combination of plant nutrients necessary to maximise the yield of hay. The answer was in fact known within a few years – 150 kilograms of nitrogen per hectare per year, 35 kilograms of phosphorus and 225 kilograms of potassium, plus 135 kilograms of silica and 63 kilograms of sodium. What was totally unexpected was the effect the fertilisers had on the diversity of plant species. The initially herb-rich grassland of forty-four species was, within a few years, converted into a species-poor sward dominated by just three tall grass species, with virtually no wildflowers. The addition of fertilisers had caused competitive exclusion of both the herbs and the nitrogen-fixating legumes, along with all their pollinators and insect herbivores, such as ants. Exactly the same thing happens in an artificially nutrient-loaded garden lawn: the reduction of the plant community to just a few species of grass.

When we eliminate fertiliser and mimic the grazing effect and intensity of herbivores in the wild, we think of a lawn in an entirely different way. It becomes a species-rich grassland mosaic where varying heights of grass benefit different specialist species. The combination of habitats provides protective cover and breeding habitat, as well as food for wildlife with complex requirements over their life cycles.[34] We can think of different mowing tools as performing the function of different animals – a strimmer, for example, providing a rough cut on longer grasses, like a cow, which, in the absence of upper front teeth, uses its tongue to rip and tear. This is a different action from that of a rotary lawnmower, which has an effect more like a horse, whose full set of front teeth enables it to nip and cut with more precision, or a cylinder mower, which produces a tight lawn similar to that produced by cropping sheep, geese or rabbits.

The conservation charity Plantlife promotes 'No Mow May' every year to give wildflowers a chance to bloom.[35] This, in essence, mimics the spring flush in the wild, when grass and wildflowers grow so fast they outrun the activities of animals that would normally eat them. For the rest of the growing season, to maintain a short grass lawn, the charity recommends mowing once every four weeks – a loose grazing effect – to stimulate the continued flowering of the short grass species. Of course, mowing also has a huge carbon cost. Plantlife estimates that if you currently spend an hour a week cutting your lawn, changing to a monthly cut would reduce your carbon footprint by 293 kilograms per year and save twenty-four hours of mowing time.

Plantlife also promotes the 'mohican', a mowing approach in which patches of longer grass are allowed within the shorter sward or at the edges of the lawn. You might mow more tightly close to a patio, for example, and less frequently further away. This benefits a wider range of flowers that like longer grass, such as ox-eye daisy, field scabious, knapweed and orchids. This two-tone approach maximises both the diversity of flowers and the abundance of nectar, and provides cover for insects, birds and small mammals – including hedgehogs – as well as frogs, newts and lizards. The longer grass will also provide a source of seed later in the year. It can be scythed or mown in September, and the plant matter removed so as to keep the soil low in nutrients and benefit native plants.

This former high-mainte-nance grass tennis court in East Sussex has been trans-formed into a wildflower lawn thrumming with life.

In its citizen-science survey *Every Flower Counts* of 2020, in which quadrats (1-metre square frames) were observed in over 9,000 lawns across the UK, Plantlife found more than 200 species of flower, including rarities such as meadow saxifrage, knotted clover and eyebright, as well as 97 species of pollinator, including 26 butterflies and 21 different bees. The survey also found that making these simple changes to the mowing regime can create 'superlawns' with enough nectar for more than ten times the number of bees and other pollinators, compared to that of a 'normal', reguarly mown lawn. Matt Rees-Warren's book *The Ecological Gardener* (2021) offers useful information about how to turn a lawn into a wildflower meadow.[36]

How much of the lawn you decide to keep closely mown will depend on what you use it for. If you want space for ball games, you will have to cut and roll the grass regularly, and if you (or your children) are likely to be running around in bare feet, you will want to remove thistles – something that can be done easily by hand. But there is no ecological defence for fostering a monoculture of grass and eradicating all grassland flowers and plants as if you're competing with Wembley Stadium or Centre Court at Wimbledon. This is a perverse aesthetic cult that nature-friendly gardeners and rewilders can challenge. Just as with agriculture or forestry, every effort a gardener makes to favour a monoculture diminishes biodiversity and the function of the ecosystem. Moreover, it leaves the favoured plant species – in this case, grass – more susceptible to pests, drought and waterlogging. Applying chemical lawn feed to encourage grass at the expense of other plants is not only carbon intensive, but also destroys life in the soil. Even spot applications of herbicide to kill plants such as plantain and thistles can, over time, destroy soil micro-organisms and affect earthworms.[37] Raking up moss reduces the lawn's ability to retain moisture, deprives birds and mice of a favoured material for their nests, and removes protective cover for beetles, spiders and other invertebrates.

A long-tailed tit uses moss scavenged from a biodiverse lawn and linchen from old branches for its nest.

Can we learn to love an old-fashioned mossy, flower-rich, worm-rich, insect-rich lawn that feels comfortable and spongy underfoot and – if camomile is part of the mix – smells delicious? Or will we continue to prefer the monotonous, high-maintenance green deserts of the 1970s? Aesthetic sensibilities run deep and give rise to strong emotions. If you relax your hold over your lawn, your neighbours may protest. Peer pressure is powerful. No one wants to incur the raised eyebrows, let alone the wrath, of neighbours.

An initiative by the Blue Campaign in the UK provides a way for gardeners to defend the rewilding of their garden and deflect criticism for laziness or inattention.[38] By planting a blue heart on a stick in the middle of an unmown lawn or by an unkempt hedge, a gardener declares that the shaggy look is intentional, to benefit biodiversity. It tells the neighbours, 'I know what I'm doing' and, by inference, suggests that they should be doing the same. If others in the neighbourhood can be persuaded to rewild and plant a blue heart of their own, the peer pressure may start working in the opposite direction. Gardeners without one may soon begin to look old-fashioned.

At Knepp, now that our children are grown up and we rarely kick a ball around, we mow our garden lawn just once a year, at the end of summer. We also keep margins of longer grass and wildflowers around the edges, which we scythe or top randomly at the end of each summer. We're amazed at the species – including pyramidal and early purple orchids, and various new species of fungus – that have manifested around us. Sitting on the lawn outside the house on a spring or summer evening with our ankles in wildflowers and the surround-sound of insects and birds, we wonder at what we've been missing all these years. While we were busy rewilding the farmland, we had still been unthinkingly mowing our lawn.

From two-dimensional to three-dimensional

We think of lawns as needing to be billiard-table flat but, of course, in the wild this is rarely the case. If you want a lawn for playing ball games, pitching tents and eating out on, you'll naturally want a level playing field. But if you can tolerate lumps and bumps, such as molehills and ant mounds, a lawn can become a multidimensional space with a much greater surface area, different aspects and even different microclimates, benefiting a multitude of species.

As we mentioned above, Tom Stuart-Smith and James Hitchmough are collaborating with us inside the walled garden at Knepp on an experiment to convert a high-maintenance, water- and fertiliser-thirsty, monoculture grass croquet lawn into what we hope will become a complex, multilayered, three-dimensional mosaic of biodiverse plant communities.[39] We're using a variety of locally available materials, including crushed concrete and brick from building sites on the estate and sand from a local quarry, to reduce the fertility of our rich clay soil. The porous, poor substrate will also improve drainage and grow a much wider range of slower-growing, drought-tolerant plants that are more capable of surviving climate change. By varying the amount of the substrate materials (including leaving the fertile soil in some areas) and shaping them into mounds and hollows to create different slope aspects, we hope to produce different combinations of productivity and growing-season temperatures. The plants that favour the hot south and west slopes will be different from those that prefer the cool north and east slopes. The aim is to create a wide range of habitats across varied topography – and using a much higher surface area – that has far greater potential for biodiversity than a flat, evenly fertile lawn. It is a more plant-friendly, lower-resource approach to creating beauty in a garden.

We've established the initial vegetation through a combination of seeding and planting. In most areas the objective is complete cover of vegetation in the summer months, although in some of the less fertile areas, cover will be sparser. To kick-start the mosaic, on west-facing slopes in the unproductive crushed concrete mulches, we've planted species that come from the hot alkaline soils of southern Europe and Colorado. On the north side of the topsoil slopes we have plants that prefer moist soils in a temperate climate, typically those of mountain meadows. We've seeded some species across all habitat gradients to see what strategies they come up with to adapt to the different conditions. To be sure, some species will fail and others will succeed, but death will not be seen as failure. The end points will be fluid but, of course, within the framework of a garden there are still 'limits of acceptable change'.

Once the initial plants are fully established, we'll act as selective grazers to steer the development of vegetation by weeding – removing species that we decide are excessive, unwanted or becoming unexpectedly dominant. In some cases, where we decide to add different plants to an area of established vegetation it may be necessary to dig up and plant anew. But the overall objective is to manage the vegetation simply, through cutting and limited weeding. We want to develop multilayered, structurally complex, relatively stable, beautiful and biodiverse vegetation that needs minimal human intervention and develops in ways that are not wholly predictable and that can respond to changes in climate. Naturally, we expect the diversity of plants to result in an increase in the diversity of insects and other species. Every five years we will conduct wildlife surveys to compare with the baseline we recorded before starting the intervention.

What sort of plants?

Agriculture has transformed the way we think about gardening. Until relatively recently, gardeners have nearly always regarded cultivation as about minimising plant stress to maximise plant productivity: watering and fertilising to maximise the above-ground biomass produced each year. Fast, big, wet plants are seen as 'good'. This makes sense when we're looking at annual crops that we want to eat, but plant diversity is largely incompatible with productivity. The bigger the plants, and the more fertile the conditions, the fewer plants will fit into a given space. When you picture how many footballs you could fit into a bathtub compared to how many golf balls, you begin to see how these phenomena create ecological rules. Conditions of low productivity favour vegetative golf balls; high productivity produces footballs.

Traditional gardeners find a way of increasing biodiversity in conditions of high productivity by applying very intense selective disturbances, such as cutting back, physically moving plants, staking floppy plants that need support, increasing spacing and removing weeds. This restricts the biomass of faster-growing species so that they don't crowd out the slower-growing ones. Because gardeners can choose their plants from a vast range of possibilities, as opposed to what is locally circumscribed in nature, they have also been able to use phenology (what grows when) and the shade tolerance of individual species to increase diversity – by stacking species in layers, one on top of another. So gardening, in the main, has involved the creation of hyper-productive plant communities by adding nutrients and water. These are then managed and directed by intense selective disturbance so that the garden does not simply become an arena of just the few, most productive species.

However, the world is running out of resources, including water, and artificial fertilisers are now recognised as damagingly unsustainable, because of the high carbon cost of producing them and their long-term detrimental impact on soil function and biota. In the past, disturbances were achieved by hand in the garden. Now we rely on garden machinery such as lawnmowers, edgers and aerators, strimmers and brushcutters, leaf-blowers and garden vacs, trimmers and chainsaws, flame guns and floor-sweepers – all of them high-carbon both to manufacture and to run. At the same time, our climate is subject to more disturbances in the form of extreme weather events such as flooding and drought, and we are becoming more interested in supporting animal life within our cultural realm. Our changing world and world view begin to make the formula of maximum productivity + intense selective disturbance less compelling. The environmental emergency is driving a revolution in land management at every level, and the parallel between a shift of approach in the garden and the movement from industrial into regenerative agriculture is clear to see.

Designing a garden that is suited to your soil and local conditions, yet with in-built resilience to rising temperatures and water shortages, is crucial for sustainability, and for your garden's long-term survival. Moving away from highly productive plants that invariably require watering and fertiliser; building drought resistance and floral diversity into a lawn; taking a relaxed attitude to beneficial or attractive 'weeds' (or native wildflowers); creating low-maintenance, low-carbon alternatives, such as replacing high-maintenance, thirsty grass pathways with gravel paths planted with low-growing dry-loving herbs: all these steps can make a garden much more sustainable.

Native versus non-native

One of the biggest debates concerning nature-friendly gardening is about native as opposed to non-native plants. Clearly a garden containing only native plants would be a challenge to most gardeners, who often choose plants for their very exoticism. The majority of plants in most gardens are non-native, and this is hardly surprising, given the astonishing selection of tempting species and cultivars available from all over the world. More than 80,000 varieties of 14,000 different plant species can be bought in the UK.

Many pollinating insects are generalists and take easily to non-native plants, especially if their flowers are accessible and/or similar to those of native species. Early-flowering non-natives, such as the flowering currant from North America (*Ribes sanguineum*), grape hyacinths from Eurasia, rockcress from southern Europe (which is also resistant to deer) and Japanese mahonia, and late-flowering species, such as baneberry or bugbane (*Actaea simplex*), colchicum, Japanese anemones, salvias, sedum, tansy-leaf aster from northern Mexico, and Argentinian vervain, can all feed pollinating insects, especially at times of the year when native wildflowers are scarce.

However, native plants, having co-evolved with native fauna, are used by many more species than are non-native plants, notably as food plants more than as sources of nectar. Many native insects depend on just a few species of plant, or even just one. In general, when gardening for biodiversity, greater attention should be given to native plants, and non-natives should be chosen with an appreciation for the benefits they might provide for native insects.

Many cultivars (plants that have been selected and cultivated by humans) are of no use to insects at all, having had the traits that make them hospitable to pollinators bred out of them. Surprisingly little research has been done on this species by species, so it's a case of taking advice from ecologists or simply trying to work out for yourself whether a cultivar still has a relationship with pollinators. Perhaps the easiest way to do this is through observation, if you already have the plant in your garden. There are a few general rules to follow, however.[40] Double blooms prevent pollinators from accessing the pollen or nectar, and are almost always sterile. If a species has been cultivated to change the flower colour, the chances are it will be less attractive to insects than the original. If a cultivar has been bred to be more compact, with larger and/or more flowers, but is otherwise little changed from the original species, it will probably still be beneficial to pollinators. Some cultivars may be less resilient to flooding and drought than wild, uncultivated species, making them vulnerable to climate change and extreme weather events; while others might be selections of ecotypes that are actually a better fit for some conditions, or at least no less vulnerable than wild populations.[41] Many cultivars offer reduced nutritional benefits, and some cultivars bred for foliage colour have become toxic to insects.

Garden trees and shrubs

Vegetation structure – the layering of plants of different heights – is hugely important for biodiversity and can make the most of garden space, allowing the play between light and shade. Trees and shrubs can help create microclimates and provide both nesting habitat and protection (a place to escape predators) for a multitude of species, as well as a wealth of food. A flowering tree can produce thousands of nectar-bearing flowers, while a herbaceous plant might produce just a dozen or so.[42] If those flowers then turn into fruit with a seed or seeds, they will feed wasps, flies, mammals and birds. A mature oak tree, the most biodiverse native tree in the UK, will support more than 280 different species of insect as well as countless birds, bats, lichens and fungi, and provide plentiful acorns for small and large mammals.[43] Silver birch – a popular garden tree – hosts more than 300 insect species and is the best tree for moth larvae.[44] Sallows and willows, now relatively scarce in the landscape of the UK, are also superlative hosts for native insects.

Conifers, on the other hand, which largely evolved before broadleaf trees and do not produce true flowers, offer less food, although they do produce cones containing seeds, such as pine nuts, that can be very nutritious. They also provide evergreen cover that offers useful protection for wildlife, especially in the winter.

It is important, though, not to be too purist. Some non-native tree species, such as horse chestnut, sycamore, mulberry, walnut, cherry laurel and snowy mespilus, provide wonderful opportunities for wildlife, including pollinators and insect- and fruit-eating birds.[45]

Trees don't have to be huge, and many, even oaks, can be grown in a mixed-species hedge – the proxy for thorny scrub in the wilder landscape. In the UK, such a hedge will provide nesting habitat for over thirty British birds, and nectar, berries and seeds for birds, insects and small mammals.[46] Domestic gardens contain just under a quarter of the 123 million trees that live outside woodland in Britain. On average, 54 per cent of gardens contain one or more trees taller than 3 metres.

However, gardens are getting smaller. Edwardian and pre-war suburban properties had much more generous gardens than late twentieth-century developments, which pack in housing to maximise profit, and 'garden-grabbing' for housing continues to erode existing gardens. In the Chilterns, 71 per cent of new homes were built on gardens in 2010, compared with 22 per cent in 1997 and 55 per cent in 2005, meaning they are less likely to contain hedges, tall shrubs and trees.[47] Clearly, we need houses to live in, but cities must also provide healthy and sympathetic space for life. The decline of green space, both public and private, and in particular of trees, which contribute to climate and temperature regulation as well as filtering pollution out of the air, is deeply concerning.

Urban gardens have shrunk dramatically in size. Compared with Edwardian-sized gardens in the village of Cholsey, Oxfordshire (top), the space given over to gardens in modern housing in the town of Didcot, five miles away, is minimal (below).

Both images © Google Earth

Biosecurity

It seems we can be relatively relaxed about having non-native plants in our gardens.[48] Surprisingly few 'exotics' have escaped from gardens, arboretums and parks into the wider UK countryside, and those that have cause far less ecological damage than the alarmist tabloid headlines would have us believe.

We must remember, too, the relative paucity of floral species in the UK. What we consider to be native plants are those that have made it to Britain without human agency in the last 10,000 years, that is to say, since the last ice age (a relatively short time in geological terms). They are but a fraction of the plants that would have been present in the landscape 2.6 million years ago, when Britain was connected to Europe and its flora was pan-Eurasian, including species that today are found only in China and other pre-Pleistocene floral refuges, such as the forests of the Caucasus, the Carpathian mountains, the shores of the Caspian Sea, and Turkey. Compared with continental Europe, Britain has markedly few plant species: 3,354 vascular plants (plants other than mosses and liverworts), of which 2,297 are native and 1,057 have been introduced. Europe has around 11,500 vascular plant species, of which 3,600 are alien taxa that are thought to be naturalised.[49]

If we take a longer evolutionary view and consider humans and the other 'native' species that have colonised Britain in the last 10,000 years as the new arrivals, and species that were here millions of years ago as fundamentally indigenous, too, we open the window on to a whole new world of possibilities. In the interests of biodiversity and climate-change survival, there is a strong argument for being more hospitable to native plants from continental Europe, especially if they were part of Britain's pre-ice ecology.[50] The relationships that co-evolved between flora and fauna are unlikely to have switched off. Insects, for example, are likely to retain the chemical 'wiring' that enabled them to interact with plant species even hundreds of thousands or millions of years ago.

More concerning for biosecurity than plants themselves, perhaps, are the diseases and faunal introductions that come with them. Imports of ornamental plants to the UK are worth over £1 billion annually, and the RHS records two or three new established invertebrates in the UK every year. Some of them are quite large. New Zealand flatworms first appeared in gardens in Belfast in 1963 and quickly spread to Scotland. They hide in plant roots and cling to the undersides of plant pots and garden ornaments. In warmer parts of Britain the Australian flatworm has made itself at home, and the Australian landhopper or 'lawn shrimp', a native of New South Wales, has spread from gardens in the Scilly Isles to mainland Britain. Three species of New Zealand stick insect are now established in gardens in Devon, Cornwall and the Scilly Isles, most likely imported on tree ferns and other ornamentals.

The RHS has set up systems, in collaboration with the Department for Environment, Food and Rural Affairs (DEFRA) and commercial growers, to improve the regulation, inspection and testing of imported plants through the Horticultural Round Table. But hitch-hikers on plants can be extremely difficult to spot, and pathogens can emerge in mysterious ways.

It's not only the horticultural industry, of course, that is responsible for new arrivals. Numerous species are stowaways in cargo – such as the European spider and European yellow-tailed scorpion that have been found close to ports in the UK. Some arrive from countries other than the one of their own origin; the Asian harlequin lady-bird, for example, now common throughout the south of England, arrived from the USA, possibly via Europe, in about 2004. Unlike the forty-six native British ladybirds, which eat exclusively aphids, harle-quins can switch to other insects, such as caterpillars and even other ladybirds. Studies have found significant declines in seven of the eight commonest ladybirds in the UK since the harlequin arrived.[51] If its trajectory in North America is anything to go by, it will soon be the UK's commonest ladybird.

How much these faunal colonisations will affect our ecosystems in the long term is hard to tell. It will almost certainly depend, to some degree, on the health and dynamism of the environments that receive them. This makes another argument for the need to encourage biodi-versity, soil health and ecosystem function in gardens.

Garden ponds

To encourage garden wildlife, create a pond. Some 70 per cent of the 1.25 million ponds rural Britain had in 1890 have been lost. Of the 400,000 that remain, four-fifths are polluted with agricultural run-off or other contaminants, such as salt from roads. Gardens have a vital contribution to make to aquatic habitat. Although garden ponds tend to be tiny, averaging about 1 square metre, there are around 3.5 million of them. About 16 per cent of household gardens have ponds, many of them in urban areas.[52]

Studies often assume that urban ponds have lower biodiversity than rural ponds.[53] But this is most likely a result of bias in the data collection. Most ecological surveys focus on rural ponds. The formal sunken pond in the garden at Great Dixter is the most biodiverse pond their ecological surveyor had ever encountered.

Garden ponds are vital oases for wildlife, providing drinking and bathing water for garden birds and declining hedgehog populations, as well as habitat for threatened species such as the great crested newt, common frog and common toad, and aquatic invertebrates, which provide food for other species.[54] Most pond animals are excellent colonists. Anyone who has created a pond in their garden, even a small 'bucket pond' (no pond is too small), can attest to the speed with which aquatic plants, insects and amphibians move in, providing joy and interest to amateurs and naturalists of any age. Night visits with a strong light can reveal underwater life that is difficult to see during the day: dragonfly larvae waiting to pounce on their next prey, and juvenile newts swimming in the shallows, their gills fanned out. Research into the wildlife value of garden and urban ponds is scarce, except for the benefits of amphibians, but the development of 'blue' infrastructure, including the trend for creating roof ponds in many European cities, is stimulating great interest.

Water quality is obviously key. But the design of a pond – variation in water depth, banks and margins, the amount of light and shade – can make a world of difference to the value of the habitat. Many books on nature-friendly gardening contain extremely useful information on how to create garden ponds that maximise the potential for wildlife. One of the best is *Rewild Your Garden* by Frances Tophill.[55] The pond section on pages 126–37 of this book explains more about the dynamics and ecological requirements of natural ponds.

Anxious parents and grandparents are often tempted to fill in a pond. But this will not only eradicate the pond's inhabitants, it will also deprive small children of the joy and fascination of pond life – often the start of a naturalist's passion. There are many websites offering excellent advice on how to child-proof a garden pond.[56]

Natural swimming ponds

It is possible to convert a conventional outdoor swimming pool into a natural swimming pond, or create one from scratch.[57] The value for wildlife varies according to the naturalness of the system you choose. The key is to reduce nutrients, preventing the growth of algae, so the water remains clear and swimmable. Some natural swimming pond companies pump the water through shingle, which acts as a filter, while others use large aquatic plants such as water lilies to absorb nutrients, micro-organisms and pathogens.

This is a wonderful option to consider if you have the space for it and do not already have a pond in your garden, since it will provide both aquatic habitat and a place for swimming. Many nature-friendly, wild-swimming gardeners are delighted with them. But they can require a lot of maintenance. The main problem is ensuring the water remains low in nutrients. Run-off of soil or fertiliser into the pond from the surrounding garden during heavy rain creates problems by adding to the nutrients in the water. So does leaf litter, which must be cleared from the water regularly – and it may then be tempting to remove surrounding trees and shrubs. Ducks and geese will also raise nutrient levels, and in some places – as in our walled garden at Knepp – it can be difficult to keep them away.

Obviously, a natural swimming pond cannot be heated without affecting the plants and other aquatic life, so you'll need to be happy with swimming in seasonal water temperatures. Not heating a pool saves a considerable amount of carbon, unless it runs on sustainable energy.

Providing habitat and food

The rising interest in attracting wildlife into our gardens is a boon for manufacturers and a welcome source of income for conservation charities. A visit to the local garden centre or the online shops of the Royal Society for the Protection of Birds (RSPB) or the National Trust, for example, will present you with an array of bird, bat, dormouse and hedgehog boxes, insect towers and bug hotels, and even frog and toad houses for your pond. But the question a rewilder should ask is why we need these artificial constructs in the first place.

The answer is generally lack of suitable habitat. Most garden birds will nest in brambles, dense foliage, creepers against a wall (both ivy and wisteria provide wonderful cover), thorny shrubs and hedges, where they can find food – insects, seeds and berries – protection from predators, and a suitable, aerated microclimate.

Infrastructure, such as a dry-stone wall, can help generate biodiversity in a garden. The flow of air over and around it, and pockets of heat and frost, create varied conditions for plants and insects.

Microclimates in gardens

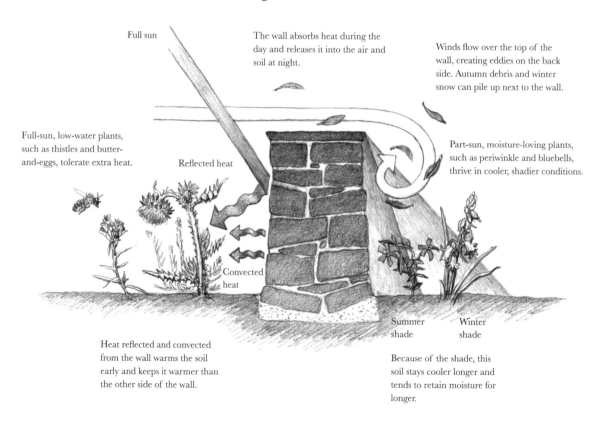

Full sun

The wall absorbs heat during the day and releases it into the air and soil at night.

Winds flow over the top of the wall, creating eddies on the back side. Autumn debris and winter snow can pile up next to the wall.

Full-sun, low-water plants, such as thistles and butter-and-eggs, tolerate extra heat.

Reflected heat

Part-sun, moisture-loving plants, such as periwinkle and bluebells, thrive in cooler, shadier conditions.

Convected heat

Summer shade

Winter shade

Heat reflected and convected from the wall warms the soil early and keeps it warmer than the other side of the wall.

Because of the shade, this soil stays cooler longer and tends to retain moisture for longer.

This is what we should be seeking to supply in the rewilded garden: organic vegetal habitat that also provides food. Manufactured habitats almost invariably fall short of what nature itself provides. Nesting boxes are often inadequately designed and can encourage mites and other pests. Wooden bird boxes can be a honeypot for great spotted woodpeckers, which drill through them to predate on eggs and chicks. Most nesting boxes designed specifically for treecreepers have never been known to be used by that species. It is generally the more common tit species that take up residence in them. Without knowing it, we may be favouring a few species over others, and actually reducing diversity.

Where bird boxes can be helpful, however, is for long-distance summer migrants such as swifts, house martins and swallows, which are currently suffering massive declines and in need of specific habitat. Before buildings, swifts and house martins nested mainly on rock faces, and swallows in caves. Swifts will also nest in old trees, as they still do in northern Europe and remnants of the Caledonian forest in Scotland. The arrival of human-made buildings quite possibly boosted populations of these birds by providing them with alternative habitat. Houses in the past would also often have been associated with livestock, even if just a house cow, pig and chickens, and would have provided a feast of insects, too. And swallows (like barn owls) love diving into the cavernous opening of a barn.

In many cultures, including in the Mediterranean, housebuilding traditions include the provision of nesting sites (often, simply, missing bricks) and ledges for swifts, swallows and house martins, as well as bats. It is still considered good luck in many countries to have birds and bats sharing one's house, perhaps partly because of their ability to keep down flies and mosquitoes.

But modern buildings rarely provide for birds. Architects and designers tend to think of the mess they generate, rather than the blessings they bring, and nooks and crannies are seen as antithetical to today's taste for clean, smooth surfaces and lines. Installing nesting boxes for these birds, which are already challenged by stupendous migrations and the loss of insects and habitat en route, provides a welcome homecoming. Not all swift, swallow and house-martin boxes on the market are up to the job, however, and care must be taken in their positioning. We recommend Peak Boxes for swifts.[58] The charity House Martin Conservation UK & Ireland has a list of recommended suppliers for house-martin boxes.[59] Swallow nesting boxes or cups are available to buy, but in fact this species is often happy with just a ledge to nest on.[60] Far better is to integrate niches for wildlife into the design of buildings. Bats can benefit, too. Many bat species need ancient trees to roost in, and their numbers have declined in parallel with the decline in ancient trees. The Bat Conservation Trust offers advice for developers and householders on how to create bat niches as part of the design of a building.[61]

Birdfeeders are another staple of garden centres. Of course, it is a joy to see garden birds at close range from one's window. About half of households with gardens – around 12.6 million – provide extra food for birds, and the UK public spends around £200 million a year on bird food.[62] But we must be aware that such generosity may have unforeseen consequences. Feeding birds during the breeding season makes it five times more likely that birds' nests near the feeders will be raided by grey squirrels, magpies, jays and other predators.[63] Chicks can choke on fragments of peanuts, so these shouldn't be put out during nesting season. Providing an excess of food in one place, which – in the case of birds – is profligately thrown around the place, also tends to attract mice and rats to the vicinity of the house, which may then need to be controlled with traps or poison.

Providing natural resources for wildlife is a far more effective way of maximising biodiversity in a garden. Leaving seed heads standing rather than deadheading and tidying up at the end of summer provides food for birds, and protects the soil beneath the plants from autumn and winter weather. Native rowan, ivy, hawthorn, blackthorn, honeysuckle, bramble, spindle, holly, elder, wild plum, sea buckthorn and dog rose, as well as exotics such as pyracanthus, Oregon grape, cotoneaster, mulberry and berberis, provide them with wonderful supplies of berries and help to extend the garden habitat.[64] Leaving fallen fruit from apple, pear, quince, medlar and plum trees and so on also provides sugar-rich calories for birds, as well as for insects, small mammals and even badgers and foxes in the run-up to winter.

Many nature-friendly gardeners put out fat balls for birds in winter. Essentially, this mimics the fat once supplied by carcasses in the landscape – a vital resource for wildlife, especially in the depths of winter. Providing winter supplements of this kind for birds in gardens will probably always aid their survival. But leaving out cooking fat and meat juices from the Sunday roast is not the answer. Cooking fat is prone to smearing and can damage the waterproofing and insulating qualities of birds' feathers, while salt is toxic to birds in high quantities, affecting their nervous system. Only pure fats such as lard and suet, which solidify and are less hospitable to bacteria, should be used to make home-made fat balls for birds.

Leaving dead wood in a garden, whether in the form of fallen branches or even a dead tree – if you can do so without it being a safety problem – provides the habitat you would be supplying by buying a bug hotel. A pile of logs left to decay slowly will do the same job. Our obsession with tidying up and generally burning dead wood, not just in parks and gardens but also in the wider landscape, has deprived a vast array of birds, bats and saproxylic (deadwood-loving) fungi and insects of their natural habitat. One of the rarest is the impressive stag beetle, a globally threatened species, and the UK's largest beetle. One of its few remaining hotspots is, perhaps surprisingly, in south and west London, including Richmond Park. The London Wildlife Trust encourages gardeners to retain as much deadwood – logs and stumps with underground roots – as possible on site, preferably some of it in shade, to avoid it drying out.

Likewise, providing a pile of hedgehogs' favoured nesting habitat, fallen leaves and sticks, in a corner will help other animals too. An enclosed box for hedgehogs alone will not do the same job.

What should we be buying for the rewilded garden?

A garden centre can be a source of inspiration, but it can also be a fount of unnecessary – and expensive – distractions, particularly when it comes to rewilding. Many of the products we're exhorted to buy to help wildlife are wholly unnecessary. Consumerism has been one of the main drivers of environmental problems, and continues to undermine efforts to heal the planet. We should be questioning every manufactured product we buy. What is it made from? How has it been made? What will happen to it at the end of its life? Do I actually need it?

It can be difficult to identify when things manufacturers encourage us to believe are necessary are actually pointless or excessive. Remember, manufacturers and retailers are in the business of selling things. They skilfully manipulate our sense of our garden as a benign and joyful space, and our wish to help wildlife. Pesticides, herbicides and artificial fertilisers are promoted in ways that suggest they are fundamentally harmless or even contribute to a healthy or natural environment. Even 'nature-friendly' products, such as bird boxes and plants for pollinating insects – as explained earlier – are not as nature-friendly as they seem. Most are costly in terms of raw materials and the energy used to produce and transport them, when nature can often provide them for free. And some products, such as peat compost, contribute significantly to habitat loss and climate change.[65]

Generally, manufacturers and governments dislike restricting commercial opportunities, preferring to place the onus on consumers to be 'green'. Under the justification of 'consumer choice' harmful, unnecessary and high-carbon products remain on the market, often heavily disguised, and it is left to the consumer to wade through the morass and work out which products it is safe or ethical to buy. Few of us have the time or information necessary to make these choices. In any case, it is debatable whether freedom of choice should be championed over the protection of nature.

Especially once your rewilded garden is established, stay away from the garden centre as much as possible. Avoid the temptations on offer. The less time you spend with a shopping trolley, and the fewer products you buy, the better it will be for your garden and the environment in general. This will also compel you to think creatively about how to source things in another way – such as plant swapping, buying seed online from organic producers, making your own compost, retaining deadwood, allowing plants and woody shrubs to provide habitat and food for wildlife – and to manage with a lighter touch, such as by mowing and strimming less often, selling the leaf-blower and practising 'no-dig' in the vegetable garden.

And, of course, all this will save you money. Indeed, one could argue that spending less is a defining characteristic of rewilding and nature-friendly gardening.

Domestic cats: An unnatural predator

Humans' love of cats is taking a terrible toll on wildlife across the planet. A study in 2013 estimated that domesticated cats kill 1.3–4 billion birds, 6.2–22.3 billion mammals, 258–822 million reptiles and 95–299 million amphibians every year in the United States alone.[66] Indeed, around the world, domestic cats have contributed to the extinction of sixty-three species of vertebrate, most of them birds.[67] In Australia, feral and pet cats together kill a million birds a day.[68]

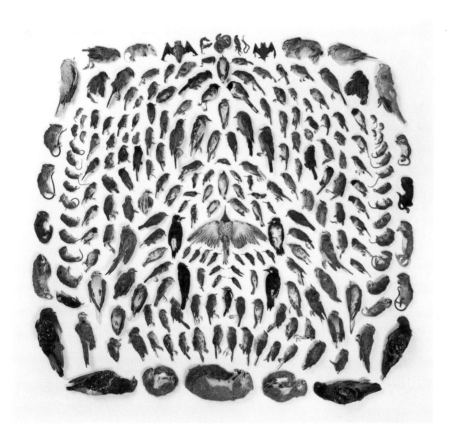

Jak Wonderly's *Caught by Cats*, depicting 232 species killed by house cats in one year in the United States. The photo would need to be multiplied 10 million times to come close to showing the billions of creatures killed by domestic cats each year in the US. In 2020, a cropped version won first place in the BigPicture Natural World Photography Competition, in the Human/ Nature category.[69]

In the UK, the population of house cats rose from 7.5 million in 2019–20 to 12.2 million in 2020–21 as people got cats during the COVID-19 lockdown.[70] Not all of these will remain attached to households throughout their lifetime. Cats are not restricted by the same roaming regulations as livestock or dogs, and owners are not legally responsible for their behaviour.[71] We have come to accept that cats are a law unto themselves and have a right to roam and even become feral. Around a million cats live in a feral or semi-wild state in rural areas in the UK, and a further unknown number of cats have only loose associations with domestic households in urban areas.[72]

Domestic cats are by far the most abundant carnivores in the UK. They catch up to 100 million prey items in the UK over spring and summer alone, of which 27 million are birds. In fact, they catch far more, since these numbers are only the prey that was brought home and counted.[73] Only a fraction of hunted prey – between 10 and 23 per cent – is brought back to the house or farm.[74] Most of the birds caught are common species, such as blue tits and blackbirds, whose populations are artificially supported by birdfeeders. But cats also regularly kill house sparrows and starlings, species that have declined dramatically in the UK in recent years. By far the largest proportion of prey killed by cats in the UK (around 69 per cent), however, are small mammals. Cats kill about twenty species, including mice, voles, shrews, squirrels, stoats, weasels, rabbits and bats. They also bring home four species of reptile, three amphibians and even some invertebrates, such as butterflies and dragonflies.[75]

A cats' impact extends further than the species it actually kills. The mere appearance or scent of a cat can increase the stress responses in prey species. This, in turn, influences the foraging behaviour, energy, body condition and reproductive ability of these species, and increases their vulnerability to other predators. Even briefly confronting blackbirds with a taxidermied cat near their nest, for example, has been shown to reduce the feeding of their young by a third, and significantly increased the risk of subsequent predation of the nest by corvids (crows and related birds).[76] Another indirect impact is competition for food; a mouse eaten or taken home by a cat, for example, cannot be eaten by a hawk.

The species we have as pets (*Felis catus*), which is thought to have been first domesticated in the Near East around 7,500 BC,[77] is distinct from the UK's native species of wildcat (*F. silvestris*), which is currently clinging on in remote reaches of Scotland. Domestic cats devastate wildlife largely because we feed, shelter and nurse them while allowing them to come and go freely. Cats like killing, irrespective of whether they are hungry or not, and domestic cats are enabled by lack of competition and natural enemies. In the wild, their natural predators would be wolves, lynx and wildcat.

Most domestic cats don't stray far from home. A six-year study tracking 925 pet cats from six European countries identified small home ranges (between 3.6 and 5.6 hectares).[78] Only three cats ranged further than 1 square kilometre. However, this means a cat will hunt in its own garden and/or those of the neighbours.

If you're serious about encouraging wildlife, not having a cat is one of the most positive actions you can take. If you have a cat already, keeping it indoors – especially around sunrise and sunset – is an effective solution. Using a brightly coloured 'Birdsbesafe' collar or attaching a bell to an existing collar is thought to improve the chances of escape for prey species by about 50 per cent.[79] Most effective, though, in reducing a cat's propensity to kill, is interactive play indoors. Throwing catnip balls or encouraging a cat to chase a toy or laser light around the house – even for just fifteen minutes a day – can help to satisfy its hunting instincts.[80]

Dogs also have an impact on wildlife. They often chase or kill birds, small mammals, and even deer and sheep – even when they are on the lead. They scare wildlife, too. On small nature reserves, just the presence of dogs can cause a 40 per cent reduction in bird species across the whole reserve.[81] The same will happen in a garden. Your dog may not be catching anything, but, by its very presence, it will be preventing birds and small mammals from establishing in the first place, or frightening them away from feeding their young.

Beekeeping: To bee or not to bee

Keeping bees is often considered positive for the nature-friendly garden. There are an estimated 274,000 honeybee hives in the UK, most of which are kept by about 44,000 amateur beekeepers. Fuelled by media coverage of catastrophic bee declines and campaigns trumpeting honeybees' role as pollinators, the craze for beekeeping has taken off, even in cities. London's bee population tripled in the 2010s to around 5,500 registered hives – the densest of any city in Europe.[82] There are even companies in London that, for a fee, will install beehives in your garden and maintain them for you, giving you a cut of the honey.

However, to help bees in this way is almost certainly counterproductive. The spread of the European honeybee around the world is threatening local biodiversity, and beekeeping is now unsustainable in many cities and threatening other native bee species.[83]

There are about 270 species of bee in the UK, most of which are in sharp decline.[84] Of those species, 250 are solitary bees, many of them specialists, feeding on the pollen and nectar of just one or a few specific plants. Honeybees, in contrast, are generalists. When kept domestically and sustained by supplementary feeding they outcompete other species, monopolising flowers. Beekeeping is essentially farming, and, just as when livestock is kept in concentrated and artificial spaces, hive bees are particularly vulnerable to parasites and diseases, which can spread to wild populations.

In April 2021 the high-street chain Marks & Spencer faced a backlash from conservationists when it announced plans to establish 1,000 beehives on twenty-five farms as part of its five-year Farming with Nature programme.[85] According to ecologists and groups such as Buglife and the Bumblebee Conservation Trust, releasing 30 million honeybees into the British countryside and calling it 'good for the environment' is 'horribly wrong'. Saturating the landscape with honeybees puts enormous pressure on wild pollinators. Far better, in the opinion of these conservation groups, would be to spend the money on restoring native habitat, so as to provide sources of nectar and pollen for native bees and other insects across the board.

Contrary to popular belief, honeybees are not even the best pollinators. Many other insects, including solitary bees, hoverflies and night-flying moths, are considerably more efficient at pollinating. Tame bees are used as pollinators in agriculture because, as a domesticated species, we can transport them by the lorry-load from crop to crop. This practice (which is also very stressful for the honeybees themselves) is now common in the USA, where wild insect populations have declined so drastically that they can no longer be relied on to do the job.[86]

This is an important consideration for both nature-friendly gardening and larger-scale rewilding. Supporting one or a few species over others affects general biodiversity, as we have seen in conventional nature conservation and on sporting estates. At Knepp we have twenty-five hives in the rewilding project, from which a local apiarist produces honey. Any more and we jeopardise the sixty-two species of native bee that are present on our site, including rare specialists such as the red bartsia bee, the rough-backed blood bee and the yellow loosestrife bee.

The best way to help wild honeybees is not by capturing them and farming them for honey, but by providing the habitat that is, thanks to human intervention, now missing in the wider landscape. This is the factor that has contributed significantly to their decline.

A wild honeybee colony requires a cavity with a volume of between 35 and 50 litres that is also long enough (about 80 centimetres) to form honeycomb. In nature, this is provided by large, standing dead or hollow trees. But over the past two centuries, with the advent of mechanised tools, we have removed such trees from the landscape with manic efficiency. Wild bees have been forced instead to seek out domestic and industrial roof space, usually with disastrous results. Leaving dead trees standing provides wild bees with habitat again. Ash trees die back particularly quickly, and already at Knepp we are finding wild bees colonising the hollowed trunks of ash that have been killed by ash dieback disease.

Matt Somerville's Bee Kind Hive provides the habitat of a hollow tree trunk for a wild bee colony in Knepp's rewilded walled garden.

But dead standing trees can be dangerous, and will not generally be welcome in a garden. Bee Kind Hives, which are made from hollowed-out logs and carefully designed for the welfare of the bees, provide an alternative.[87]

Once in the wild, honeybees naturally revert to a darker colour and, under the influence of natural selection, swiftly adapt to local conditions. Knepp now has dozens of wild honeybee colonies in the rewilding project, as well as one in the rewilded walled garden and one in the orchard. Wild honeybees are known to be healthier and more resilient than farmed populations. Their reproduction rate is slower, mediated by the natural availability of food. In bad weather, they simply wait it out. Providing habitat for wild honeybees helps to bolster the genetic diversity of native populations, a factor that is key for the survival of the honeybee in the face of climate change and disease, without skewing the balance against other bees and insects.

Urban gardens may not be the best place for a Bee Kind Hive, given the existing burden of domestic bees in many towns and cities. But in rural gardens, orchards and even nature-friendly cemeteries and graveyards, providing habitat for wild bees, where they can do their thing without intervention, is a fascinating exercise.

Light pollution

The impact of artificial light on wildlife has been largely overlooked. A rewilded garden should, wherever possible, have only natural light. Our passion for illuminating gardens artificially at night has produced a dazzling array of commercial products from porch and outdoor wall lights, floodlights, spotlights and uplighting for trees to lamp-post and bollard lights for the garden path, tile and brick lights for the garden wall, ground-level and deck lights, hanging lanterns, and fairy and festoon lights.

This has a big impact on wildlife.[88] Nocturnal insects, including many moths, navigate by the moon. They can become disorientated by or positively attracted to artificial light, wasting their energy and often dying if they cannot draw themselves away. Many bats avoid artificial light altogether, and those that hunt the insects that are magnetically attracted to lights can be put at risk of predation themselves by domestic cats. Sensor-driven security lights that burst on at night can stimulate garden birds, particularly robins, to start to sing or even feed. Some species, such as frogs, may be temporarily blinded by lights, and owls find lit areas much harder to hunt in.

Light pollution can have an even more disruptive effect on the breeding of nocturnal species, such as glow-worms, whose numbers declined by around 75 per cent in the UK between 2001 and 2018 largely, it is thought, because of artificial light.[89] Female glow-worms (which are actually flightless bioluminescent beetles) emit a dim green light from the tip of their abdomens to attract mates. Rather than moving away from artificial light, a study found that many females remained in the same location but simply stopped glowing, significantly reducing their chances of attracting a mate. Even distant light sources can disrupt the dispersal and life cycle of these fascinating insects that were once common in British gardens.

It seems we have forgotten how magical gardens are without artificial light, and how many nights in a month the moon is out. Do we really need lighting in our gardens? Of course, it may be impossible to avoid artificial light if your garden is affected by street lighting. But you may be able to persuade your local council to adopt a switch-off scheme. Even changing from full lighting at night to partial lighting reduces the effect on the behaviour of moths. And light pollution, we are beginning to understand, may well negatively affect people as well.[90]

If you do need lights in specific places in your garden, there are things you can do to minimise the impact: switching them off when they're not in use; using low-intensity (preferably solar-powered) lights with warmer white, yellow or amber hues that are less likely to affect wildlife (coloured solar lights seem to attract and confuse glow-worms); positioning them as low as possible; and using hoods to direct the light downwards. Best of all is to relish the tranquillity and subtlety of darkness and make do with portable solar-charged or candle lanterns that you can take out into the garden as and when you need them.

Threats to garden space and habitat

One of the biggest threats to nature in the garden is the car. Hard surfacing over gardens for parking has become the norm in British towns. A study of London gardens found a 12 per cent drop in vegetation cover in the years between 1998 and 2008.[91] Across London, the total area of paving in front gardens is estimated to be 32 square kilometres, the equivalent of twenty-two Hyde Parks.[92] Most of this has been lost to car-parking, particularly in older, high-density areas where garages were not part of the original design. And even where garages were provided, most are not used for parking. By 2006, nearly half of all front gardens in northeastern England, and just under a third in Scotland, southwestern England and eastern England, were at least 75 per cent paved over.[93]

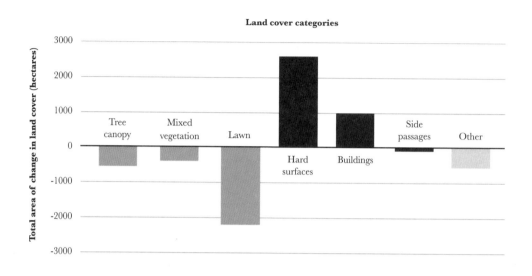

Land cover categories

Total area of change in land cover (hectares)

From green to grey: gardens everywhere are being built on and paved over. This table for the period 1998–2008 shows a dramatic loss of vegetation in London gardens to hard surfacing such as patios, decking, parking areas, sheds and greenhouses.

Patios and decking are also to blame. Losing soil and greenery to hard surfaces dramatically increases surface water run-off, increasing the risk of flooding.[94] Over the course of the present century, rainfall will become more intense, more frequent and more variable.[95] Drilling up hardcore, lifting decking and returning these areas to garden once again reduces flood risk, stores carbon and returns habitat for wildlife. Car-sharing schemes and good urban transport – allowing people to give up the car (or at least reduce the number they have) – is crucial.

As we have seen, gardens are getting smaller, thanks largely to developers, who pack as many houses on to one site as they can. But there is also the problem of turnover.[96] The average private rental of a house is just a year. Even owner-occupiers remain in residence for an average of only eleven years, moving house four times after buying their first property.[97] Generally, every time a new owner or renter takes over, the garden receives a makeover. Everyone has different taste, and fashions for plants change, too. A gardener with a passion for alpine plants may be supplanted by an enthusiastic grower of vegetables. Ponds may be removed for the safety of children (see page 439); wonky wooden garden gates replaced with automatic metal driveway gates; trees, beloved of the previous occupant, considered a nuisance by the new.

The constant disruption of starting a garden from scratch over and over again affects both plant and animal communities. The older a garden is, the more likely it is to contain rarer species of invertebrates that have made their home in old walls, timbers, gates and sheds. Replacing old brick, stone and wooden infrastructure can destroy habitat that is almost impossible to replicate.

A tree in a front garden in Waterthorpe, Sheffield, was cut in half, following a row about pigeons roosting in the tree and dirtying the driveway.[98]

Monitoring and surveying

Recording the species that are present in your garden is great fun. It is even more rewarding if you can establish a baseline before you begin changing your management practices, so that you can identify any increase in biodiversity. You may be able to elicit the help of volunteers or consultant ecologists, but you can also teach yourself. The Pan-Species Listing website explains how to compile a list of species if you're starting from scratch, and how to make a database of all your records if you're a more experienced naturalist.[99] See Chapter 9 for more on monitoring and surveying.

You may prefer to focus on just one group, such as birds, butterflies or bees. During the COVID-19 lockdown in 2020–21 the filmmaker Martin Dohrn filmed sixty bee species in his urban garden in Bristol.[100] Or you could home in on a particular patch of ground. The square metre project, started by the ecologist Dr Patrick Roper, has inspired many gardeners to tune in to the astonishing amount of life that can exist in a tiny area.[101] Roper began a minimum-intervention study of a single square metre of his garden in Sedlescombe, near Hastings, East Sussex, in September 2003. By April 2016 he had clocked up over 1,000 species. Half the area is managed as a miniature meadow, with Roper hand-plucking the grass, imitating herbivory. The other half has taller vegetation bordering shrubs. There is even a miniature pond, no bigger than his hand. Alongside woodlice, grasshoppers, spiders and ants, he has found nationally rare invertebrates and microfungi, as well as protected species such as slow-worms and common lizards.[102]

Roper has since added numerous other square, cube and circle studies of his garden. His window box, started with sterile soil in November 2005, has now naturally rewilded over a period of more than ten years. It has been colonised by goat willow, birch, hazel and yew saplings, as well as mosses, rushes, willowherb, hairy tare, white clover, tutsan and ragwort – a plant community that completely defies the National Vegetation Classification. The plants continue to thrive, despite being only ever supplied with water when necessary, and Roper has counted over 200 species of fauna in the window box.

Rewilding schools

We can apply the principles of rewilding a garden to other small spaces, public and private. Cemeteries, churchyards, supermarket car parks, industrial and housing estates can all benefit from rewilding, adding to the habitat our wildlife needs, becoming part of the living network of nature and providing ecosystem services for human beings. But arguably the most vital of all are school grounds.

There is clearly a huge need to connect children with nature. Just a generation ago, 40 per cent of children still played regularly in natural areas. Today, only 10 per cent do. Some 40 per cent of children never play outdoors at all. The loss of nature from our children's lives

– the 'extinction of experience' – is reflected in changes to the *Oxford Junior Dictionary*. In 2012 it replaced 'acorn', 'buttercup' and 'conker' with 'attachment', 'blog' and 'chat room'.[103] Catkin, cauliflower, chestnut, clover, heron, herring, kingfisher, lark, leopard, lobster, magpie, minnow, mussel, newt, otter, ox, oyster and panther have also been dropped.

Surprisingly, perhaps, the percentage of the population living in urban areas (80 per cent) is the same as it was in the 1950s. It's nature that has disappeared. It is now much harder for children, particularly in cities, to find green space to explore. And adults, who themselves have grown up without nature, are more fearful about letting their children play freely in it when they do find it.

The benefits of nature for children are enormous, however, and fundamental to their development. Evidence shows that if children have access to nature – whether it's countryside, parks or gardens – they are healthier, physically fitter and better adjusted, and their behaviour and schoolwork improve.[104] Impulsive behaviour, aggression, hyperactivity, inattention and low levels of self-discipline all improve through contact with nature. Studies of children who were being bullied, punished or relocated, or suffering from family strife, all showed that they benefited from closeness to nature, in levels of both stress and self-worth.[105]

We need to think, too, about where our future ambassadors for conservation will come from. Children who spend time in green spaces between the ages of seven and twelve tend to think of nature as magical. They are most likely, as adults, to want to protect it. Those who have no such experience tend to regard nature as hostile or irrelevant, and are indifferent to its loss.

The not-for-profit organisation Operation Future Hope has launched the remarkable Rewilding Schools initiative. Any teacher in the UK can sign up on its website to become a rewilding school. This gives access to an excellent manual on how to rewild school grounds, from identifying existing bad practice (such as the routine use of glyphosate to keep down weeds) to establishing 'hedges and edges', such as species-rich hedgerows for wildlife, wide grass margins and native wildflower strips around playing fields, and orchards. It explains how to map, how to monitor habitats and species, and how to identify opportunities to connect with green space beyond the school grounds. There's information about how rewilding reduces maintenance costs, and how to source trees, plants and wildflower and hedgerow seed mixes for free. And there are plenty of ideas for schools with limited space: among them thinking vertically (growing ivy and other native climbers up buildings and over fencing), creating mini-meadows and hedgerows in tiny spaces, and making mini orchards (apple and cherry trees in planters).

Operation Future Hope also runs Rewilding Apprenticeships from its online rewilding education centre. A course of thirty-plus lessons designed for all age groups uses case studies and video clips from around the world to teach about all aspects of rewilding, including its benefits for mental health. It can be timetabled into lessons, used as part of the curriculum in preparation for the new natural-history GCSE, or form an after-school club. It can also include field trips to rewilding sites within the UK.

Rewilding the Chelsea Flower Show

Whether you're rewilding your garden, schoolyard, churchyard or roadside verge, providing space for nature is largely about changing practices that we have adopted, often unthinkingly. It's about questioning our aesthetic sensibilities, asking ourselves what we really consider beautiful, and why. It can be discomfiting to dig into our deep-rooted attachment to particular ways of doing things, especially if nostalgia and a sense of cultural identity bind us to them. Much of rewilding is about letting go, not just practically, but also psychologically.

It can be done, however. Nowhere was this more evident than in that bastion of horticultural convention and neat-freakery, the RHS Chelsea Flower Show, in 2022. Among the stands of orderly flower beds, decking kits and pristine lawns, and the explosions of gladioli and dahlias in the Floral Marquee, one garden stood out from all the rest. Sponsored by the charitable organisation Rewilding Britain, the Somerset-based garden designers Lulu Urquhart and Adam Hunt created a garden around a beaver dam. The dam was made from sticks gnawed by real beavers, and held back a pool of water with rivulets trickling around it. A grove of sallow, hawthorn, hazel and alder shaded a riverside wetland of ferns, kingcups, greater tussock sedge, marsh valerian, marsh helleborine, devil's bit scabious and marsh thistle, mingling in managed disorder with grasses, last year's dead foliage and seed heads. By the second week of the show this temporary habitat – in the middle of London – had attracted birds and scores of insects, including butterflies, dragonflies and solitary bees.

The garden certainly divided opinion. The *Daily Mail* described it as 'a ramshackle plot featuring … a pile of sticks'. Despite calling it 'a brilliant show garden', celebrity gardener Monty Don said 'rewilding is a really unhelpful term when it comes to gardening'. Others were moved and entranced – including the show's judges. Defying the odds, 'A Rewilding Britain Landscape' claimed the prestigious award for Best in Show.

A beaver garden designed by Lulu Urquhart and Adam Hunt, with a leaky dam of beaver-gnawed sticks and native bog plants, rocked the 2022 Chelsea Flower Show when it won the Best in Show award. Even in the UK's bastion of conventional horticulture, it seems, an appetite for wildness in our gardens is growing.

12

Urban Rewilding

Bringing nature back into our cities

With tolerance and imagination, our cities and their surroundings could be thronging with wildlife. Greening the grey and learning to share our urban space with animals great and small would not only bring joy, it would also transform the health of our cities.

The tiny 330-hectare Kraansvlak nature reserve, overlooked by the beach resort town of Zandvoort in the Netherlands, supports a herd of free-roaming bison, one of which can be seen here. The Kraansvlaak's 4-kilometre Bison Trail is a hugely popular local walk for city-dwellers.

Cities may seem unlikely places for rewilding, but they have enormous potential for nature and a vital role to play in tackling climate change and the loss of biodiversity. It is now the case that 55 per cent of the world's population live in cities.[1] While cities cover less than 2 per cent of the Earth's surface, they are responsible for 60 per cent of the world's carbon emissions and use 78 per cent of its energy.[2]

In the UK, urbanisation is even higher than the global average, with 84 per cent of the population currently living in cities and towns. The population of London is projected to reach 10 million by 2035.[3] This concentration of energy consumption in cities is, itself, an opportunity. Rethinking the design and management of cities is fundamental in the drive to address climate change. Technology will play an important role in providing alternative energy and energy efficiency, but the greening of urban areas can also contribute by providing nature-based solutions to energy use. By absorbing heat from the streets, trees and vegetation reduce the need for air conditioning, and clean the air.[4] Green space and rewilded river systems help to store and purify water, and mitigate against flooding, which causes millions of pounds' worth of damage to infrastructure every year. Gardens and allotments can contribute significantly to the provision of sustainable food, and, of course, trees, soil and vegetation sequester carbon.[5]

Good health is generally framed, quite correctly, as a fundamental human right. But one might also express it as indicative of a healthy planet and integral to the fight against climate change. Ill health is costly not only in terms of human happiness and prosperity, but also in terms of the environment and energy. The health-related effects of urbanisation include mental, chronic and heat- and pollution-related health problems, as well as infectious diseases, violence and road-traffic injuries.[6] Increasing nature in cities is proven to mitigate many of these health problems, including reducing allergies, asthma, cardiovascular disease and depression, improving cognitive ability and even reducing crime and violence.[7] If every household in England had access to good-quality green space, £2.1 billion could be saved every year in averted healthcare costs.[8] Access to green space for children is also vitally important for their cognitive development.[9]

Progressive cities are already experimenting with new ways of increasing green space: planting trees along streets and roads, creating inner-city nature reserves, growing plants on the walls and roofs of buildings, and converting manicured civic plant beds with bare soil to continuous wildflower cover. One might think that this is where it ends; that cities, being by definition human constructs, can never move beyond the realm of intensive human management: we may improve the urban environment, greening up the grey, but cities seem unlikely contenders for rewilding, dominated as they are by immutable buildings and roads, divorced from functioning ecosystems and the influence of natural processes.

But a rewilding approach is possible, even in a huge area of concrete. Thinking holistically, in terms of the city as a living organism, with a pulse and arteries, brain, lungs, kidneys, a digestive system and heart, muscles, limbs, nerves, sinew and skin, can bring about a different approach to urban ecology. It can conjure up opportunities that would seem impossible if we restricted ourselves to the view of the pavement beneath our feet and the buildings rising up around us.

The idea of the city as organism is not new. Thinkers as chronologically diverse as Plato, Vitruvius and Leonardo da Vinci, modern economists, social reformers and politicians, and architects Joseph Allen Stein, Le Corbusier and Frank Lloyd Wright, have all envisaged the city as an entity with its own metabolism and pathology, with energy and a collective consciousness, and the capacity to react, absorb and reject.[10]

A city with green buildings, rewilded waterways and connected natural habitats would be a far healthier, happier place to live, and would dramatically assist in the reduction of energy use. It would also provide wonderful opportunities for wildlife. Here in Europe, eagle owls might take up residence, subsisting off urban fox cubs and rats, and spoonbills might trawl for aquatic snails and insects in a roof-garden pond. Red kites, a familiar sight in London in the eighteenth century, would be back in city skies, and pine martens would find their niche, predating largely on non-native grey squirrels.

The city is capable of attaining a state of balance, but it can also become sick and dysfunctional. The English urban planner and social reformer Ebenezer Howard was one of the first in modern times to recognise the malaise of overcrowded, industrial housing and the importance of green space for the well-being of residents and the harmonious functioning of a city. He designed the Garden Cities of Letchworth (1903) and Welwyn (1920) to improve the quality of urban life, each with a garden at the centre and extensive parks and avenues radiating from it, and each ring-fenced with a rural belt that could not be built on. His concepts – in particular his emphasis on green-belt areas, restricting urban sprawl and controlling population density – have become an integral part of suburban and city planning in the UK.[11]

But we must think about long-term sustainability as well as human well-being. We must focus not just on new urban design (which is always under pressure from developers to minimise costs and reduce green space), but also on redesigning and retrofitting existing towns and cities so they can cope with declining resources and the climate crisis. The ambition should be to turn the boat around entirely, so that urban areas begin to contribute positively to water storage, alternative energy, carbon sequestration and biodiversity.[12] This requires integrated thinking, long-term goals, vision and political determination. Vishal Mehta, an expert on energy, waste and water resources in Bangalore and watersheds in India and the USA, talks of 'urban metabolic mapping', of 'securing the biophysical foundation of cities'.[13]

Many of the approaches that can help achieve this are common to rewilding. We can create corridors and connected habitats to enable a flow of wildlife and natural processes between urban areas and countryside. We can increase blue (water) as well as green space in cities, using nature-based solutions to address rising temperatures, flooding and drought in urban areas. And we can allow the evolution of novel ecosystems and new communities of species, including – where they don't pose a problem – non-natives. We can even consider introducing large herbivores, such as sheep, goats, ponies, cattle and llamas, into urban parks, and also keystone species. The pair of beavers introduced into a 60,000-square metre enclosure in Enfield, north London, in 2022 will, it is hoped, help restore biodiversity and mitigate the impact of flooding in the area (see page 472).[14]

Wandering livestock were a common sight in urban areas in the UK until the 1990s. This cow was photographed in Wanstead High Street in east London in 2008.

Even apex predators are a possibility. Mountain lions in Los Angeles, leopards in Mumbai, black bears in Seattle and hyenas in the medieval city of Harar in Ethiopia have all learned to coexist with humans in the urban environment, sometimes with unexpected benefits.[15] The leopards in Mumbai predate on stray dogs, reducing the prevalence of rabies. In Seattle, mountain lion, black bears and coyotes regulate populations of raccoons, opossums, skunks and rodents, and this has had a positive impact on native birds.[16] In Europe, wolves, bears and lynx are on the rise, thanks to legal protection and growing public tolerance. Germany now has more than a hundred packs of wolves, which can sometimes be spotted passing through the suburbs. They can pose a threat to domestic cats and dogs, but most urban residents are relaxed about the occasional visit. It may be some time before carnivores are sanctioned in urban areas in the UK – the first missing apex predator is yet to be returned to Britain – but birds of prey, such as peregrine falcons and sparrowhawks, create a landscape of fear for urban pigeons and gulls. This helps control their numbers, saving buildings from the corrosive effect of guano and preventing the spread of disease.[17]

As always, a shift in mindset will be required for urban rewilding to happen: challenging cultural norms and embracing a different aesthetic, and allowing natural processes, including natural hydrology, predation, herbivory and dynamic, 'messy' habitats, the space to perform. Seattle, Vancouver, Sydney, Auckland and Wellington, among other cities, recognise that the 'biophysical foundation' of future urban areas must be rooted in deep ecological understanding. Knowing the ecosystems that existed before cities were born can, like rewilding in the wider landscape, point the way to ecosystems of the future. The knowledge of indigenous communities has helped these cities develop a new and closer relationship with nature, inspiring citizens to cherish life other than just their own.[18] A wonderful example was set in April 2020 by the mayor of Curridabat, an urban suburb of San José in Costa Rica, who passed a motion recognising local wildlife, including every bee, bat and tree, as a citizen of the town.

Green belts

An important step in the rewilding of towns and cities is to
reconnect them with the wider landscape. Since the Industrial
Revolution cities have become increasingly ghettoised. Once, they
were permeable to livestock, with green corridors for the passage
of animals and commons for grazing. These arteries have gradually
become blocked, stifling the potential for nature in cities and physically distancing urban populations from the countryside.[19] We saw
how, under the travel restrictions of the COVID pandemic, urban
communities felt particularly starved of nature.

In the UK, green belts can help address this problem. A product
of planning laws from the 1930s and 1940s, green belts were originally
designed to contain urban sprawl. To date, about 1.6 million hectares
have been designated green belt, representing about 12.5 per cent of
the area of England.[20] Of the fourteen green-belt designations, four
encompass England's largest metropolitan areas: London, Liverpool
and Manchester, Sheffield and Leeds, and Birmingham. But the
remaining ten, which surround smaller cities, are substantially smaller.
Many cities, including Leicester (population 300,000), don't have
a green belt at all. There is clearly enormous scope for increasing
green-belt designations. Just under 3 million hectares of land – 22 per
cent of the area of England – consists of countryside within 5 kilometres of towns and cities with populations greater than 100,000.[21]

Green belts were also heralded as buffers that would provide
leisure amenities for urban populations and improve air quality in
cities. However, this potential has never been fully realised. There has
never been a directive to encourage land use in the green belt that
is specifically positive for nature. Much green belt, as Friends of the
Earth explains, 'consists of conventional agriculture, scattered with
golf courses and the occasional Country Park. Just 13.6 per cent of
Green Belt land in England comprises broadleaf and mixed woodland – barely higher than the UK average.'[22]

Rewilding green belt and creating new green-belt zones around
cities that have none would dramatically improve the quality of life
for urban residents and bring enormous cost benefits. In 2015 the
Natural Capital Committee, the independent body that previously
advised the UK government on protecting and improving natural
resources, estimated that planting 250,000 hectares of woodland close

to urban centres in Britain would result in a net economic benefit of nearly £550 million in recreation and carbon sequestration.[23] If this new woodland was achieved through natural regeneration the gains would be even higher, especially in terms of biodiversity. It would also improve air quality. In 2010 the cost of poor air quality in the UK equated roughly to 5 per cent of GDP. The environmental law charity ClientEarth has successfully sued the UK government on three occasions since 2014 for breaking national and international law by continuing to allow harmful levels of air pollution in its towns and cities.[24] In 2021 thirty-three out of forty-three reporting zones, including Greater London, the West Midlands, North East, South East, South West, North Wales and Central Scotland, charted levels of nitrogen dioxide pollution above legal limits.[25] Air pollution, primarily from major roads and in cities, causes 29,000 deaths a year, and costs the UK between £9 and £19 billion a year, far more than the health and economic impact of obesity.[26] Creating functioning areas of wetland upstream of major towns and cities would also help prevent flooding, as well as improving biodiversity and providing additional space for recreation.

In the Netherlands, one of the most densely populated countries in Europe, nature reserves are squeezed into every available space. Here in Slikken van de Heen, part of Krammer-Volkerak, a Natura 2000 site, a herd of bison wade through wetlands on a major canal route between Rotterdam and Antwerp.

The move to rewild green belt has been sluggish, though, and these strategic buffer zones are coming under increasing pressure from development. Indeed, a rising percentage have been removed from classification so that houses can be built on them.[27] In 2020 the countryside charity, CPRE (formerly the Campaign to Protect Rural

England), launched a policy document calling for the green-belt policy to have more teeth, for local plans to prioritise brownfield over greenfield development, and for new green-belt designations.[28] In the same year, the Wildlife Trusts launched its 'Wildbelt' campaign, concerned that the government promotes housing and development growth over nature recovery, and so seeks to dismantle the green belt.[29] It has called for the 'rewilding of the planning system', to ensure that nature recovery is integrated into the planning process.[30] As part of its ambition to improve 30 per cent of land and seas for nature by 2030, the Wildlife Trusts seeks to secure protection for 500,000 hectares of land on the outskirts of cities, as well as slivers of habitat within cities themselves that are not currently under conservation management. Friends of the Earth has joined the fray, calling for a rewilding 'incubator': a community-led rewilding pilot on land close to where people live, that would increase demand for rewilding near cities and showcase the potential of a 'wildly different green belt'.[31]

The question is how to unlock green-belt land for rewilding. At present, the two sides pitching for control of green belt are intransigent: those who believe the land is ripe for development against those who consider it sacrosanct as green space. Many owners of green-belt land are holding out for the windfall of development money, and will be reluctant to surrender it for rewilding. The most pragmatic solution may be to release, say, 10 per cent of green-belt land for housing on the condition that 90 per cent returns to nature. As we discuss in the funding section of Chapter 8 (pages 338–48), the American easement system could assist here. The public purse and/or the developer pays the landowner a capital sum to surrender in perpetuity the right to develop the land and, instead, manage it for nature. This would be a payment for 'public goods' – in other words, in the interest of the general public. Those with an interest in securing space for nature can pressure government to rewild green belt by holding events to imagine, with artistic impressions and (CGI) computer-generated images, how the surroundings of their local town or city should look.

Establishing rewilding sites and nature reserves in green-belt areas around cities throughout the UK will be crucial for raising ambitions and trumpeting the benefits of access to nature to urban residents and policymakers alike. One example is Gloucester City Council's hugely popular Alney Island nature reserve, sandwiched between two strands of the river Severn less than a kilometre from the city centre, and dissected by a busy road. The 32 hectares of flood meadows, which lie a stone's throw from a busy retail park and flow around a transformer station and an electricity substation, are managed with rare-breed cattle. They host orchids, waterfowl, raptors, amphibians, wading birds and many species of insect, including dragonflies and butterflies. Four long-distance footpaths and two cycle routes pass through the reserve.

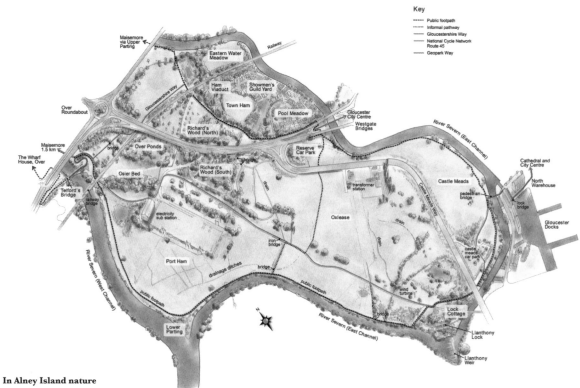

In Alney Island nature reserve, less than a kilometre from Gloucester city centre, flower-rich flood meadows grazed by free-roaming cattle flow around an electricity substation and busy retail park, providing stunning walking and cycle routes for city residents.

Rewilding urban rivers and waterways

Most cities are built beside rivers. The alluvial soils of floodplains originally provided rich, productive farmland, and the river transported goods before trains and modern roads. As cities have grown, much of the original water system has been transformed. Wetlands have been drained for development and tributaries either canalised or driven underground into conduits, storm drains and sewers. Streams and wetlands often live on in name only. In London, the tributaries of the river Thames are immortalised in the place names of Westbourne Park, Holborn, Kilburn, Fleet Street,

Bayswater, Stamford Brook, Knightsbridge, Deptford, Brentford, Mortlake, Shoreditch and Marylebone (originally 'St Mary by the bourne'), even if the rivers are no longer remembered.

About 70 per cent of the total 600 kilometres of river tributaries in London today flow through culverts or concrete/steel channels, most of them underground. This devastates habitats and the ability of species to survive extreme weather events, such as flooding. The challenge for cities, now, is similar to that in the rest of the countryside: to reverse the confinement of natural water systems brought about by heavy drainage engineering, and shift the idea that rivers are an environmental hazard and water must be got rid of as quickly as possible. In London much of the hydrological engineering was undertaken by the Victorians. They solved the considerable problems of sewage, effluence, stink and disease. But with climate change, cities are becoming vulnerable to extreme weather events. Flash floods can quickly overwhelm the system. Just as on agricultural land, efforts are now being made to slow the flow by renaturalising rivers and streams, returning meanders, removing culverts and weirs, and restoring riparian vegetation and floodplains wherever possible. Obviously, this has additional benefits, such as improving water quality, wildlife habitat and quality of life for city residents.

Most of London's rivers, tributaries of the Thames, are now buried as sewers, but their memory survives above ground in district and street names, providing inspiration for rewilding projects in which stretches of lost rivers are being brought to the surface again.

As with any river restoration project, a whole-catchment approach is needed, from identifying habitats upstream, outside the city area, that can absorb rainfall as a first defence, to improving habitat and blue space within the city itself to increase water absorption, and allowing space for the tidal functioning of the river-mouth or estuary. However, the urban environment heaps up challenges for river restoration. Land tenure and management control can be complex in cities. The river may be owned by one body but managed by another, and the land surrounding it may be owned by numerous others. And, of course, the physical challenges for river restoration in a built-up area are far greater than in a rural setting. But the very density, energy and creativity of urban populations can also work to a project's advantage.

In London, roughly 39 kilometres of river – around 6.5 per cent of the total length of the Thames, excluding the tidal section – have been restored since 2000, and this is set to reach 20 per cent by 2050.[33] Much of this has been instigated and carried out by local communities.[34]

Operation Kingfisher is a particularly successful example of a community-led river restoration scheme. In 1994 the Quaggy Waterways Action Group formulated its own flood-prevention plan to restore sections of the river Quaggy running through the boroughs of Bromley, Greenwich and Lewisham in south London. In 2002, with the Environment Agency, concrete channelling 300 metres long was removed at Chinbrook Meadows, and natural meanders and a small floodplain restored. Between 2002 and 2004 a previously buried stretch of the Quaggy was 'daylighted' in Sutcliffe Park in Eltham and a section of the park lowered to re-create a floodplain. This award-winning project featuring a lake, ponds, meadows and reed beds has created one of London's finest nature reserves, where you can now see dragonflies and damselflies, herons, little egrets, king-fishers and reed warblers.[35] The number of people visiting the park specifically for wildlife rose from 2 per cent before the improvements to 47 per cent afterwards.[36]

A previously buried stretch of the river Quaggy, a tributary of the Thames, has been 'daylighted' in Sutcliffe Park, Eltham, and connected with its old floodplain. Visitor numbers have rocketed since its restoration.

Where streams and rivers cannot be brought back to the surface, other interventions can mitigate flood risk, and the presence – or even absence – of a river can be a catalyst for change by capturing the public imagination. London Wildlife Trust has been working closely with local communities in Herne Hill, Loughborough Junction and Brixton on the Lost Effra project, creating rain gardens and living roofs, and de-paving hard surfaces in the catchment of the 'lost' river Effra, an area of south London that is particularly prone to flooding.

At the same time, however, developers often fight river restoration strategies. A £5 billion mega-theme park, the London Resort, is still planned on the Swanscombe Peninsula in Kent, despite this being a vital natural flood-mitigation site for the Thames estuary. Its many rare species – including water voles, cuckoos, nightingales, black red-starts, scarce plants such as man orchids and more than 2,000 insect species (250 of them of conservation concern, including the critically endangered jumping spider) – have qualified it as a Site of Special Scientific Interest (SSSI).

Species introductions

Charismatic species can inspire the public and ignite catchment-scale thinking. Kingfishers, grey wagtails and Daubenton's bats have become flagship species for London's river restoration projects, increasing in numbers since 2000.[37] But nothing captures the imagination like species reintroductions. While cities may not seem the obvious place for wildlife releases, they can be as galvanising as introductions anywhere else, and potentially draw even greater attention and support.

The small and charismatic water vole has impassioned support in Kingston upon Thames, southwest London. Once populous on the river but not seen there since 2017, this small mammal has become the focus for the community-led restoration of the Hogsmill River.[38] Spearheaded by Citizen Zoo, an organisation dedicated to engaging communities with rewilding and citizen science, the project has involved 350 volunteers carrying out habitat surveys, restoration days and crowdfunding. They have planted more than 1,000 native river plants, and engaged with dog-walkers to explain the project.[39] They released their first cohort of 100 water voles across a 2.5 kilometre section of the river in 2022, and aim to continue releases in the coming years until they have built up a viable population. Inevitably, the reintroduction of even a small creature is a complex undertaking in the urban environment; this one involved the removal of a weir and the integration of numerous management, ownership and vested interests in the river and surrounding buildings. The eradication of non-native American mink, one of the biggest threats to the survival of the water vole, may also be a challenge for London residents less comfortable with the concept of culling. But the discussions conducted around the process itself are helping to raise public awareness of nature and the issues at stake.

In Scotland, the Glasgow Water Vole Project has adopted a plan to protect voles that are already living in the city. They were recently discovered in a wide variety of grassland sites, including parks, road verges and derelict land more than a kilometre from water. The project is raising awareness (particularly of the fact that these creatures are not rats), persuading developers to provide suitable space and habitat for voles at the start of new building projects, and plotting habitat connectivity for the voles using geographic information system (GIS) mapping.[40]

Beavers are the flagship species of rewilding par excellence, and one might think that, programmed as they are for radical ecosystem engineering, there might be no space for them in cities. But they have a remarkable ability to adapt, and – when we have an open and tolerant mind – so do we humans. Beavers live in cities in Canada, the USA and Europe. In Massachusetts, which now has more than 70,000 beavers, 'beaver deceivers' (structures that lower water levels and prevent the blocking by beavers of crucial drains and culverts) are widely used to prevent flooding in urban areas; culling is a solution of last resort. In Munich, Germany's third-largest city with a population of more than 1.4 million people, beavers are broadly tolerated and riverside parklands in new housing developments are designed to accommodate them, with beaver-proof protection around specific trees. A pair of beavers reside in a patch of wet woodland just minutes from the arrivals hall of Munich airport.

Following the popular introduction of beavers in the London Borough of Enfield in 2022, Citizen Zoo, in collaboration with the Beaver Trust, is directing the London Beaver Working Group to investigate other potential sites for beaver establishment in the capital. Hackney Marshes, the London Wetland Centre in Barnes and the Olympic Park in east London all have promising habitat for beavers. But none would be more fitting than the historic Beverley Brook, which flows through Wimbledon, Putney Commons and Richmond Park – a 14.3-kilometre stretch of river being restored by the Royal Parks charity. Its name derives from 'Beaver's Ley', a 'place where beavers live'.[41] At the time of writing these plans – in line with government policy for beavers in England – are for releases into enclosures only.

That beavers are living free and harmoniously in cities elsewhere in the world is largely thanks to successful communication campaigns. When beavers began to recolonise Vancouver, including the Olympic Village, fears about flooding and damage to trees were alleviated by common-sense such as solutions as wrapping the base of selected trees with chicken wire and using beaver deceivers to lower water levels. Vancouver's beavers have since captivated the local population, inspiring art installations and school projects. Interestingly, as public tolerance has risen, signs of aggression in these urban beavers, such as tail-slapping, have greatly reduced.

Unleashing the power of urban parks

Many prominent cities around the world have large parks. Some of the oldest and largest cities have expanded to such an extent that they have engulfed the smaller towns around them, incorporating existing common land and heathland into the patchwork.

Oslo is said to be the greenest capital city in the world, with 72 per cent of its public space made up of parks and gardens.[42] Other cities with a high proportion of green space are Singapore, Sydney, Vienna, Shenzhen and Moscow. London scores highly, too: as much as 47 per cent of the city is identified as green, comprised of 3,000 parks totalling 14,200 hectares, 3.8 million gardens, 30,000 allotments and 8.3 million trees − almost one tree per inhabitant.[43]

An obvious first step for cities is to increase habitat for wildlife within existing parks, communal gardens and urban commons. Over the past century, as in the countryside, parks and commons have been tidied up and sanitised to satisfy an aesthetic dictated by the efficiency of modern machinery, an obsession with tight management and a distaste for 'weeds', thorny scrub, wild grasses and dead wood. As with the UK's landscape-scale national parks, the emphasis has been on parks as an amenity for human recreation, rather than nature itself, as if the two are mutually incompatible.

Now, however, some change is happening. Rougher areas of grass and wildflower meadows are beginning to appear in some urban parks, and the occasional fallen tree is left in situ. The City of Edinburgh Council had by 2015 created seventy-four new wildflower meadows and 105 hectares of more naturalised grassland. This has resulted in a significant increase in biodiversity and public engagement, as well as financial savings in terms of management and mowing.[44] In London, wildflower areas have recently been established in Manor Park in Kingston, Burgess Park in Camberwell and Beckenham Place Park in Lewisham, among others. In Derby, a redundant 129-hectare golf course owned by the city council at Allestree Park is being rewilded after a remarkable unanimous council decision to do so in 2021.

But this is not the story for most city parks. According to the charity Butterfly Conservation, all urban parks could support at least half of all the UK's butterfly species if they were better managed. Yet butterflies are declining more rapidly in urban areas than in the countryside, primarily because of loss of habitat.[45] Often it is council staff who are the hardest to convince. The 'keeping grass short' and civic flower-bed culture is deeply ingrained, and less hands-on management is sometimes perceived as a threat to jobs.

When COVID-19 struck, town and city councils were forced to furlough gardeners and maintenance staff, resulting in the spontaneous rewilding of parks, gardens, roundabouts and roadside verges. This suspension of mowing through the spring and summer showed how urban areas could look, and did much to shift public sensibilities. Even some of the quads and courts of Oxford and Cambridge colleges, once upheld as the most pristine lawns in the world, have been transformed into shaggy meadows.[46] Councils may one day understand that intensive mowing is bad for biodiversity and carbon emissions, and an unaffordable drain on the public purse. Wouldn't we rather spend that money on care for the elderly or school lunches?

In 2020 the conventional 'billiard table' lawn that had been carefully maintained in front of the chapel at King's College, Cambridge, since the 1720s, was transformed into a wildflower meadow. Towards the end of summer, gardeners rake up the freshly mown hay replete with harebells, buttercups, poppies and cornflowers.

Urban scrub and the nightingale umbrella

The ambition for habitats in parks needs to be far greater than just laying off the mowing. Let's rewind to a time, not so long ago, when urban parks, commons and heaths were much wilder, with greater vegetation complexity. In 1819 there were still enough thickets on Hampstead Heath to invite the nightingale that inspired the poet John Keats, who was living nearby, to write his famous ode. (Unfortunately, the nightingale singing in Berkeley Square in the famous wartime song was a romantic figment of the lyricist's imagination.)

Urban parks can be this wild again. The nightingale is a useful gauge of habitat complexity, nesting, as it does, close to the ground and requiring thorny scrub, bramble and stinging nettles to protect it from urban predators, such as dogs, foxes, cats and other birds. In Berlin there are now between 1,300 and 1,700 nesting pairs of nightingales, their population having increased in the city by 6 per cent every year from 2006 to 2016.[47] The reason for this rise is that the city's parks have not been managed for years, and brownfield sites are plentiful. The habitat in these areas is now robust enough to protect nightingales from predation, even though Berlin also has one of the highest concentrations of goshawks (which eat them) in Europe.

Berlin boasts a population of around 1,500 nesting pairs of nightingales, thanks to plentiful brownfield sites and scrubland habitat in its parks. Following Berlin's example by encouraging thorny scrub and brambles in all our urban parks, commons and heaths would provide nesting habitat for numerous birds and small mammals, and protection from predation by foxes, dogs and cats.

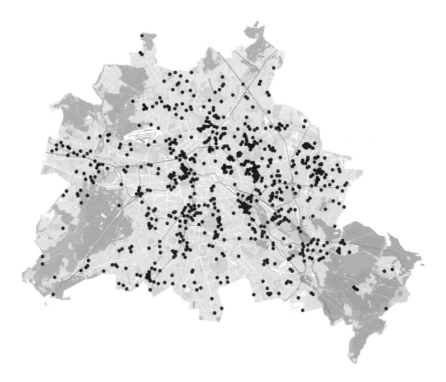

In the UK nightingales are known as famously shy creatures, almost impossible to see. But that is because their numbers have declined by more than 90 per cent since the 1960s. With critical mass, these birds become much more confident. One nightingale in Berlin has taken up his territorial singing position on a traffic light. Another nightingale hotspot is on the brightly lit Strasse des 17. Juni, just metres from the Brandenburg Gate, defying the theory of some ecologists that nightingales may be adversely affected by light pollution.[48] The nightingales are not only loved by Berliners. They have drawn international attention with a book, a film and an entire album of musical nightingale collaborations compiled by the American jazz musician David Rothenberg.[49]

Berlin has moved the baseline. We can once again think of the nightingale as a bird that could live successfully today in London and other UK cities. It is also an important indicator, or 'umbrella', species. Providing thorny habitat for nightingales in urban parks and communal gardens will attract a host of other creatures, including other songbirds and many moth species feeding on the scrub and flowers. And if nightingales can return themselves to London in high enough densities, then, one day, a male may after all be bold enough to stake out its territory in Berkeley Square.

Grazing animals

Barely a century ago sheep and cattle were a familiar sight in cities. The pasture of urban parks did not go to waste; it provided grazing for livestock, and livestock kept the grass down, their dung and urine fertilising the soil, and their browsing, grazing, seed-spreading and trampling sustaining floral meadow complexity and a rich diversity of pollinating insects, dung beetles and other invertebrates. It was a sustainable cycle. In the UK, the practice fell away in the 1950s with the advent of tractor mowers, although Freemen of the City of London retain the right to drive livestock across London Bridge. The tradition of urban grazing lives on in cities such as Newcastle, where cattle have summer-grazed Leazes Park and Town Moor since the thirteenth century, and Beverley in East Yorkshire, where cattle still have the right to roam the streets. Old English longhorns graze Christ Church meadows in Oxford, and on Midsummer Common in the centre of Cambridge longhorns are managed by a 'pinder', who is traditionally charged with penning any strays and extracting a fee from their owners to reclaim them.[50]

In 2017 the Royal Parks, the charity that looks after 2,000 hectares of parks in London, began an experiment to return grazing sheep and cows to Green Park in the centre of the city. The animals were taken early every morning from Mudchute Park and Farm on the Isle of Dogs by the farm's manager to spend the day grazing a fenced wildflower meadow at the north end of Green Park, then returned to the farm at night – supposedly for their own safety. The animals drew large crowds and proved so popular that, during the COVID-19 outbreak, the experiment was suspended because it was considered a threat to social distancing.[51]

A Dexter cow and rare-breed sheep graze in Green Park in central London as part of a conservation project aiming to increase floral complexity in the Royal Parks.

Invertebrate surveys conducted from the start of the Green Park grazing trial will record the effects of grazing in the meadow on plant and insect biodiversity and, it is hoped, provide an impetus for conservation grazing to be rolled out to other parks and heaths. Presumably, when cattle and sheep become a familiar sight in the capital they will draw less attention and be left overnight without concern, as they are in cities elsewhere.

Even pigs can have a role to play in park maintenance as an eco-friendly, low-cost alternative to tractors, ploughs, herbicides and man-power. In 2010 and 2011 six British saddlebacks spent two months in Holland Park, west London.[52] Their rootling stimulated wildflowers in the park's arboretum meadow and oak enclosure, excavated brambles, nettles and deep-rooted thistles and docks, aerated the soil, and increased the number of butterflies, grasshoppers and bees. Crowds of people came to watch them. The city of Melbourne in Australia is using goats in Royal Park to graze down non-native weeds and provide basking spots for White's skink lizards.[53]

Of course, grazing animals already live in London, in the form of deer. At 1,000 hectares, Richmond Park – a royal hunting ground enclosed by Charles I in 1637 from farmland and small commons – is Britain's largest urban walled park, containing about 345 red deer and 315 fallow. Deer have shaped the landscape there continuously since King Charles's day, maintaining species-rich acid grassland – a rare and fragile habitat. Forty-nine species of grass, rush and sedge thrive there, supporting 144 species of bird (including 63 breeding species, several of them ground-nesting birds, such as skylarks), 139 spider species, around 750 different butterflies and moths, more than 1,350 beetle species (including one specialist that lives on deer dung) and 150 solitary bees and wasps. The 1,200 ancient trees of Richmond Park, many of which pre-date the park's enclosure, host stag and saproxylic beetles, 400 species of fungus and 11 species of bat.

Stags clash antlers during the autumn rut in Richmond Park. Red deer have been grazing this former royal hunting ground since the time of King Charles I, and their presence has contributed to the park's extraordinary variety of plants, insects, fungi and birds.

There are also small numbers of red and fallow deer in nearby Bushy Park, once a hunting ground for Henry VIII, and in Greenwich Park, which was enclosed in 1433.[54] Elsewhere in the UK, Birmingham's 970-hectare Sutton Park, the UK's second-largest urban walled park, was denuded of deer in 1528, but tiny non-native muntjac have squeezed in through the railings to replace them.[55] The deer park in Magdalen College, Oxford, grazed by fallow since 1700, must be the UK's smallest urban example, covering about 12 hectares.

The role of herbivores in urban parks is just the same as in the wider landscape, with the same ecological benefits and logistical challenges. Keeping deer in the park and away from roads and urban gardens is more challenging than managing domesticated livestock, requiring either a high brick wall, which is expensive to build, or a 2-metre deer fence of high-tensile wire. Neither is likely to be popular in parks. Cattle, ponies, pigs, geese and sheep, or possibly such exotics as llamas and alpacas, which can be managed with simple, temporary electric fencing or 'no fence' collars, are likely to be the species of choice for conservation grazing in urban areas.

The very presence of deer and livestock in city green space demonstrates one of the most fundamental principles of rewilding: that of the relationship between vegetation, herbivory and biodiversity. It's once again the question of balancing animal numbers. Too few herbivores and the flower-rich meadow of an urban park will become species-poor, thatchy grassland that is gradually colonised by scrub and trees, with increasing shade diminishing biodiversity. The grass will require mowing again. Too many herbivores, on the other hand, and diversity is compromised by intensive grazing, the soil becomes compacted and excessively fertilised, and the animals themselves run out of food. The need to control the number of herbivores becomes a tool for education about animal welfare and conservation.[56] Although the presence of deer, cattle, sheep and other animals in cities may be hugely popular, urban populations may not be familiar or comfortable with the idea of culling. This must be addressed openly and as part of the bigger ecological picture.[57] Meat from conservation-grazing projects in the capital could have as much cachet as wild-range meat anywhere, perhaps greater.

While city parks, deer parks, commons and heaths generally get more visitors than most sites in the broader countryside, they still offer opportunities for creating wilder spaces. Introducing other species of herbivore to deer parks, and lowering the deer population to accommodate them, is likely to increase floral and vegetation complexity. Randomising those introductions over time, varying the numbers of cattle, ponies or sheep and the seasons and length of time they spend in a given area, should have the same effect, and excluding some areas completely from herbivory for a time may allow thorny scrub, brambles, nettles and young trees to establish. No-fence collars, or virtual fencing, once again, will prove an indispensable tool for creating habitat mosaics within confined spaces.

Green connections

Some ecologists argue London's designation in 2019 as a National Park City demonstrates old-school thinking, with too much focus on individual, high-value but isolated sites and not enough on connectivity, and where the sum of the parts does not add up to the whole.[58] In 2020 twenty-five European capital cities were ranked according to their 'ecological integrity': in other words, the extent of their green infrastructure and connection with ecologically functioning landscapes in the surrounding areas. London scored very low indeed, second from the bottom, above only Brussels. London may contain a number of nature reserves, some impressively large parks and a high percentage of green space.[59] But these green areas are islands in a sea of grey, with little flow or connection between them.

Maps (left) and ranking (right) of European capitals according to their ecological integrity. The colour gradient represents the areas with different ecological integrity values, with the lowest integrity in brown and the highest in dark green.

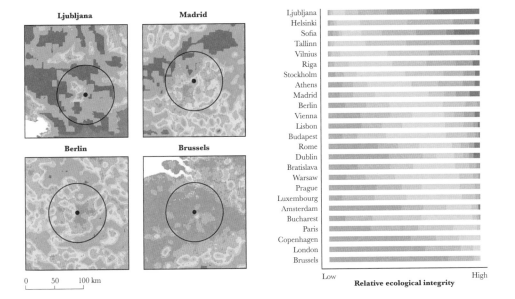

Ljubljana: Europe's greenest capital

Ljubljana in Slovenia is ranked number one for ecological integrity, with 'good ecological connectivity to its hinterland and a relatively high functional diversity of species'.[60] Of course, Ljubljana (population 270,000) is far smaller than London (9.4 million) or Brussels (2 million), but its overall ambition and holistic planning provide lessons for all urban centres.

In 2007 a new Ecological Zone in the city centre shifted the emphasis from cars and other vehicles to pedestrians and cyclists.[61] This reduced carbon dioxide emissions by 30 per cent, black carbon concentrates by 70 per cent and noise levels in the city centre by 6 decibels. The Ljubljanica River and its corridor have been restored, removing weirs, reconnecting a Natura 2000 site upstream of the city with floodplains downstream of Ljubljana's urban areas, and transforming the riverbanks into parkland and walking and cycle routes for city residents.[62]

Ljubljana's Ecological Zone today covers an area of 120 square kilometres and continues to expand as the city creates more green space, plants thousands more trees and converts brownfield sites into parks, pocket parks, playgrounds, cycle routes and horse-riding trails. Now that 75 per cent of the city area is considered to be green, Ljubljana has about 542 square metres of public green areas per resident.[63] In 2020 the city council declared that it would reach zero waste by 2025. No other capital in Europe collects as much separated waste as Ljubljana. According to Mayor Zoran Janković, residents have changed their habits radically. They are now very sensitive to actions that are disrespectful to the natural environment, such as fly-tipping, littering and vandalism, and have become active participants in keeping the city clean and green.[64]

Railways

If managed well, railways can provide corridors for wildlife flowing between urban and rural areas (see Chapter 2), and connect habitat within larger cities. In contrast to roads, with their constant heavy traffic, the tracks themselves are more permeable to wildlife – unless they are electrified. Embankments and sidings provide important brownfield and early successional habitat, too. They can become arteries, much like naturalised rivers.

Disused railways are even more valuable for nature and can act as a unifying principle for reconnection. Sydenham Hill Wood in south-east London is a remnant of the Great North Wood, a wood pasture that once stretched between Deptford, Streatham and Selhurst, providing commons grazing as well as timber, charcoal and firewood for London. The wood was divided and largely sold off in the late eighteenth and nineteenth centuries. Now, London Wildlife Trust hopes to reunite and restore twenty fragments of the Great North Wood, including Sydenham Hill Wood on the old Nunhead–Crystal Palace railway line. It originally served the Great Exhibition pavilion after it was relocated in 1854 to Crystal Palace Park, but since the line's closure in 1954, it has effectively rewilded itself. A highlight is the old railway tunnel to the south of the site, now an 'urban cave' for a range of species and a roosting site for brown long-eared bats.

Nearby, four fragments of woodland form the Sydenham to New Cross Railway Cutting, one of London's finest examples of railside habitat supporting a diverse range of species. They form part of a chain of six nature reserves and green spaces along nearly 9 kilometres of railway line between Sydenham and New Cross. All are managed by local 'Friends of' groups, volunteers and organisations such as London Wildlife Trust. Of course, green corridors like these facilitate the movement of predators, too. Foxes are regularly spotted wandering down the railway line, cutting through gardens into the reserves. Sparrowhawks, which in recent years have roosted at Garthorne Road Nature Reserve in Honor Oak, cross the railway line into Devonshire Road Nature Reserve to hunt. What matters is the quality of habitat in these corridors and in the reserves themselves. Plenty of cover, including trees, brambles and thorny scrub, will provide prey species with protection from predation.

Highlines

New York's High Line – inspired by the 4.6-kilometre Promenade Planteé in Paris, a nineteenth-century viaduct converted to an elevated park walkway in 1993 – is now the world's most famous ambassador for turning disused urban railways into green space. Its 2.3 kilometres of freight railway, elevated above the streets on Manhattan's West Side in the 1900s, shut down in 1980. It was threatened with demolition in 1999, but by then the railway had rewilded itself as plants and trees colonised it, and it became beloved by New Yorkers. A Friends of the High Line campaign successfully fought off the demolition order and raised funds to preserve the structure and convert it into space for people and wildlife. Today this linear park, designed by the famous Dutch landscape designer Piet Oudolf, consists of a spectrum of micro-habitats from woodland and wood edge to meadows and grassland. Although actively managed, the plants are inspired by those that colonised naturally during the line's years of disuse.

New York's High Line along the lower west side of Manhattan is planted with species inspired by those that naturally colonised the elevated railway line after it fell into disuse in 1980.

The High Line is now one of New York's most popular attractions for both residents and tourists, clocking up around eight million visitors a year, and it provides an important resource for wildlife among the skyscrapers. A study in 2017 found thirty-three species of bee, the majority of them cavity nesters using the hollow stems of plants for their nest chambers. The findings inspired the High Line's horticulture team to leave plant stems standing later into the spring to allow nesting bees to emerge, and to keep a few plant species as standing dead vegetation throughout the park to provide additional bee habitat.[65] Ten species of plant that are rare for New York State have colonised the line, including purple milkweed, which thrives at Tenth Avenue Square and is the food plant for monarch butterfly caterpillars. In 2012 a pair of peregrine falcons nested on the side of the Drug Enforcement Administration headquarters at West 17th Street and Tenth Avenue, where they could be seen swooping over the High Line searching for prey – an indication of the booming bird population.

Yet even there, the temptation to over-garden and create conventionally tidy spaces seems irresistible. The complex habitats of reclaimed brownfield sites can swiftly lose their value for nature through the gardening mindset and excessive care and attention. Happily, during the COVID lockdowns, the lawns on the High Line were spared the attentions of mowers and allowed to rewild.[66]

Seoullo 7017, a sky garden designed by the Dutch urban-design firm MVRDV in 2017 on a former inner-city highway in Seoul, was inspired by the High Line.[67] It shows what can happen when a 'green corridor' falls victim to design architects and almost completely loses sight of nature. The designers were dealing with a load-bearing challenge. Intended for lighter vehicular loads than New York's High Line, the abandoned motorway viaduct could accommodate only an estimated 100 millimetres of soil – not deep enough, it was thought, for continuous plant cover. But many plants thrive in shallow, poor soils. Indeed, many need hardly any soil at all, just a loose covering of gravel, crushed brick and/or sand, which is why brownfield sites are so successful for plants. The concept of plants in this case was, unlike for the New York High Line, dictated by the garden/agricultural mindset, which favours highly productive, tall, leafy species that require deep soil, regular watering and fertiliser.[68]

The Seoullo 7017 project in Seoul, South Korea, is an example of how not to design a green corridor. The former inner-city highway, described in publicity as a high-line 'forest', consists of a concrete walkway and isolated planters, and has almost no value for nature.

To get around the weight limitations for continuous soil cover, the designers installed 645 cylindrical concrete planters – archipelagos of green in a walkway of concrete. While the 'plant library' of 24,000 plants representing 228 species and subspecies contained in the pots may sound impressive, the strategy is to uproot trees when they get to a certain size and move them to other areas of the city.[69] This continual transplanting of vegetation is both energy-demanding and disturbing for insects, birds and other life that may have found refuge there.

Although it was described in publicity as a high-line 'forest', the ecological credentials of this high line are clearly limited. It serves a sustainable purpose by encouraging pedestrian rather than vehicular traffic across the city. But its potential as a nature corridor that might connect the few green spaces of inner-city Seoul with surrounding natural landscapes, such as Namsan Mountain, has not been realised.

In London, the Camden Highline plans to convert a disused 1.2-kilometre stretch of railway viaduct into a new elevated park and walking route from Camden Gardens to York Way at King's Cross. The project, costing an estimated £35 million, is scheduled to open in 2024.[70] Like the New York High Line, it will serve as a recreational space for arts, cafés and charitable activities, bringing new green space to 20,000 local residents and habitat for a range of species, including butterflies and birds.

In 2022, the disused Victorian Castlefield Viaduct in Manchester was transformed for the National Trust by architectural firm Twelve Architects into an urban 'sky park'.

A similar initiative in Manchester has transformed a disused viaduct 330 metres long into an urban 'sky park' planted with trees and shrubs, including Manchester poplars, one of the few trees resilient enough to withstand the pollution of the Industrial Revolution.[71] Sheffield's 'Grey to Green' project, completed in 2020, has turned a 1.5-kilometre stretch of dual carriageway – part of a former ring road – into a green corridor, planted with drought- and flood-resistant trees and shrubs. It contributes to the city's flood defences as well as connecting long-distance walking and cycling routes.

As always, the value for nature of such projects depends on the quality of the habitat and the ability of designers to resist the compulsion to over-manage and assign too much hardstanding for cafés, exhibition space, walkways and playgrounds. The projects should also be planned with a view to connect with other areas of nature, rather than as a stand-alone amenity solely for human recreation. It should be possible – perhaps especially in cities – to keep some abandoned railway areas exclusive to wildlife, thereby minimising disturbance from humans and dogs and enabling the free movement of species between areas of high nature value.

Urban trees

Trees in cities have never been more crucial. As temperatures rise with climate change, and heat radiates from roads, pavements and buildings, trees cool us. Tree cover of 40 per cent or more can lower the summer daytime temperature in towns and cities by as much as 5.5°C.[72] This is more trees than are currently in most cities; in UK cities, for example, the average is 16 per cent tree cover.[73] Trees are needed even more in poorer suburbs, where people live, than in central parks and high-profile avenues. For every 1°C increase in temperature above 21°C, heat-related deaths increase by 3 per cent.[74] An increase of 10 per cent in urban green cover in high-density residential areas in Greater Manchester, for example, would decrease the expected maximum surface temperature in the 2080s by about 2.5°C (and up to 4°C). Conversely, removing 10 per cent of green cover would increase that surface temperature by 7°C.[75]

The lowering of street temperatures reduces the need for air conditioning, thus reducing emissions of greenhouse gases. One mature tree can give off up to 450 litres of moisture a day. That is a cooling equivalent of five room-sized air conditioners left on for nineteen hours.[76]

Trees also reduce air pollution. Those with bigger canopies and larger leaves are better at trapping the pollutants generated by fossil fuel-burning vehicles, tyre dust, factories and construction sites. Rough, rugged and hairy leaves, such as those of silver birch, yew and elder, act as the best filters for removing these harmful minute particles from the air.[77] The sticky honeydew created by aphids feeding on lime leaves is also very effective at trapping particulates. Globally, 8.9 million deaths a year are thought to be attributable to the inhalation of fine particulates.

And yet we continue to get rid of trees in our cities. In the USA, 36 million trees are cut down in urban areas each year, and 67,500 hectares of hardstanding added.[78] The loss of public benefits caused by tree removal is estimated – very conservatively – to cost the country $96 million a year.

The discrepancy between green proclamations and the reality on the ground is nowhere more striking than in the plight of urban trees. In 2016 the Mayor of London, Sadiq Khan, announced plans to plant 2 million trees in the capital. It has taken six years to plant around 430,000 trees, including two new woodlands in London's Green Belt.[79] Meanwhile, London councils cut down almost 50,000 mature trees in the five years to 2017, an increase of 20 per cent on the period 2003–7.[80]

In 2014 the city of Sheffield began a controversial mass felling of a targeted 17,500 trees, ostensibly to reduce the cost of road maintenance and resurfacing. By the time public outcry had put a stop to the felling, in 2018, the council had cut down 5,500 mature trees, replacing them with saplings.

Middlefield Road, Bessacarr, Doncaster, before and after its avenue of sixty-four healthy lime trees were felled by the council because of 'tree root damage' to pavements.

The removal of 'problem' trees – those inconvenient to developers – are often falsely justified on 'health-and-safety' grounds. Residents also complain to councils about trees blocking their view, leaves and debris being shed into gardens, tree roots creating uneven pavements, and bird guano and sticky honeydew from aphids in lime trees dropping on to car roofs and windscreens. The vested interests of contractors in the work of cutting down trees and replacing them is also a significant driver. Clearly, the message about the importance of mature trees for quality of life and sustainability in cities needs to be more clearly understood, and permeate all levels of urban planning and management. Residents themselves must learn to love trees for what they provide, and live with the associated minor inconveniences, such as falling leaves.

In 2011 the Forestry Commission published excellent advice for councils in *Common Sense Risk Management of Trees*.[81] Replacing old trees is never easy. Soil quality and compaction, air pollution, increasingly unpredictable weather patterns and disease pose huge threats to urban trees. Trees that reach maturity are miracles of survival. Clearly, also, big trees sequester more carbon, trap more pollution, provide greater flood protection, further reduce noise pollution, give more shade, moisture and temperature control, and produce more pollen and nectar as forage for insects. Every effort must be made to save them.

We must also plant many more trees, and ensure that those new trees have the best possible chances of survival. Tender new saplings, generally imported from commercial nurseries hundreds of miles away, or even from overseas, invariably have a mountain of challenges to overcome in cities: pollution, overheating, nutrient overload, over- and under-irrigation, inadequate drainage, excessive compost, soil compaction, lack of mycorrhizal associations, lack of soil volume under paving and hardstanding, tight spacing and restrictions to the flare of the trunk.[82] The landscape architects and contractors who plant trees often know little about how to look after them, and local planning authorities fail to enforce maintenance and aftercare.[83] On average, 30 per cent of newly planted trees in towns and cities in the UK die within twelve months of being planted.[84]

Local planning authorities have a duty to focus not on tree planting as a box-ticking exercise, counting the number of saplings put into the ground, but instead on the survival of the trees, including improving soil and site conditions and monitoring aftercare and maintenance. One exciting new contribution to urban tree health is biochar (the residue from burning biomass).[85] Enriched with worm casts, seaweed and mycorrhizal fungi, a blend of biochar called Carbon Gold, produced in the UK, significantly reduced losses in 2,000 mature urban trees in Stockholm, and increased the survival rate of newly planted saplings by improving the aeration of the soil and its water-holding capacity, and introducing supportive micro-organisms.

It is also important to choose the right species of tree for urban conditions, and ensure enough genetic variety to spread the risk of disease.[86] Only a handful of genera make up urban tree cover around the world. In the UK, native species such as silver birch, ash and English oak and exotics such as sycamore and London plane are the most common.[87] These five alone account for a third of trees in London's parks, gardens, playing fields and streets. Genetic diversity within these species is often low, since they have originated from limited sources or are clones propagated in commercial nurseries.

The arrival of new diseases is one of the biggest threats to trees. They can be imported through global trade or transmitted via fungal spores and insects travelling great distances on their own. The best protection we can give our trees is genetic and species diversity. In cities, where natural regeneration is restricted and conditions are affected by climate change, we should choose drought-tolerant and disease-resistant genetic stock. It is also important to plant less traditional, even rare, trees. Botanic gardens and nurseries, in particular, need to collect new source stock from the wild and develop cultivars that are suitable for urban environments.[88]

In 2018 the first World Forum on Urban Trees, held in Mantua, Italy, set out a vision of how trees could be used to make cities greener, healthier and happier places to live in. In 2020 the Food and Agriculture Organisation of the United Nations announced a list of the first fifty-nine Tree Cities of the World, among them Ljubljana, Paris, New York, San Francisco and Toronto, the Italian cities of Mantua, Turin and Milan, and small towns such as Thunder Bay in Ontario, Tempe in Arizona and Bradford in the UK. The aim is to encourage other cities to follow suit. So far, a hundred or so have pledged (note, only pledged) to meet the standards for qualification. This is a drop in the ocean, however, given the 10,000 cities across the globe, half of which have come into existence within the last forty years without having been designed with trees in mind.

The new Treepedia app, launched in 2017 by MIT's Senseable City Lab, quantifies both overall tree coverage and individual street coverage within a city.[89] Using Google Street View rather than satellite images, it gauges 'human perception of the environment from street level'. This throws up some surprising results. Of the twenty cities currently listed on its Green View Index, New York scores higher for street trees than London. Singapore has the highest median score (29.3 per cent) and Paris the lowest (8.8 per cent). This is largely because public parks (of which Paris has many) are excluded from the score, but it also shows most of Paris's tree coverage clustered in the southern sectors and around its borders.[90] By mapping the distribution of trees down to street level, the app identifies areas that would benefit from tree planting.

Greening Paris

Paris has now clearly focused on its need to increase tree cover. As part of its aim to be a carbon-neutral city by 2050, the Mayor of Paris, Anne Hidalgo, announced plans in 2019 for an 'urban forest' of 170,000 new trees, including in big new garden schemes around the historic sites of the Place de l'Hôtel de Ville, Gare de Lyon, Palais Garnier and along the banks of the river Seine. She also plans for 50 per cent of the city's surfaces to be vegetated and permeable, including 100 hectares of newly planted roofs and façades, by 2030.

Making space for trees will be a driver for rethinking Paris's purpose and structure, largely by reducing the area previously devoted to cars. Hidalgo, who was elected in 2014, has created about 1,450 kilometres of bicycle lanes across the city; major roads, including the *quais* along the Seine, have been shut to traffic entirely; and grand squares such as Place de la République, Place de la Madeleine, Place de la Nation and Place de la Bastille have been made pedestrian-friendly.[91] All diesel cars will be banned from the city from 2024, and all petrol cars from 2030. During the first COVID-19 lockdown in 2020, some 65 kilometres of road, including the Rue de Rivoli, were converted into bicycle lanes, or *corona pistes*. The measure inspired an ambition for the first four *arrondissements* (5.6 square kilometres, or 7 per cent of the city) to be permanently closed to transit traffic – vehicles that pass through without stopping – by 2024.[92] The goal is to reduce traffic in the heart of Paris by at least half. In preparation for this permanent closure the city centre has, since 2022, become a car-free zone on the first Sunday of every month.

Implementing Hidalgo's green vision was not easy at first. Protesting motorists and taxi drivers filled the streets of Paris. In 2018 a group of 128 municipalities joined the conservative president of the Ile-de-

By adding trees and reducing traffic on the Champs-Élysées, the Mayor of Paris promises a dramatic green makeover for the French capital's iconic boulevard. This interpretation by Philippe Chiambaretta Architecte shows how the greening of Place Charles de Gaulle, around the Arc de Triomphe at the western end of the boulevard, could look.

France region, which includes Paris and its suburbs, in filing legal action against Hidalgo's project to turn the 3.3-kilometre stretch of the Pompidou Expressway along the Seine into a park for pedestrians and cyclists – a case she won on appeal. In 2020, however, she was soundly re-elected by Parisians supporting her car-reduction policy.[93]

No other mayor has been so determined in pursuit of the greening of a capital city. Indeed, Hidalgo has set the bar for city mayors around the globe.[94] In Paris, now, every new building is required to have rainwater-collection systems installed on its roof for use in washing machines and dishwashers, and for flushing lavatories. Water from the Seine – rather than municipal drinking water, as in most cities – is piped to streets for street-cleaning. The city is now building a vast subterranean water tank to store stormwater and prevent sewage spilling into the Seine, with the aim of making the river swimmable in time for the 2024 Summer Olympics.[95] As a way of encouraging community interest in trees and green spaces, citizens can download the Arbres Remarquables de Paris app to identify and learn about the history of the city's 300,000 trees. They can also apply for a *permis de végétaliser* to plant flowers and vegetation in public spaces.

Trees for connectivity

Urban tree canopy is critical for wildlife, too. In the urban industrial sector of Belo Horizonte, a city of nearly 3 million people in southeastern Brazil, the streets with more trees – particularly native, well-established trees – have a greater diversity of native birds and other organisms.[96]

Trees aid the movement of birds, increasing their reach for habitat and food. Many bird species are reluctant to fly across roads, possibly through fear of predation and/or the creation of territorial boundaries where the vegetation ends at roadsides.[97] Streets with tall trees on both sides, whose canopies touch or nearly touch, hugely increase the permeability of the urban environment for birds and, presumably, other species, including insects and tree-living mammals.

We need to think more imaginatively and ambitiously about tree cover in cities, and not restrict ourselves to isolated trees, parks and avenues. In the 1970s the Japanese botanist Akira Miyawaki pioneered a technique that has since been developed to create 'micro-forests' in cities.[98] Originally designed to fast-track natural forest regeneration on degraded land, the Miyawaki technique has proved hugely successful in cities around the world, with all kinds of soil and climate, from Japan, India, Indonesia and Brazil to France, Switzerland, Sicily and the Netherlands.[99] Taking inspiration from nature and the cooperative relationships between trees and shrubs, it involves planting between fifteen and thirty native species densely together in small areas (usually about 200 square metres). Particular care is taken with the seedlings in the first few years, and no chemicals or fertiliser are used. The resulting tree cover, even in arid places, establishes astonishingly quickly. These micro-forests are structurally more complex and diverse than conventional plantations of just a few species, creating far better habitat for wildlife.

In the UK in 2021 the environmental charity Earthwatch adopted the Miyawaki technique to create a micro-forest in Bristol, on what was once a car park and a magnet for fly-tipping next to the river Trym in Southmead.[100] Planted with native species from oaks to birch, elder, blackthorn and guelder rose, it is part of a bigger regeneration project to bring trout back to the river and rare black poplar to the riverbanks. A micro-forest can attract more than 500 animal and plant species in its first three years. Like pearls on a necklace, these tiny city forests become biodiversity hotspots from which wildlife can flow down avenues, railway embankments and other green corridors and flyways to the next hotspot. Using armies of volunteers for planting, and citizen scientists for monitoring species and collecting data, Earthwatch aims to create 150 more micro-forests in Birmingham, Wolverhampton, Leicester and Glasgow, among other cities.[101]

Biophilic design

Unregulated cities lack parks and gardens, as well as streets wide enough for trees. A more tolerant approach towards urban trees will find space for them where civic authorities have never considered them before. This might involve the closure of roads or the redirection of land to new urban parks and spaces. But we need not think of trees simply at ground level, and in conventional green space; buildings themselves can create the greenery. Architects and builders across the world are now thinking in terms of biophilic design.[102] E. O. Wilson coined the term 'biophilia' in 1984, referring to humans' innate love of living things and our biologically inherited need to associate with nature (see page 384). Increasing trees and vegetation both outside and inside buildings helps to connect the occupants with nature, improving their general well-being and their performance at work. It also provides the important ecosystem services we've been discussing in this chapter, such as cooling, air purification and water storage. Green buildings can replace the green space that has been lost because of them. Singapore, for example, requires new buildings to replace 1:1 the nature lost at ground level by providing nature on roofs and walls.

Also key to effective biophilic design is creating nature-friendly aspects within a building, such as natural light, natural ventilation, views, vegetation, and natural forms and shapes, then connecting them with nature on the outside of the building, and the trees, shrubs, plants, parks and rivers in the vicinity. In this way, the building's occupants feel part of a wider, living landscape. The designers of today's hospitals and psychiatric facilities, particularly, recognise these benefits. Biophilic design can be used to great effect to accelerate the recovery of patients.

The essential aspects of biophilic design – increasing vegetation and tree cover on buildings – can become a vital part of the green infrastructure of a city. Green walkways and bridges can be used to link buildings above ground, increasing connectivity for wildlife and plants, and allowing people to walk around off the street.

Green façades

An obvious way to increase greenery and habitat for wildlife in urban areas is to train climbing plants up the walls of buildings. In effect, this mimics the way climbing plants establish themselves on rocks and trees in the wild. Common ivy, a natural tree scrambler that is native to Britain, is one of the easiest to grow, and, since it attaches itself to surfaces using climbing stems with specialised adhesive hairs, needs no structure or wires. It can grow to a height of 30 metres. Its dense foliage provides nesting habitat for birds and bats, and food for caterpillars, and it plays host to a fascinating bright-yellow parasitic flower, ivy broomrape, which looks like an orchid. Because ivy is one of the last plants to flower each year, in mid-autumn, it is an important late resource of pollen and nectar for insects before they hibernate, and it produces its berries in the winter, providing blackbirds, thrushes, wood pigeons, blackcaps and other birds with high-fat food when insects are hard to find.

Many homeowners and property managers don't adopt this simple measure because they believe that climbers – and ivy, in particular – damage brickwork and cause damp. However, studies have shown that masonry is rarely affected by ivy, whose aerial roots penetrate only existing cracks.[103] Indeed, its evergreen foliage insulates walls, protecting them from the damaging contraction and expansion of freeze and thaw.[104] The dense leaf cover and water-absorbing aerial roots of climbers may even help to keep walls drier.

Of course, gardeners have long been enjoying climbers. Climbing hydrangea is another 'self-clinging' plant, but others are more demanding. Honeysuckle and clematis need a trellis, mesh frames or wires to twine around, and rambling roses and fire thorn need initial training and tying in to a structure. Other climbers that are wonderful for wildlife are hops, passion flower and star jasmine. There is certainly a degree of maintenance involved in creating a green façade; many climbers need regular pruning to increase vigour and flowering, and all must be kept clear of loose fittings, such as guttering, windows and roof tiles. This may require a tall ladder and a head for heights, or even the cost of hiring a cherry picker and/or a contractor to do the work. But the benefits in terms of reducing indoor temperatures in the summer, keeping buildings warmer in the winter, cleaning the air, storing carbon and providing habitat for wildlife are huge. Arguably, a building looks naked and unloved if it is *not* clothed in green.

Stand-alone structures using scaffolding or permeable screens can also support plants, creating green façades, and trays, planter tiles or flexible bags can be vessels for rooting plants. These innovative designs can create green screens to block undesirable views or noise, such as around a construction site. They can mark the boundary of a garden or set of buildings, where a solid fence or barrier might have been used in the past. They can improve the aesthetics and habitat value of streets and pedestrian walkways that are unsuitable for large trees or where access is difficult. And they can add a vertical dimension to smaller parks and gardens.

Living walls

Living walls take the concept of a green façade further, reaching heights that climbers on their own cannot scale. Using hydroponics or a growth medium such as soil, substitute substrate or hydroculture felt, plants are grown directly on the vertical face of a building. Although the technique was patented by the University of Illinois in 1938, its popularisation is generally credited to a French botanist, Patrick Blanc, who worked with the architect Adrien Fainsilber and the engineer Peter Rice to create the first successful *mur végétal*, at the Cité des Sciences et de l'Industrie in Paris in 1986.

Since then, green walls – whether living or made up of climbers – have been shown to reduce the temperature of external walls in the heat and insulate buildings in the cold, reducing energy consumption and costs.[105] In the UK climate, green walls reduce the temperature of external walls by as much as 12°C in the hottest months.[106] In hotter countries the reduction is even greater.

Living walls now grace the façades of numerous high-profile buildings around the world, such as Edmonton International Airport and Semiahmoo Public Library in Canada, Universidad del Claustro de Sor Juana in Mexico City, CaixaForum in Madrid and Il Fiordaliso shopping centre in Milan. One of the largest, covering 5,300 square metres, is at the Institute of Technical Education in Singapore.[107] In Santalaia, Bogotá, Colombia, plants cover over 3,000 square metres of a nine-storey residential building. The world's largest green wall (as of 2020) covers 7,000 square metres of an elevated road interchange in Khalifa Avenue, Qatar. Notable sites in the UK include Westfield shopping centre in west London, the corner of Edgware and Marylebone roads in central London, and the Sir David Attenborough Building in Cambridge. But green walls are also increasingly popular on a smaller scale and in private spaces. They can even be indoors, such as the 560 square-metre living artwork inspired by the painter Jackson Pollock in the Old Tabacalera building in Santander, Spain; and smaller ones too, such as the two-storey green wall of Columbian coffee plants in Devoción Coffee Shop in Brooklyn, New York.

The 15-metre-high living wall designed by ANS Global and the Green Infrastructure Consultancy is a striking feature of the internal atrium of the David Attenborough Building, the centre for the Cambridge Conservation Initiative, a global conservation hub at the University of Cambridge.

Numerous companies now supply green walls, and the technology is advancing all the time. New, lightweight geotextile membranes, for example, can accommodate extensive root structures, allowing them to spread across the face of the building. The membrane is strong enough to support shrubs and even some lighter trees, such as birch and rowan.[108] A huge range of plants are suitable for establishing on vertical walls, depending on the conditions, from ferns and orchids to hebes and hostas, from yuccas and solanum to blueberries and box.[109] As with any form of gardening, the look can be formal and orderly or relaxed and wild. Living walls can take on a life of their own if mosses are allowed to take over and wildflowers, such as wild sorrel, that arrive with birds are allowed to establish.

The system of watering and maintaining a living wall, which for large buildings often includes computer-controlled irrigation and nutrient systems, is inevitably more complex than for a green façade, where the plants grow up from the natural soil base or containers on the ground. Poorly installed and maintained living walls with dying plants – often through inadequate or excessive irrigation – are obviously a self-defeating eyesore. There is as yet no certification requirement for providers of living walls, and experts in the industry are pushing for the development of a trade association to set up a standards process.

As technology for green walls and roofs advances, nature-based solutions are being developed to sustain them. These include using rainwater harvested on rooftops to irrigate vegetation passively by wicking, without the need for pumps or pipes. These have the benefit of not requiring power to operate, and are usually less visually intrusive, as well.

Vertical forests

One step on from the living wall is the concept of the 'vertical forest' – a vast structure that can support a huge complexity and mass of plants, even full-grown trees. Milan's Bosco Verticale, designed by the architect Stefano Boeri and established on two residential towers in 2014, was the world's first. Alternating cantilevered balconies provide the strength and space for trees to climb upwards, providing around 30,000 square metres of 'woodland' on a footprint of just 3,000 square metres. To date, twenty species of bird have nested among the towers' trees and shrubs, including pale, or pallid, swifts, redstarts and house martins.[110] The vegetation reduces humidity, and it produces 19 tonnes of oxygen and absorbs 80 kilograms of particulates and 30 tonnes of carbon dioxide from the air every year. It shields the flats' balconies and interiors from noise pollution and lowers the temperature of the 'heat island' caused by sunlight reflecting from glass façades by as much as 30°C in summer. The trees and shrubs are irrigated with groundwater pulled by solar-powered roof pumps to all the pots in the building. The water is transpired by the vegetation, returning purified water vapour to the atmosphere, which also helps to extract heat from the surrounding environment. Three times a year a team of 'flying gardeners' – expert arborist-climbers – carry out pruning and maintenance from ropes and pulleys.[111]

This astonishing vertical forest, the Milan Bosco Verticale, created by Stefano Boeri Architetti in 2014 on two residential tower blocks in Milan, Italy, was the first of its kind. Alternating cantilevered balconies support 30,000 square metres of woodland on a footprint of just 3,000 square metres. The vegetation produces 19 tonnes of oxygen and absorbs 80 kilograms of particulates and 30 tonnes of carbon dioxide from the air every year.

The Easyhome Huanggang Vertical Forest City Complex in the city of Huanggang, 70 kilometres east of Wuhan in Hubei province, China, was completed by Stefano Boeri Architetti in 2022. It spans 4.5 hectares and includes two residential and three commercial towers. The vegetation of native species, predominantly *Ginkgo biloba*, is expected to absorb around 20 tonnes of carbon dioxide and emit 10 tonnes of oxygen a year.

In 2016 Boeri Architects was commissioned to create two vertical forest buildings along the same lines in the Chinese city of Nanjing. There, 600 tall trees and 500 medium-sized trees of 23 native species, and 2,500 plants and shrubs have been grown, equating to 6,000 square metres of urban 'woodland'. In the words of Boeri, this intervention is the architectural equivalent of a skin graft.[112]

Clearly it will take more than two vertical forests to tackle China's pollution crisis, but the Nanjing project has been the springboard for a much more ambitious vision to create a series of sustainable mini-forest cities, the first just north of Liuzhou, a city of 1.5 million residents in the southern province of Guangxi. Housing up to 30,000 inhabitants, the project, which began in 2020, will provide an environment for 40,000 trees and 1 million plants of more than 100 species along its streets and in parks and gardens, as well as on the sides of buildings. A second project is planned in Shijiazhuang, an industrial city in the north, ranked among China's ten most polluted cities.[113]

Living pillars

Infrastructure and architecture can become a proxy for living trees. Singapore's Gardens by the Bay claims to be the most sustainable and ambitious urban green space in the world. Central to the design are twelve artificial 'Supertrees' (the tallest as high as a sixteen-storey building), which provide space for vertical gardens containing around 165,000 plants of more than 200 species. Surreal and futuristic, they featured as the conclusion to Sir David Attenborough's *Planet Earth II* series in 2016. Seven of the Supertrees are designed to harvest solar energy. On a more accessible scale, ordinary lamp posts can be transformed into 'living pillars': multidimensional habitats to support birds and insects.

Clearly these innovators, exciting as they are, should not replace trees themselves. Hard infrastructure comes with a carbon cost. The greening of buildings is one thing, but replacing carbon-sequestering live trees with artificial tree 'sculptures' is hard to fathom.

Green roofs

Green and garden roofs have enormous potential. German studies dating back to the 1980s have shown that green roofs mitigate storm flooding. Roofs with soil between 60 and 100 millimetres deep can intercept 50 per cent of annual rainfall, and roof gardens, with soil typically 500 millimetres deep, can intercept 90 per cent.[114] Like living walls, they are also very effective at insulating buildings – saving energy and money.[115]

In the Dutch city of Utrecht – a greening pioneer – the 'no roofs unused' policy is part of a drive to create a patchwork of connected nature through the concrete cityscape and to improve its inhabitants' quality of life. A total of 316 bus stops now have easy-maintenance sedum roofs, providing food for bees, butterflies and other insects as well as catching rainwater, absorbing pollutants and decreasing street temperatures in the summer.[116] Citizens of Utrecht are incentivised through subsidies to green their own roofs.

Some cities in the UK, including London and Brighton, have adopted green roofs on some bus stops, but the uptake has been small. The city of Leicester is leading the way, with 30 green bus stops installed since 2021. However, Clear Channel, which manages

30,000 commercial bus shelters on behalf of councils, is now working with the Wildlife Trusts to create at least 1,000 'bee buzz stops' in the UK, planted with native flowers such as kidney vetch, thyme, selfheal and wild marjoram.[117]

In Germany, where the technology has been used since the 1970s, the number of green roofs has increased dramatically since the turn of the millennium. Some cities, such as Münster, Munich and Cologne, incentivise people to green their roofs by reducing their municipal water fees.[118] Stuttgart pays its residents up to 50 per cent of the cost of installing a green roof, to a maximum of €10,000, and Berlin includes green roofs in its mandatory targets for green space.[119] Seen from the air, Berlin is now a sea of green.

Many cities around the world have developed policies to encourage green roofs. But France is the only country that has – since 2016 – a national law requiring all new commercial developments across the country to 'integrate green roofs or solar panels'. Around one million square metres of green roofs were installed in France in 2017 alone.[120]

Most green roofs are, naturally, beyond street view and useable only by people with access to the building, but this isn't always the case. In Porto, Portugal, a green roof has transformed the historic square of Praça de Lisboa. The old, semi-derelict covered market in the centre of the square was crowded, unhygienic and a magnet for crime. A new building, designed in 2007 as both a partially covered commercial street and a public garden, features a roof supporting 4,500 square metres of grass and fifty olive trees – a wood pasture within a city. The roof garden, now a popular tourist attraction and public leisure space known as the Jardim das Oliveiras, is visible and accessible from street level and has helped rehabilitate this part of the city. Other spectacular green roofs are at the Vancouver Convention Centre, California Academy of Sciences in San Francisco, and Emporia in Hyllie, Malmö, Sweden.[121]

In the UK, 42 per cent of green roofs are in London.[122] Central London currently has 700 green roofs, many of them private or corporate, covering an area of around 175,000 square metres.[123] Greater London has over 1.5 million square metres of green roofs, a considerable achievement considering the capital has had a green-roof policy only since 2011 and that green roofs do not receive government subsidy.

This extensive, biodiverse 'dry meadow' roof on a building in Bishopsgate, in the City of London, has shallow soil favouring low-growing, drought-resistant plants. It is low maintenance and does not need watering.

The Bauder BioSOLAR system of photovoltaic panels installed on the flat roof of Noah's Ark Children's Hospice in Barnet, north London, allows growing room for wildflowers without blocking light to the panels. The green roof helps sustain an optimal temperature around the panels, enabling them to work more efficiently.

The designs and purposes of green roofs, roof gardens and green walls are gloriously varied. They can provide space for growing food, including fresh herbs and salads for restaurants. Some – usually those with soil deeper than 200 millimetres – require irrigation and annual maintenance, but more can be grown on them. Others have shallow soils, and low-growing, drought-resistant species such as sedum. They do not need watering, and are extremely low-maintenance. 'Biodiverse green roofs', with relatively deep soil and a wide variety of plants, including shrubs, provide habitat for wildlife. They are only 'semi-intensive', needing little or no irrigation and maintenance. 'Blue green roofs' store rainwater as part of a sustainable drainage system, and 'biosolar roofs' have frames for a photovoltaic array over a green-roof substrate.

Buglife publishes advice on how green roofs can be modified to encourage wild bees, beetles, spiders and other invertebrates.[124] Planting sedum with drought-tolerant native wildflowers and incorporating such features as deadwood, sandy banks, pools and stones increases habitat for wildlife. If done well, green roofs and their associated vegetation provide stepping stones for insects and birds to travel through the cityscape, especially if they are near trees and green or living walls. Cascading plants from roofs, imitating plants scrambling down ravines and cliff-faces in the wild, can establish green walls from above. In London, green roofs have enabled rare black redstarts to spread from derelict brownfield sites in the Docklands to breed on rooftops in Soho. Shrill carder bees burrowing into the shale sediments of a roof garden now feed these birds in Canary Wharf.[125] Green roofs may be perfect for ground-nesting birds, since people rarely visit them and cats may not be able to reach them.

Greening the space above our heads can eliminate a city's heavy, negative grey footprint. Physically, this can replace the gardens, meadows and other green areas that have been incrementally built on over the centuries. Of course, green roofs are inaccessible to most, so they do not increase public access to green space. But they can recover some of the public services these lost spaces would once have provided, such as encouraging wildlife, preventing flooding, storing water and cleaning the air.

Generally, the shallowness of the substrate (growing medium) on rooftops restricts the potential for carbon sequestration. An extensive green roof typically sequesters 375 grams of carbon per square metre both below and above ground. However, research is being carried out into how the uptake of carbon dioxide on green roofs could be improved by including charcoal into the substrate and inoculating the soil with mycorrhizal fungi. Green roofs have long been recognised as superb insulators for buildings, and for their ability to regulate the outside temperature by reducing the 'heat island effect'. This is the amount of heat and energy buildings absorb from the sun and reflect back into the atmosphere (temperatures in London are expected to rise by 5–6°C by the 2050s). They also absorb sound. Substrate just 100 millimetres deep on a roof can reduce noise – such as that from passing aeroplanes – by up to 10 decibels.

Sustainable drainage systems

Many cities have antiquated sewage systems designed for a much smaller, more permeable city, and are ill equipped to cope with the severe storm events of today. As cities have grown, a higher percentage of their surface area has become impermeable, preventing water from soaking into the ground. At the same time, building more and bigger storm drains and sewers into conventional underground piped networks has become prohibitively complex and expensive. Hard revetments (sloped reinforcements against flooding and erosion), such as retaining walls and barricades, are costly in terms of carbon and still run the risk of being overwhelmed by increasingly violent storms.

Sustainable drainage systems (SuDS) encompass various nature-based solutions for problems with drainage, water storage and flooding in cities. They include green roofs and walls and rainwater-collection systems. As we have seen, providing greater areas of blue space, such as ponds and wetlands, and renaturalising rivers helps to absorb stormwater and increase water storage for times of drought. Another integral part of any sustainable drainage plan is reducing hard surfaces and introducing road swales – broad, shallow, plant-filled roadside channels – and permeable paving, which can include pervious concrete, porous asphalt, paving stones and inter-locking paving.[126] The Mayor of London has provided a *Grey to Green* guide for communities that want to run their own de-paving projects in the city.[127]

Portland, Oregon, was one of the first cities to establish a sustainable drainage system, in the 1990s. Other cities – among them Philadelphia, New York and Melbourne – have adopted a nature-based approach since then. In Europe, Copenhagen made a 'Cloudburst Plan' to combat flash floods after a heavy storm dropped 150 millimetres of rain on the city in less than two hours in 2011, submerging streets under a metre of water and causing more than €800 million of damage. Now that extreme weather events are more common and sea levels in the harbour are expected to rise by a metre by 2110, Copenhagen's authorities estimated that the cost of doing nothing would amount to approximately €55–80 million a year.

A holistic solution

The greening of a city addresses numerous pressing problems at once. It provides sustainable drainage for the city, habitat for wildlife, carbon sequestration, and improved quality of life for residents.

The Lost Effra project in London (see page 471) is a good example. Its driving aim was to prevent flooding in areas such as Herne Hill, where, in 2004, half an hour of heavy rain overwhelmed sewers and waste water flowed into local shops and businesses. In 2012 a burst mains pipe leaking for just ten minutes caused similar devastation. The fact that the project is managed by the London Wildlife Trust indicates how important it is ecologically. Green roofs and rain gardens have been installed, and key areas de-paved within the river catchment. Concrete on the Lowden Road traffic island has been replaced with a rain garden. A total of 110 square metres of concrete have been removed from Oborne Close and replaced with hugely popular wildlife-friendly community gardens, with free water butts and a green roof-building workshop for local members. Air pollution has decreased, biodiversity has improved, the local community's enjoyment of the surroundings has dramatically risen, and the new green space is contributing to local food supplies.

Food production

Increasingly, attention is turning to growing crops in cities. Reducing transport miles and the cost of storage, packaging and refrigeration could, it is thought, make food cheaper and more accessible. This could be conventional intensive agriculture; around 100,000 tonnes of fresh food are produced annually in Dar es Salaam, Tanzania, and most of the milk and eggs consumed in Shanghai are now produced within the city limits.[128] But new-tech hydroponic systems and grow tents are also making headway, such as Gotham Greens rooftop farms in New York and Lufa Farms in Montreal.[129] In Guangzhou, China, one 150-square-metre rooftop produces hundreds of kilograms of vegetables a year from fourteen hydroponic tanks.[130]

However, greenhouses, vertical farms and hydroponics have their own carbon construction and running costs. Compared to soil-based roof gardens, these systems, especially if enclosed, are unlikely to provide ecosystem services to the same degree as other green space. They will not reduce the heat-island effect in the way that open-air green roof gardens do. They can't reduce and reuse stormwater run-off in the same way as soil-based gardens and living walls can, and they clearly don't provide habitat for wildlife.

One system of urban food production that does all this is generally overlooked, presumably because it doesn't register on the radar of big commerce. Garden allotments produce up to 35 tonnes of food per hectare, compared to 8 tonnes of wheat or 3.5 tonnes of oilseed rape per hectare under industrial farming.[131] This is because allotment gardens are labour-intensive and pack in a huge variety of crops in close proximity. The emphasis is on healthy soil and good compost and manure, generating several harvests per year. Good soil also sequesters carbon.

Because allotment crops are mostly grown for home consumption, gardeners tend to be more cautious with chemicals and have a greater understanding of the need for insects for pollination and pest control. Indeed, allotments have the highest insect diversity of any urban habitat, including gardens, parks and even city nature reserves.

An aerial view of a community garden with its footpath in London shows how complex and productive these tiny parcels of land can be when assiduously managed by green-fingered allotmenteers. Garden allotments produce up to 35 tonnes of food per hectare, compared to 8 tonnes of wheat or 3.5 tonnes of oilseed rape per hectare under industrial farming.

Gardening – particularly growing food – has tremendous benefits for physical and mental health, too.[132] A study in the Netherlands found that people with allotments tend to be healthier than the general population, especially in old age.[133]

In theory, careful, labour-intensive, small-scale horticulture of this kind could almost entirely ensure the UK's food security, reducing food miles and the nutrient drain involved in importing food from abroad. The UK currently consumes about 6.9 million tonnes of fruit and vegetables per year, of which 77 per cent is imported at a cost of £9.2 billion. According to entomologist Dave Goulson, under allotment-style management all the UK's current fruit and vegetable requirements could be produced from just 200,000 hectares of land

– the equivalent of 40 per cent of the area that is currently under gardens, and just 2 per cent of the area that is under industrial agriculture. Just a million people engaged in small-scale horticulture, in belts around cities, he believes, would provide sustainable food for the entire urban population. But we would have to learn to be content with a much higher proportion of seasonal fruit and vegetables, using imports from earlier growing seasons in warmer countries, such as Spain and France, as luxuries and to cover the 'hungry gap' (usually April to early June) between the end of winter brassicas and the beginning of spring crops.[134]

There is no shortage of people wanting to garden for their own food. Currently 120,000 people are on waiting lists for allotments in the UK.[135] Freeing up green belt and providing rooftop gardens could address this need. But more allotments could also provide food for wider distribution, enabling gardeners to make money out of their surplus crops. The pioneering REKO-ring movement (REKO stands for 'Real Consumption'), begun in Finland in 2013, enables producers to connect with consumers via Facebook groups. Producers put up lists of available products and consumers place their orders in advance (thereby reducing food waste); they then meet somewhere like a local car park to make the transaction. REKO-ring has spread through Scandinavia and now 800,000 people in Sweden (nearly 10 per cent of the population) have joined the movement. Prices are not fixed – they are set by each producer – and payments are made via SWISH, phone to phone. The system is highly profitable for small farmers and allotment gardeners, and customers pay less than they would in shops, while getting fresh local produce from people they trust.

Small-scale horticultural enterprises within and on the outskirts of cities could work in the same way, and perhaps encourage a new generation of small farmers and horticulturalists. In the UK, the post-Brexit shake-up of farming subsidies aims to pay farmers for public goods such as carbon sequestration, soil restoration, water storage and purification, and biodiversity, rather than food production alone. Under the previous basic farm payment scheme subsidies were not available to farms under 5 hectares. Community Supported Agriculture (CSA) and others are lobbying DEFRA to include small-scale farmers and fruit and vegetable growers in the new payment schemes.[136]

Some communities have used food production as the primary driver for greening and revitalising their town. The Incredible

Edible Todmorden project was established in 2008 by residents of this former mill town in Calderdale, West Yorkshire.[137] Vegetable beds, herb beds and fruit trees have been planted all over the town, providing free food for anyone who wants to pick it. Food markets, gardening and agricultural workshops, harvest festivals and cookery demonstrations are also run as part of the project. Its influence has been global. There are now 120 official Incredible Edible groups in the UK and over 700 worldwide.

This mural by the Rochdale Canal celebrates the Incredible Edible project in the town of Todmorden, West Yorkshire. Residents are encouraged to help themselves to free food growing in plant beds around the town.

In California in 1995 the world-renowned chef Alice Waters pioneered the Edible Schoolyard project. This non-profit organisation uses organic school gardens and kitchens to teach children how to grow and eat healthy food, and also as classrooms in which to learn about the connections between food, the economy, health and the environment. By 2019 the Edible Schoolyard had a network of over 9,500 kitchen garden programmes across the USA.

Joining the dots

Connecting green space in cities, creating an intricate, all-encompassing ecological web, not only increases the amount and viability of urban wildlife, but also allows urban nature to become resilient and self-sustaining. It helps to generate a flow of natural processes through the built-up environment, increasing the availability of ecosystem services to the public – including health and happiness.[138] The Greenspace Information for Greater London centre collects and shares data about wildlife, green space and green infrastructure in the capital, and provides advice and training for communities that want to create their own projects. The site has collated nearly 6 million species records and 13,000 open-space records, mostly thanks to citizen science. Its visual mapping tool identifies obvious areas where greenspace is missing, and can be used to model ecological corridors.[139] Clearly, every town and city needs its own plan for easy-to-navigate green-space data collection and green networks.

We've discussed the potential of railways, disused tracks, flyovers and tunnels, canals, streams, wetlands, rivers and floodplains, parks and avenues to function as corridors for urban wildlife, and for rooftops to act as ecological stepping stones. These are the arteries that can enable the flow of life through the city. Much of this greening will be down to the projects and policy decisions undertaken by municipalities, local councils and corporations. As voters, customers, philanthropists and shareholders, we all have a role to play in ensuring the environment and rewilding are at the top of their agendas, and that they provide for the long-term maintenance and protection of green space.

How this network is intensified, supplying the finer capillaries that can reach into every corner and cul-de-sac, is largely down to the actions of private individuals and local communities. We can all design and manage our gardens with nature at the forefront. We can collaborate and connect with our neighbours. We can promote and oversee the management of other green spaces over which we have influence, such as communal parks and gardens, and, crucially, advocate the greening of space to connect them.

Everywhere we turn we should be looking for space for nature. Cemeteries and graveyards can be fantastic wildlife habitat, as well as peaceful places for meditation and remembrance. St Mary's churchyard in Walthamstow Village, east London, is a wonderful example. An ancient graveyard respectful of people visiting their loved ones, it is also a working garden with a pond, cold frames and compost heaps, bat boxes and beehives. It hosts garden courses and nature workshops, seed swaps and a small flower market. Local gardeners can adopt an old, forgotten grave, tending and planting around it. As this example demonstrates, the Church can play a leading role in educating communities about the climate and biodiversity crisis and – in their words – be part of repairing God's creation. The UK charity Caring for God's Acre, established in 2000, provides advice from qualified ecologists and conservationists for those interested in managing their local burial ground for nature.[140]

Plenty of advice is available on how to increase biodiversity on our urban doorsteps (see Further Resources, pages 533–5). Plantlife, for example, publishes an excellent guide on how to transform roadside verges into wildlife havens.[141] But translating advice into action is another matter. It takes imagination to see the potential for flowers and greenery in places where there is currently none. Often councils, dominated by the older generations, are wilfully blind. 'Guerrilla gardening' – a movement that began in New York in the 1970s and was taken up by activists such as Ron Finley in Los Angeles and Richard Reynolds in London – can expand the vision, inspiring local residents and challenging management practices carried out by local authorities (such as spraying toxins and excessive mowing and pruning) that are hostile or positively harmful to wildlife.[142] Although controversial, guerrilla gardening has a light touch. By its very nature, it tends to win hearts and minds more than it antagonises. Reynolds, author of *On Guerrilla Gardening: A Handbook for Gardening without Boundaries* (2008), sneaks out at night to plant seeds, plants and occasionally trees in car parks, on roundabouts, in playgrounds, in cracks in pavements, at the base of street trees and in the bare earth between insect-hostile civic planting. They may not be the plants everyone would choose – often intended to make a statement, they can seem somewhat incongruous – but the sudden burst of a sunflower or the ripening of a hidden gooseberry bush can help to break the mould and reclaim the streets for nature and human enjoyment.

Pesticide-free towns and cities

Increasing green space in urban areas for human health and bio-diversity won't help if pesticides continue to be used. Some cities still routinely fog urban areas with pesticides in an effort to eradicate 'invasive' species. Sacramento in California recently waged war on the Japanese beetle, for example, and Miami and New Orleans spray seasonally to get rid of mosquitoes. Pesticides (see Chapter 11) are blanket killers of insect species across the board, harm human health, can make the soil toxic for years to come and, ironically, are rarely successful in eradicating the target species.[143] Many gardeners, however, continue to consider insects in general so undesirable that they employ contractors to routinely drench their gardens.

The town of Hudson in the Canadian province of Quebec led the way in 1991 by introducing a local law banning the use of all pesticides within town limits. Now more than 170 towns and cities in Canada, including Vancouver and Toronto, are pesticide-free (this includes herbicides and fungicides). Similarly, France has 900 *villes sans pesticides*, including Paris, which has been pesticide-free for more than fifteen years.

The UK has been notably slow on the uptake, with the exception of Glastonbury town council in Somerset, which banned herbicides in 2016, and the London Borough of Hammersmith and Fulham, which went pesticide-free in 2016.[144] At the time of writing, however, forty-six towns and cities around the UK, including Brighton & Hove, Sunderland, Worthing, Hexham, Reading, Chelmsford and Balerno on the outskirts of Edinburgh, have passed motions to ban glyphosate and/or phase out pesticides entirely.[145] Pesticide Action Network UK, which launched its Pesticide-Free Towns campaign in 2015, provides advice on how to lobby town councils to go pesticide-free.[146] In 2021 Dave Goulson sent a petition with 53,000 signatures to the UK government to ban urban and garden pesticides in order to protect bees, other wildlife and human health.[147] He received a bland response. The UK government stands by its position that pesticides are well regulated and harmless.

Putting cities at the heart of rewilding

It is tempting to think of rewilding purely in terms of rural land-scapes. But cities are where nearly all of us live. The city landscape is something we have to get right. Nature-based solutions answer almost every serious problem our cities face. Putting cities at the heart of the rewilding vision is one of the most important steps we can take for the future. Thinking holistically, as rivers help us to do, brings cities into context with mountains, farmland, green belt, wetlands and forests. Creating flow between a city and its hinterland, turning grey into green and blue, connects the solar plexus of human life with the rest of its physical body.

The economic incentives for urban rewilding are abundantly clear. Properties on tree-lined streets – just one example – have a higher value than those on predominantly grey streets. The trees themselves save the taxpayer money in terms of public health, reduce the costs of air conditioning and heating, and help in national and global efforts to mitigate climate change.

But what rewilding brings to society in terms of fairness and inclusivity is often overlooked. Urban rewilding brings all demographics together. The more green spaces there are in cities, the more accessible nature and its benefits are to all levels of society. One of the greatest inequalities in society is health. Rewilding can improve living conditions and mental and physical health for everyone, but particularly for those living in deprived areas.

Rewilding can create jobs and volunteering, and build communities. In Medellín, Colombia, the city's low-income citizens are given training to become gardeners and ecologists to look after their green corridors. Free Town in Sierra Leone pays its citizens to plant trees. In Paris, residents actively participate in the greening of their city. Just as in the countryside, rewilding can regenerate degraded areas, bringing life back in. In Metropolitan Detroit, abandoned, depressed, post-industrial areas along the river have been transformed into a river walk described as a 'beautiful, exciting, safe, accessible, world-class gathering place for all'. Nearly three million visitors now use it every year, strolling between the city and the Great Lakes.[148]

Education is one of the most important roles rewilding can play in the urban environment, bringing an appreciation for nature into the lives of city residents. This is, arguably, where a shift in thinking can reap the most rewards. Places where populations are densest can create a groundswell of support for nature recovery and action for climate change. They can exert pressure on governments and councils, driving an unstoppable force for change. No single city is providing all the solutions. But many are leading lights, exploring courageous, imaginative new ways to transform our cities. Rewilding can bring all these ideas together, holistically. Embracing a new vision for sustainable life in our cities will radiate hope for nature's recovery in every corner of the globe.

Eighteen colossal solar-powered 'Supertrees', designed by UK landscape architects Grant Associates in the Bay South garden (part of the 100-hectare Gardens by the Bay project in Singapore), act as vertical gardens of tropical flowers and climbing ferns, while also generating solar power, acting as air-venting ducts for nearby green conservatories, and collecting rainwater. They show how nature and hi-tech infrastructure can come together to address climate change and increase biodiversity in the cities of the future.[149]

According to scientists, the negative impacts of climate change are mounting much faster than predicted and are already hitting the world's most vulnerable populations in devastating ways.[150] UN secretary-general António Guterres calls this a 'damning indictment of failed climate leadership'. On the current trajectory, the impacts will soon be irreversible. It is now or never, according to scientists, to limit global warming, conserve global biodiversity and set in place strategies for the sustainable use of natural resources. Systemic change is needed right now.

Politicians are proving slow to act. The kind of change that is needed can only be achieved by an upwelling of global public action. Most pivotal moments in history, the radical changes that have advanced the course of human thinking and enlightened our collective behaviour, have happened as a result of an irresistible surge from grassroots. As citizens of the earth, it is up to us to drive the agenda, to change the mindset of the institutions that govern us.

At the same time, there are things we can do ourselves. The late, great American biologist E. O. Wilson told us that, if we are to save earth's life-support systems on which all species – including our own – depend, we must dedicate half of all terrestrial land to nature. How on earth do we do that, when so much of our land is depleted and fragmented? Rewilding is one of the most positive and exciting answers to that question. Across the world, rewilding projects have proved that nature is ready to respond, even in the unlikeliest of places. Rewilding shows how to ignite the natural processes and reinstate the creatures that restore ecosystems. If we do it in the right way and let nature take the lead, humans can be a keystone species, drivers of biodiversity, creators of habitat, repairers of damage.

Scale is important. But still more important is connectivity. The tiniest piece of land can make a vital difference if it relates to other patches of nature, helps to join the dots. Whatever scrap of soil we have agency over, whether it's a field, a grass verge or a window box, can make a difference. Opening our hearts to a wilder world, letting nature in to our backyards, increasing the potential for life on every inch of soil, is the key to our future. At Knepp, as we watch the rewilding movement grow, it feels as though the tendrils of change are spreading like mycorrhizal fungi fizzing with connections, fuelling the determination of individuals and communities from mountaintop to sea, from village green to city street. A wilder, more resilient world is within our reach.

How Wild Are You?

Is nature delicately balanced and fragile, or wild, unpredictable and dangerous? Should we restore ecosystems as they were in the past, protect landscapes and species as they are now, or let them change into the future? Are humans part of nature, or separate from it? How should we measure the value of nature, or the success of conservation?

Questions like these are central to debates about nature and rewilding. People hold different views about conservation – indeed, about nature itself. This is one of the reasons that rewilding divides opinion, but we don't think it needs to be divisive. For rewilding to be successful, it's important to understand clearly our motivations, consider different perspectives and find common ground.

This quiz is designed to help you explore your views about nature and rewilding. It's based on research (that we took part in) by social scientist Dr Benedict Dempsey, who did some of his PhD fieldwork at Knepp.[1] Ben identified four different perspectives on conservation, and the quiz will show you which of these best represents your views, as well as offering suggestions for how to develop your thinking further.

There is no 'right' answer. People's opinions are subjective, and will probably be made up of a mixture of the four viewpoints. A conservationist responsible for a precious nature reserve is likely to have a different view from someone managing land with little biodiversity value, as Knepp was when we started our rewilding project. This quiz will help you discover what blend of views makes up your own unique perspective.

How to take the quiz

For each question, choose which answer you most closely agree with (A, B, C or D) and write it down. You may well agree with more than one answer, but pick the one that resonates with you the most. At the end, count up how many of each letter you have chosen to see which perspective you most agree with, and read about it on the next page.

1. Which of these statements best represents your view of nature as a whole?

A) Nature is always changing. We should help it thrive by managing that change with a range of techniques, such as protected areas combined with habitat restoration.

B) Nature is under huge pressure from humans. We must protect vulnerable species and habitats in nature reserves.

C) Nature is dynamic and unpredictable. We should restore ecosystems – including by reintroducing lost species – then take a hands-off approach.

D) Nature should be allowed to be wild. We should try not to interfere with it at all, but give it space to recover and thrive.

2. How do you feel about attempts to restore ecosystems to the way they were in the past?

A) We shouldn't try to recreate nature from the past, because nature is continually changing. We need to avoid having fixed ideas about where nature should be.

B) We should try to restore habitats that existed in the last few centuries and protect the precious fragments of biodiversity that still exist.

C) We should embrace radical visions of what nature could be like in the future, even if they challenge our current ideals, not try to recreate what existed in the past.

D) We should try to recreate wild ecosystems from the deep, prehistoric past, as they were before humans started damaging them.

3. Imagine you're managing a conservation project aiming to restore natural ecosystems. How would you set goals for the project?

A) We need clear objectives, but should also accept that we may have to adapt as things change over time.

B) We should set very clear, specific objectives and try to reduce uncertainty as much as possible.

C) We should not tie ourselves too tightly to specific goals, because nature is inherently unpredictable. We should be able to experiment with different approaches.

D) We should avoid setting objectives that constrain the ability of nature to take the lead, as well as experimental interventions that might have negative consequences.

4. Imagine that you're asked to consider a proposal for introducing a species that used to exist in your area. What is your reaction?

A) Carefully managed reintroductions of missing native species, like lynx, should be welcomed where appropriate, as part of a range of approaches for restoring nature.

B) Reintroductions of native species like lynx may be acceptable, but they should take a back seat to protecting the threatened species that are already there.

C) Species introductions should be a central part of conservation and we need to be more daring. We should also consider introducing non-native or domestic 'proxy' species that can play the role of extinct species.

D) We should reintroduce native wild species to help re-establish wild ecosystems, but should not introduce non-native or domestic 'proxies' like cattle and ponies.

5. Where do you think most nature conservation and restoration activity should be targeted?

A) We should help nature everywhere – in towns and cities, farmed land and nature reserves – and connect it up in a wider network.

B) We should concentrate on protecting the places that have the most threatened species and habitats, such as nature reserves, even if those places are quite small.

C) We should concentrate on large, radical initiatives where people play an active role in restoring nature at scale.

D) We should designate large areas for wild nature to exist – including some places where people are not permitted to go.

6. What is your attitude to people actively intervening to shape ecosystems (for example by creating new habitats)?

A) Intervening is fine. We generally have enough scientific knowledge to reduce the risk of causing harm – and if necessary we can change as we go along.

B) We need to manage nature, but should be cautious about using anything but tried-and-trusted techniques. The wrong intervention may do more harm than good.

C) Radical intervention is needed because current methods are not working. We can't know the exact outcome, but we can take calculated risks to help nature recover.

D) Ecosystems are too complex and unpredictable for us to try to 'design' how they function. Our best approach is to keep out of the way and let nature take the lead.

7. Imagine you work for a national park authority where the landscape has been shaped by human activity for thousands of years. What would be your general approach?

A) Rural landscapes have always changed due to changes in farming practices, economy and climate. We need to balance protecting habitats that already exist and allowing gradual change.

B) In the UK, important habitats and their species (like hedgerows, woodland, heathland and meadows) are the result of historical human activity. We need to make sure we protect these habitats.

C) Rural communities should be incentivised to be more radical in delivering large-scale nature restoration. We need to be more innovative.

D) We should create some large areas where nature can exist free of human interference. That will probably mean letting some farmed landscapes go wild.

8. If you have a garden, how might you manage it to benefit biodiversity?

A) I want my garden to fulfil a range of functions, including as a place for people to enjoy. I want to create a mixture of habitats, including plant species that are tolerant of climate change. I will talk to my neighbours about how we can help nature across all our gardens.

B) I want to build habitats that will support vulnerable native species that are already in the area – for example by creating a nature pond. I will then carefully manage my garden to encourage and protect those species.

C) I want to try different ways of encouraging biodiversity – for example having areas of poor soil and piles of crushed bricks, and bringing in non-native species if they are good for attracting wildlife. My neighbours might think I'm unusual!

D) I want to manage my garden as little as possible and let the whole thing run wild.

9. How do you feel about the use of new genetic technology to support nature restoration? For example 'de-extinction' (attempts to bring back extinct species like mammoths, passenger pigeons or dodos)?

A) I think genetic science can be useful, but experimental approaches can be unrealistic and should not be a priority.

B) I completely reject the use of experimental genetic technology like de-extinction in conservation, because tinkering with genes may have unpredictable outcomes.

C) I think a wide range of genetic technologies should be considered as part of the more radical approach to nature recovery that we need.

D) I strongly dislike the idea of using genetic technology, because how can nature be wild if it has been genetically engineered by people?

10. How important is it to prioritise the protection of species that are native to the UK, relative to non-native species?

A) It is quite important. Protecting native species should be a significant element of what conservationists do, but in some cases it will be sensible to accept the presence of non-native species too.

B) It is very important. Protecting native species is one of the most important objectives of conservation and should include eradicating non-native species if necessary.

C) It is not especially important. The priority should be how the ecosystem functions as a whole – and that could include non-native species as well as native ones. In any case, species have always moved around, so who is to say what is 'native'?

D) It is important. We should be re-establishing wild nature as it was before humans introduced non-native species.

Results

A) Management of Changing Nature

This is a relatively moderate, mainstream position. You are comfortable with the idea that nature changes over time, and oppose attempts to recreate ecosystems from the past. You believe that people are part of nature, and that we have a role to play in managing ecosystems. You likely favour setting nature recovery targets in law – because otherwise nature will not be protected. You'll be happy to integrate conservation into wider society and the economy.

Alongside traditional conservation management techniques, you may be happy to include rewilding projects in your work as well. In your own garden you will likely be doing a range of things to balance different needs. As a conservationist, you would be comfortable managing a protected site with specific legal requirements, but you would want to allow the ecosystem to change over time. Overall, your approach is pragmatic, flexible and adaptable.

If you're considering your own nature restoration project, you might want to mix it up a bit with different approaches – some of them quite tightly controlled, others introducing a bit more uncertainty. Are there ways you can enable your project to adapt in the future, for example by encouraging species that will tolerate the changing climate? Can you identify a set of indicator species you can monitor over time? You may also want to link your project to others in your area, to give your work impact on a bigger scale.

Consider whether you have chosen the best ways to judge the success of your project. If you are focused on particular species, are you sure you haven't lost sight of the bigger, whole-ecosystem picture? Could you be more radical?

B) Protection of Threatened Nature

This is a careful perspective. You are probably focused on protecting what nature still exists, using established methods that prioritise endangered species and nature reserves. You believe that nature is fragile and favour its close management; you value traditional rural practices, heritage and human-made landscapes such as farmland. You worry that more radical conservation approaches put at risk species and habitats that conservationists have spent decades trying to protect. Your focus is likely to be on the present and recent past, and you may be suspicious of 'novel' ecosystems. Overall, your approach is careful, dedicated and diligent.

You probably want to protect something that is already precious, rather than building nature back from a low base. Which species or habitats are most important to you, and why? Are you able to tell whether they are in good or bad condition? In a garden, what can you plant to provide food or habitat for the most threatened species? Can you build special habitats for them, for example bug hotels?

Can you think of ways to allow more flexibility and uncertainty into your project, without risking the loss of the nature that already exists? Are there ways to connect your project with a wider network of sites? Do you have a long-term objective to expand nature, not just prevent decline? How will your project cope with change, especially that driven by the warming climate?

C) Innovation in Nature

Innovation in Nature is a more radical perspective. You reject the re-creation of past ecosystems and you're comfortable with the idea of novel ecosystems. You may well think of nature as being wild, unpredictable and robust. You probably don't like the idea of too much management and may get frustrated with the constraints of bureaucratic approaches. People aligned with this viewpoint think traditional conservation is not doing enough, and that more innovative approaches are needed.

You are able to embrace experimental approaches that may have unclear outcomes. You are probably in favour of intervening in nature – particularly through species introductions – and then letting natural processes take their course. You probably won't want to be constrained by legal targets and strict management frameworks. Overall, your approach is innovative, experimental and ambitious.

Can you introduce new species or features that will perform a particular role, or increase the complexity of the ecosystem? In a garden, for example, which combination of plant species will have the most transformative effect? What kind of intervention or disturbance can you implement to benefit nature, for example reducing soil fertility or creating a wet area? How does your work link up to the wider landscape, rivers or sea in your area?

Are you really comfortable with the idea that you might not know where your project is heading? Could it have negative unintended consequences? Might you meet resistance from people who are worried about what you are doing? Are there ways to mitigate the potential risks to existing nature?

D) Re-establishment of Wild Nature

This view is defined by the belief that nature should be allowed to exist unmolested by people. Rather than embracing human-influenced landscapes, you prefer to anchor conservation in the deep past, when nature existed without interference. You think of nature as wild and free, and pristine, untouched ecosystems are the best of all.

You are less enthusiastic about types of rewilding driven by human intervention like species introductions, and more enthusiastic about 'letting nature go' and allowing it to be wild. You may be willing to see some species disappear from the UK, if that meant we could return managed, human landscapes to something closer to their prehistoric condition. Overall, your approach is radical and visionary.

You want to 'manage' nature as little as possible. Think about what that might look like in practice, and how you can minimise your intervention. Think about the scale of your project; if it is a fairly small area or garden, how will letting it run wild relate to the nearby landscape? How big an area do you need for this kind of approach?

Are you sure you want to remove human involvement from nature? Could careful intervention or management ever increase the chances of nature recovering – especially in the absence of species that existed in the past? Is there any value in ecosystems that have been enmeshed with people for a long time? If you remove all management, how might your neighbours react? Are there species you care about that exist thanks to human-managed habitat – and can you accept that those species might disappear?

Notes

The notes for this book can be found at https://www.bloomsbury.com/uk/thebookofwildingnotes where they can be downloaded and read at leisure.

Further Resources

CHAPTER 1: WHAT IS REWILDING?

Books and articles

Barlow, C., *The Ghosts of Evolution*, Basic Books, 2000

Carroll, S. B., *The Serengeti Rules*, Princeton University Press, 2016

Carver, S. et al., 'Guiding principles for rewilding', *Conservation Biology*, 2021, vol. 35, issue 6, pp. 1,882–93

Flannery, T., *Europe – The first 100 million years*, Penguin, 2019

Foreman, D., *Rewilding North America*, Island Press, 2004

Fraser, C., *Rewilding the World*, Metropolitan Books, 2009

Groves, C., *Guidelines for Conserving Connectivity through Ecological Networks and Corridors*, IUCN Publication, 2020

Harris, J. and Kopecky, D., 'Ecosystem restoration: Securing biodiversity, complexity and resilience', *Environmental Scientist*, 2022, vol. 31, issue 3. Articles on ecosystem restoration, including international policy, climate and biodiversity, featuring successful examples including the National Trust's Wicken Fen and Knepp.

Hilty, J. et al., *Guidelines for conserving connectivity through ecological networks and corridors*, IUCN report, 2020

Jepson, P. and Blythe, C., *Rewilding: The radical new science of ecological recovery*, Icon Books, 2020

Jones, C. G. et al., 'Organisms as ecosystems engineers', *Oikos*, 1994, vol. 69, pp. 373–86

Kerr, M., *Wilder – How rewilding is transforming conservation and changing the world*, Bloomsbury Sigma, 2022

Kurtén, B., *Pleistocene Mammals of Europe*, Routledge, 1968

Ledger, S. E. H. et al., *Wildlife Comeback in Europe: Opportunities and challenges for species recovery*, Rewilding Europe report, Zoological Society of London, 2022

Levy, S., *Once and Future Giants: What Ice Age extinctions tell us about the fate of Earth's largest mammals*, Oxford University Press, 2011

Marris, E., *Rambunctious Garden: Saving nature in a post-wild world*, Bloomsbury, 2011

Martin, P., *Twilight of the Mammoths: Ice Age extinctions and the rewilding of America*, University of California Press, 2005

Monbiot, G., *Feral: Searching for enchantment on the frontiers of rewilding*, Allen Lane, 2013

Pereira, H. M. and Navarro, L. M. (eds), *Rewilding European Landscapes*, Springer, 2015

Quammen, D., *The Song of the Dodo: Island biogeography in an age of extinctions*, Scribner, 1996

Smit, R., *The Oostvaardersplassen: Beyond the horizon of the familiar*, Staatsbosbeheer, 2010

Thompson, K., *Where Do Camels Belong? The story and science of invasive species*, Profile Books, 2014

Wilson, E. O., *Half Earth*, Liveright, 2016

—, *The Diversity of Life*, Harvard University Press, 1992

Other sources

ARK Nature, www.ark.eu Ark Nature has pioneered

rewilding since its founding in 1989. It works mainly in the Netherlands but has projects in other European countries. It is a founding partner of Rewilding Europe.

Center for Large Landscape Conservation, largelandscapes.org Based in Montana, USA, this organisation works with over 2,000 community-based conservation projects around the world. Its main focus is on connectivity and wildlife corridors.

Citizen Zoo, www.citizenzoo.org A social enterprise that supports community rewilding projects around the UK.

Coalition for Wildlife Corridors in India, corridorcoalition.org An excellent example of a collaboration of people and organisations working to advance connectivity conservation.

Connectivity Conservation Specialist Group, conservationcorridor.org Established in 2016 under the IUCN World Commission on Protected Areas to restore ecological connectivity.

Endangered Landscapes Programme, www.endangeredlandscapes.org Funds large-scale landscape restoration projects across Europe.

EU Biodiversity Strategy for 2030, https://sdgs.un.org/partnerships/eu-biodiversity-strategy-2030-bringing-nature-back-our-lives The EU's ambitious, long-term plan to reverse the degradation of ecosystems across Europe.

European Nature Trust (TENT), theeuropeannaturetrust.com Runs conservation projects in the UK and Europe, and produces documentary films to raise awareness of its and others' work.

Nature-based Solutions Evidence Platform, www.naturebasedsolutionsevidence.info An interactive map linking nature-based solutions to climate change adaptation based on a systematic review of the peer-reviewed literature.

Rewilding Britain, 'Defining Rewilding', https://www.rewildingbritain.org.uk/explore-rewilding/what-is-rewilding/defining-rewilding Founded in 2015 with Charlie as its chair, to promote rewilding in Great Britain.

Rewilding Europe, rewildingeurope.com Based in the Netherlands, works to create rewilded landscapes throughout Europe.

Wild11, *Global Charter for Rewilding the Earth,* 11th World Wilderness Congress, 2020, Jaipur, India

Get involved

Global Rewilding Alliance, rewildingglobal.org A network of more than 125 rewilding projects around the world. Also offers advice to individuals on how to make a difference both in daily life and by joining rewilding initiatives.

Rewilding Britain, www.rewildingbritain.org.uk Its Rewilding Network provides advice and support to rewilding projects and connects rewilders with local groups and NGOs.

Rewilding Europe, https://rewildingeurope.com/join-us/become-active/ Runs a Rewilding Volunteer Database, the Rewilding Training Tourism programme and Rewilding Europe Travel. Also runs the European Rewilding Network, which helps members communicate about their rewilding work to other members and a wider European audience, and to share the best rewilding practices. https://rewildingeurope.com/european-rewilding-network/

CHAPTER 2: REWILDING IN THE UK

Rewilding works hand in hand with nature-friendly farming. Here, alongside publications on declining nature in the UK, we recommend some of our favourite reading, consultancies and events on regenerative agriculture – a movement gathering momentum around the world – which demonstrate how it reverses the environmental impacts of industrial agriculture, supplies sustainable food and keys into wider nature restoration.

Books and articles

Brown, G., *Dirt to Soil*, Chelsea Green, 2018

Fiennes, J., *Land Healer: How farming can save Britain's countryside*, BBC Books, 2022

Granstedt, A. and Thomsson, O., 'Sustainable agriculture and self-sufficiency in Sweden – Calculation of climate impact and acreage need based on ecological recycling agriculture farms', *Sustainability*, 2022, vol. 14 (10), pp. 1–23

Harvey, G., *The Carbon Fields – How our countryside can save Britain*, Grassroots, 2008

Henderson, G., *The Farming Ladder*, Vintage Farming Classics, Marcel Press, 2010. First published in 1943, a prescient exhortation to sustainable, profitable, small-scale farming.

Hetherington, D., *The Lynx and Us*, Wild Media Foundation, 2018

Lawton, J. (Chair), *Making Space for Nature – A review of England's wildlife sites and ecological network*, Independent report to the Secretary of State, the Department for Environment, Food and Rural Affairs, 2010, https://web archive.nationalarchives.gov.uk/ukgwa/20130402170324/http:/archive.defra.gov.uk/environment/biodiversity/documents/201009space-for-nature.pdf

Lymbery, P., *Sixty Harvests Left: How to reach a nature-friendly future*, Bloomsbury, 2022

Massey, C., *The Call of the Reed Warbler: A new agriculture, a new earth*, Chelsea Green, 2018

McCarthy, M., *Moth Snowstorm*, John Murray, 2016

—, *Say Goodbye to the Cuckoo*, John Murray, 2010

Mikolajczak, K. et al., 'Rewilding – The farmers' perspective. Perceptions and attitudinal support for rewilding among the English farming community', *People and Nature*, 2022, vol. 4, issue 6, pp. 1–15. https://doi.org/10.31235/osf.io/u3a5e

Mondière, A., Corson, M. S., Morel, L. and van der Werf, H. M. G., *Agricultural rewilding: A prospect for livestock systems*, 2021, https://doi.org/10.32942/osf.io/mv6dn

Montgomery, D., *Growing a Revolution: Bringing our soil back to life*, W. W. Norton & Co, 2017

Painting, A., *Regeneration: The rescue of a wild land*, Birlinn, 2021

Poux, X. and Aubert, P-M., 'An Agroecological Europe in 2050 – Multifunctional Agriculture for Healthy Eating. Findings from the Ten Years For Agroecology (TYFA) modelling exercise', *Iddri-AScA*, 2018, Study No. 09/18

Pywell, R. F. et al., 'Wildlife-friendly farming increases crop yield – Evidence for ecological intensification.' *Proceedings of the Royal Society B*, 2015, vol. 282, issue 1816, http://dx.doi.org/10.1098/rspb.2015.1740

Rebanks, J., *English Pastoral*, Penguin, 2021

Schofield, L., *Wild Fell: Fighting for nature on a Lake District hill farm*, Doubleday, 2022

Thomas, V. et al., 'Domesticating rewilding – Combining rewilding and agriculture offers environmental and human benefits', in Bruce, D. and Bruce, A. (eds), *Transforming Food Systems: Ethics, innovation and responsibility*, Wageningen Academic Publishers, 2022, Section 24, pp. 165–70

Tree, I., *Wilding: The return of nature to a British farm*, Picador, 2018

Wheeler, N. et al., *Smarter Flood Risk Management in England*, Green Alliance policy insight, 2017, https://green-alliance.org.uk/publication/smarter-flood-risk-management-in-england-investing-in-resilient-catchments/

Other sources

Farming & Wildlife Advisory Group (FWAG), www.fwag.org.uk Comprises regional groups providing independent advice to the UK farming community on the environmental value of their land and how to benefit from agri-environment funding.

Groundswell, groundswellag.com The annual Regenerative Agriculture Show and Conference, begun by John and Paul Cherry in 2015 at Lannock Manor Farm in Hertfordshire, UK, has become an influential forum for farmers, food producers, policy makers and environmentalists to explore new ideas about regenerative systems.

Linking Environment and Farming (LEAF), www.leaf.eco A membership organisation, it advises on regenerative farming and nature-based solutions for farmers and the food industry in the UK.

Natural England, *Lost Life – England's lost and threatened species*, 2010. Special report, produced to coincide with the International Year of Biodiversity, identifying nearly 500 animals and plants that have become extinct in England – almost all within the last two centuries. http://publications.naturalengland.org.uk/publication/32023

Nature Friendly Farming Network (NFFN), www.nffn.org.uk Unites farmers and the public with a passion for wildlife and sustainability in farming in the UK.

Our Planet, WWF and the World Economic Forum, Film and article about Knepp, https://www.weforum.org/agenda/2019/10/farming-agriculture-biodiversity-wildland/

Oxford Real Farming Conference, orfc.org.uk Held every January, this is the largest gathering of the agroecological movement on the planet, connecting people in the UK and around the world with the aim of transforming the food and farming system.

Rewilding Britain, *Adapting to Climate Heating: How rewilding can help save Britain's wildlife from extinction during the climate emergency*, 2020, https://bit.ly/3Rjcwty

—, *Reforesting Britain: Why natural regeneration should be our default approach to woodland expansion*, 2021, https://www.rewildingbritain.org.uk/news-and-views/research-and-reports/reforesting-britain

Soil Association, www.soilassociation.org The main certification body in the UK for organic accreditation in food production, it also campaigns within communities, schools, hospitals and government for healthier food and sustainable, nature-based agriculture.

Sustainable Food Trust, sustainablefoodtrust.org
A global organisation working to accelerate the transition to more sustainable food and farming systems.

Tompkins Conservation, *Livestock Guardian Dogs*, Wildlife Bulletin, March 2017, no. 2

Get involved

Community Forest Trust, https://englandscommunity forests.org.uk/englands-community-forests-map
This organisation often runs volunteering groups. The Restoring Hardknott Forest Project in the Lake District National Park, for example, recruits volunteers for practical tasks such as removing non-native Sitka spruce from the site, as well as for help with monitoring and research. See the list of local Community Forests on the website for events and activities.

Knepp Estate, West Sussex, knepp.co.uk
We're open from Easter to the end of October for glamping, camping, guided safaris and rewilding workshops, plus 25km of public and permissive footpaths for year-round access. We run a team of about 200 volunteers helping with tasks from public footpath management to feeding white storks.

Nattergal, www.nattergal.co.uk Two rewilding projects, at Boothby in Lincolnshire and High Fen in Norfolk, and workshops and guided tours on rewilding.

Rewilding Britain, www.rewildingbritain.org.uk/rewilding-projects Rewilding projects big and small in the UK, most of which you can visit. Some may, in time, offer courses and workshops on rewilding.

Rewilding Coombeshead, rewildingcoombeshead.co.uk
Reintroduction specialist Derek Gow runs courses on practical rewilding and species introductions from his

300-acre rewilded farm in Devon where you can also stay, camp and go on guided rewilding walks.

The Conservation Volunteers, www.tcv.org.uk
This provides information about volunteering in nature restoration projects in the UK.

Wild Ken Hill, wildkenhill.co.uk This 4,000-acre estate in Norfolk is a combination of rewilding project, conventional nature reserve and regenerative farm, offering guided tours, workshops and The Gathering, an annual nature festival.

Wildlife Trusts, https://www.wildlifetrusts.org/closer-to-nature/volunteer A federation of 46 independent wildlife conservation charities covering the whole of the UK, with more than 870,000 members. Their aim is to restore 30% of land and sea for nature by 2030. Contact the Wildlife Trust in your area for advice on nature restoration and volunteering opportunities.

CHAPTER 3: REWILDING WATER

Books and articles

Addy, S. et al., 'River Restoration and Biodiversity: Nature-based solutions for restoring rivers in the UK and Republic of Ireland', IUCN, 2016, CREW (Centre of Expertise for Waters) reference: CRW2014/10

Biggs, J. and Williams, P., *Ponds, Pools & Puddles*, New Naturalist series, HarperCollins, 2023

Brazier, R. E. et al., 'River Otter Beaver Trial: Science and Evidence Report', 2023, https://bit.ly/3kWk2ih

Campbell-Palmer, R., et al., *The Eurasian Beaver Handbook*, Pelagic Publishing, 2016

Coles, B., *Beavers in Britain's Past*, Oxbow Books, 2006

Collier, E., *Three Against the Wilderness*, Touch Wood Editions, 1959, reprinted 2007. The true story of a trapper from Yorkshire who restores wildlife in an over-hunted landholding in British Columbia by reconstructing an old beaver dam.

Elliott, M. et al., *Beavers – Nature's Water Engineers. A summary of initial findings from the Devon Beaver Projects*, Devon Wildlife Trust, 2017, https://bit.ly/3WN6ov2

Gow, D., *Bringing Back the Beaver*, Chelsea Green, 2020

Heritage, G. et al., *A Field Guide to British Rivers: Implications for restoration*, Wiley-Blackwell, 2021

Howard, J., *The Wildlife Pond Book: Create your own pond paradise for wildlife*, Bloomsbury, 2019

Lewis-Stempel, J., *Still Water: The deep life of the pond*, Black Swan, 2020

Perfect, C. et al., *The Scottish Rivers Handbook*, 2013, CREW (Centre of Expertise for Waters) reference: C203002. Available online at www.crew.ac.uk/publications. See page 14 for a fascinating analogy of a river system functioning like the human body.

Raven, P. and Holmes, N., *Rivers: A natural and not-so-natural history*, British Wildlife Collection, Bloomsbury Wildlife, 2018

Rothero, E., Lake, S. and Gowing, D. (eds), *Floodplain Meadows: Beauty and Utility: A Technical Handbook*, Milton Keynes, Floodplain Meadows Partnership, 2016

Sayer, C. D. et al., 'The role of pond management for biodiversity conservation in an agricultural landscape', *Aquatic Conservation: Marine and Freshwater Ecosystems*, 2012, vol. 22, issue 5, pp. 626–38

Sayer, C. et al., 'Restoring the ghostly and the ghastly – a new golden age for British lowland farm ponds?' *British Wildlife*, 2022, vol. 33, no. 7, pp. 477–87

Schwartz, J., *Water in Plain Sight*, St. Martin's Press, NY, 2016. A brilliant explanation of the global ecology and physics of water.

Williams, P. et al., *The Pond Book: A guide to the management and creation of ponds*, Freshwater Habitats Trust, Oxford, 2010

Zeedyk, B. and Clothier, V., *Let the Water do the Work: Induced meandering, an evolving method for restoring incised channels*, Chelsea Green, 2014

Other sources

Amber Barrier Atlas, https://amber.international/european-barrier-atlas/ Pan-European database of artificial in-stream barriers.

DEFRA, https://www.gov.uk/guidance/owning-a-watercourse

Floodplain Meadows Partnership, www.floodplainmeadows.org.uk

Rewilding Britain, *Rewilding and Flood Risk Management*, 2020, http://bit.ly/3WP7WES

Get involved

Catchment Based Approach (CaBA) Partnerships, catchmentbasedapproach.org Launched in 2013 by DEFRA, they are currently working in 100+ catchments across England and Wales with farmers, landowners, communities, local authorities, water companies, local businesses and NGOs to restore river catchments. See their website for your nearest project.

Freshwater Habitats Trust, https://freshwaterhabitats.org.uk/pond-clinic/ The website has an online Pond Clinic with advice on pond creation and management in factsheet form.

River Action, riveractionuk.com A UK campaigning charity addressing the problem of river pollution, monitoring and challenging industrial and agricultural polluters.

River Restoration Centre, https://www.therrc.co.uk/supporting-uk-trusts-partnerships-and-community-groups Advice, training and factsheets for all aspects of river restoration. Its Citizen River Habitat Survey involves river enthusiasts and citizen scientists in recording and assessing the ecology and water quality of rivers and streams.

—, 'Manual of River Restoration Techniques', 2020, www.therrc.co.uk/manual-river-restoration-techniques

Rivers Trust, theriverstrust.org The umbrella organisation for 65 member river trusts in the UK. They provide conservation expertise to farmers, government, businesses and communities to restore rivers. They offer advice on how to create the most effective campaign to clean up your local river, and volunteering opportunities in partners' projects.

CHAPTER 4: REWILDING WITH PLANTS

Books and articles

Alexander, K. N. A., 'The invertebrates of Britain's wood-pastures', *British Wildlife*, 1999, vol. 11, issue 2, pp. 108–17

Alexander, K. et al., 'The value of different tree and shrub species to wildlife', *British Wildlife*, Oct 2006, no. 18

Blakesley, D. et al., *Realising the wildlife potential of young farm woods in South East England*, East Malling Research, 2010

Broughton, R. K. et al., 'Slow development of woodland vegetation and bird communities during 33 years of passive rewilding in open farmland', *PLoS ONE*, 2022, vol. 17 (11), e0277545

Çolak, A. H. et al., (eds), *Ancient Woods, Trees & Forests*, Pelagic Publishing, 2023

Crane, E., *Woodlands for Climate and Nature*, RSPB report, 2020

Di Sacco, A. et al., 'Ten golden rules for reforestation to optimize carbon sequestration, biodiversity recovery and livelihood benefits', *Global Change Biology*, 2021, vol. 27, issue 7, pp. 1,328–48

Green, T., 'The importance of an open-grown tree – from seed to ancient', in Çolak, A. H. et al., *Ancient Woods, Trees & Forests*, Pelagic Publishing, 2023, chapter 5, pp. 91–8

Harding, P. T. and Rose, F., *Pasture-Woodlands in Lowland Britain: A review of their importance for wildlife conservation*, Institute of Terrestrial Ecology, 1986

Lewis, P., *Making a Wildflower Meadow*, Frances Lincoln, 2015

Lewis, S. L. et al., 'Restoring natural forests is the best way to remove atmospheric carbon', Comment, *Nature*, 2 April 2019 https://bit.ly/3Y8bfaZ

Lonsdale, D. (ed), *Ancient and other veteran trees: Further guidance on management*, Ancient Tree Forum, 2013

Painting, A., *Regeneration: The rescue of a wild land*, Birlinn, 2021

Peterken, G. F., *Natural Woodland: Ecology and conservation in northern temperate regions*, Cambridge University Press, 1996

—, *Meadows*, Bloomsbury Wildlife, 2018

—, 'A long-term perspective on rewilding woodland', *British Wildlife*, August 2022, vol. 33, no. 8, p. 584

Rackham, O., *Woodlands*, Collins New Naturalist Library, William Collins, 2015

Read, H., *Veteran Trees: A guide to good management*, Natural England, 2000 publications.naturalengland.org.uk/publication/75035

Sheldrake, M., *Entangled Life: How fungi make our worlds, change our minds and shape our futures*, Random House, 2020

Shrubsole, G., *The Lost Rainforests of Britain*, William Collins, 2022

Simard, S., *Finding the Mother Tree: Uncovering the wisdom and intelligence of the forest*, Penguin, 2022

Stiven, R. and Holl, K., *Wood Pasture*, Scottish Natural Heritage, 2004

Thomas, P., *Trees*, Collins New Naturalist Library, William Collins, 2022

Vera, F., *Grazing Ecology and Forest History*, CABI Publishing, 2000

Other sources

Economist, 'Climate Change: The Trouble with Trees', 2021, www.youtube.com/watch?v=EXkbdELr4EQ

Flora Locale, This website sadly no longer exists but its library of guides on wild meadow creation, managing new grasslands and sourcing plants, etc., is now managed by the Chartered Institute of Ecology and Environmental Management (CIEEM). https://cieem.net/?s=Flora+Locale

Forestry Commission, Scotland, *Management of Ancient Wood Pasture*, 2009. A practical guide. https://forestry.gov.scot/images/corporate/pdf/fcsancientwoodpasture guidance.pdf

Institute of Terrestrial Ecology, 1986, https://nora.nerc.ac.uk/id/eprint/5146/1/Pasture_woodland.pdf Ahead of its time, anticipating the work of Frans Vera, this publication identifies the enormous value of ancient wood pasture as an ecosystem.

Kneppflix, A natural regeneration cartoon under our Knepp media label. https://www.youtube.com/watch?v=0UUfxw1S8pU&ab_channel=Kneppflix

Plantlife, *Our Vanishing Flora*, 2012, https://www.plantlife.org.uk/uk/our-work/publications/our-vanishing-flora

PuRpOsE, Funded by the Biotechnology and Biological Sciences Research Council, this project has identified a staggering 2,300 species (excluding fungi, bacteria and micro-organisms) associated with the oak, https://www.actionoak.org/projects/purpose-uncovering-biodiversity-oak-trees

Rewilding Britain, 'Reforesting Britain – Why natural regeneration should be our default approach to woodland expansion', 2020, www.rewildingbritain.org.uk/reforesting-britain

Rewilding Europe, 'How living forests can mitigate our climate and biodiversity crises', https://rewildingeurope.com/wp-content/uploads/2020/02/GRAZELIFE-Herbiforests-RewildingEurope.pdf

Woodland Trust, *Trees in Historic Parks and Landscape Gardens*, Woodland Trust, 2008. Practical guidance: ancient tree guide 2, available to download at www.woodlandtrust.org.uk/publications

—, *Ancient Woodland Restoration: Introduction to the principles of restoration management*, 2020, www.woodlandtrust.org.uk/publications

Woodwise, https://www.woodlandtrust.org.uk/media/49178/woodwise-woods-in-waiting-autumn-2020.pdf Special issue of the Woodland Trust's magazine, dedicated to natural regeneration.

Get involved

Ancient Tree Forum, www.ancienttreeforum.org.uk
Events and workshops in the UK. Join the Ancient Tree
Hunt to improve the national database for ancient and
veteran trees.

Continuous Cover Forestry Group (CCFG),
www.ccfg.org.uk Runs field meetings and workshops
on commercial forestry management.

Plantlife, www.plantlife.org.uk Provides expert advice
and guidance on creating wildflower meadows in
the UK. See also the associated website Coronation
Meadows (run by Plantlife, the Wildlife Trusts and
the Rare Breeds Survival Trust): coronationmeadows.
org.uk

Trees for Life, https://treesforlife.org.uk/support/
volunteer/ A highly effective, passionate, community-
based organisation in Scotland, which connects
surviving patches of original Caledonian forest
through a combination of tree planting and rewilding.
To volunteer, see the website.

UK Wood Pasture and Parkland Network,
ptes.org/wppn/ Managed by the People's Trust for
Endangered Species, it has excellent videos and case
studies on the restoration of wood pasture and parkland,
as well as volunteering opportunities for survey work and
professional training days.

Woodland Trust, https://www.woodlandtrust.org.uk/
support-us/act/volunteer-with-us/opportunities/
The UK's largest woodland conservation charity,
offering advice on tree planting and creating community
woods. They run an army of volunteers.

CHAPTER 5: REWILDING WITH ANIMALS

Books and articles

Danell, K. et al., *Large Herbivore Ecology, Ecosystem Dynamics
and Conservation*, Cambridge University Press, 2006

Painting, A., *Regeneration*, Birlinn, 2021. Contains a useful
description of how reducing numbers of deer in
Scotland allows for natural regeneration.

Vera, F., *Grazing Ecology and Forest History*, CABI Publishing,
2000. The seminal book that inspired Knepp and
revolutionised nature conservation in Europe.

Other sources

Association of Deer Management Groups,
www.deer-management.co.uk Advice on deer management in upland Scotland, and on native woodland
expansion and peatland restoration for member groups.

British Deer Society, bds.org.uk Advice on deer
management in the UK to better protect the natural
environment, and courses and training for culling deer.

The Deer Initiative, *Deer Best Practice Guides – England &
Wales*, www.thedeerinitiative.co.uk/best_practice

CHAPTER 6: TYPES OF HERBIVORE

Books and articles

Clutton-Brock, J., *A Natural History of Domesticated Animals*,
Cambridge University Press, 1999

Grandin, T., *Animals in Translation*, Bloomsbury, 2006

—, *Humane Livestock Handling: Understanding livestock behaviour
and building facilities for healthier animals*, Storey Publishing,
2008

—, *Temple Grandin's Guide to Working with Farm Animals: Safe,
humane livestock handling practices for the small farm*, Storey
Publishing, 2017

Ingraham, C., *Animal Self-Medication: How animals
heal themselves using essential oils, herbs and minerals*,
Ingraham Trading Ltd, 2018

Ransom, J. I. and Kaczensky, P. (eds), *Wild Equids: Ecology,
management and conservation*, Johns Hopkins University
Press, 2016

Tolhurst, S. (ed), *A Guide to Animal Welfare in Nature
Conservation Grazing*, The Grazing Animals Project, 2001

Other sources

Bud Williams, Cattle handling videos at stockmanship.
com Bud Williams' (1932–2012) walked-up system of
livestock mustering has revolutionised our cattle handling at Knepp. His instructive videos are not cheap – or
brilliantly edited – but they're extremely useful. For other
free-roaming cattle mustering and handling systems,
try the books by Temple Grandin listed above.

DEFRA, https://www.gov.uk/guidance/deer-keepers-tag-
ging-deer-and-reporting-their-movements Updated 2022

Fisher Modular Construction (UK) Ltd, fisheruk.co.uk
Design of a deer larder.

Food Standards Agency Legislation, https://www.food.gov.uk/business-guidance/wild-game-guidance

Government Legislation Archive, https://www.legislation.gov.uk/ukpga/1991/54/contents Advice on deer management, selling venison and building a deer larder in the UK.

Oklin, oklininternational.com/commercial/ Knepp's recommended high-temperature aerobic composter for animal waste.

Rare Breeds Survival Trust, www.rbst.org.uk Promotes the preservation of old and rare breeds of livestock, and publishes excellent information leaflets developed by the Grazing Animals Project on naturalistic grazing and livestock management, including 'Gathering, Handling and Transporting Stock'.

Rewilding Europe, https://rewildingeurope.com/publications/ Publishes a number of excellent books and practical guides related to grazing animals for rewilding:

Aurochs: Born to be Wild

Bison Best Practice Guidelines

Bison Grazing Plan 2014–2024

Circle of Life: a new way to support Europe's scavengers

GrazeLIFE Practitioners Guide

Grazing for Life (GrazeLIFE leaflet)

How European policies – especially the Common Agricultural Policy – can better support extensive grazing

Life Bison: Layman's report

Natural Grazing: Practices in the rewilding of cattle and horses

Rewilding Horses in Europe

Supporting effective grazing systems (GrazeLIFE layman's report)

CHAPTER 7: BECOMING THE HERBIVORE

Books and articles

Broad, K., *Caring for Small Woods*, Earthscan Publications, 1998

Buckley, G. P., *Ecology and Management of Coppice Woodlands*, Chapman and Hall, 1992

Casey, D. et al., 'Opportunities for wildlife through small-scale wilding in lowland farmed landscapes', *British Wildlife*, 2020, vol. 31, no. 3, pp. 179–87

Goulson, D., *A Buzz in the Meadow*, Jonathan Cape, 2014

Green, T., 'Tree hay – The forgotten food', *Conservation Land Management*, Winter 2016. Ted Green, one of the inspirations behind Knepp, argues for the revival of pollarding, an ancient agricultural practice that contributes to both biodiversity and the health of livestock.

Hopkins, J. J., 'Scrub ecology and conservation'. *British Wildlife*, 1996, vol. 8, pp. 28–36

Tabor, R., *A Guide to Coppicing*, Eco-Logic Books, 2013

Other sources

Natural Processes for the Restoration of Drastically Disturbed Sites, https://www.ser.org/news/506071/Open-Access-Natural-Processes-for-the-Restoration-of-Drastically-Disturbed-Sites.htm Webinar presented by Dave Polster, Canadian plant ecologist and expert in ecological restoration.

CHAPTER 8: YOUR REWILDING PROJECT

Books and articles

Dasgupta, P., *The Economics of Biodiversity – The Dasgupta Review*, HM Treasury, London, 2021. Final report of the Independent Review on the Economics of Biodiversity led by Professor Sir Partha Dasgupta. Summary (only 100 pages!) also available: https://www.gov.uk/government/publications/final-report-the-economics-of-biodiversity-the-dasgupta-review

Deutz, A. et al., *Financing Nature: Closing the global biodiversity financing gap*, The Paulson Institute, the Nature Conservancy, and the Cornell Atkinson Center for Sustainability, 2020

Ducros, A. and Steele, P., *Biocredits to Finance Nature and People: Emerging lessons*, International Institute for Environment and Development, 2022, https://iied.org/21216iied

The Economist, 20 December 2022, 'Why climate change is intimately tied to biodiversity'. Discusses the financial case for investing in biodiversity.

Helm, D., *Natural Capital: Valuing the planet*, Yale University Press, 2016

—, *Green and Prosperous Land: A blueprint for rescuing the British countryside*, William Collins, 2019

McLuckie, M. et al., *Wildlife watching hides: A practical guide*, Rewilding Europe, 2014

Porras, I. and Steele, P., *Making the Market Work for Nature:*

How biocredits can protect biodiversity and reduce poverty, The International Institute for Environment and Development, 2020, https://www.iied.org/16664iied

Sutherland, W. J. (ed.), *Transforming Conservation: A practical guide to evidence and decision making*, Open Book Publishers, 2022

White, N., 'UN says new biodiversity credits can succeed where carbon offsets failed', *Bloomberg UK*, 5 December 2022, http://bit.ly/3Y57143

Other sources

Biodiversity Consultancy, *Exploring design principles for high-integrity and scalable voluntary biodiversity credits*, 2022. A technical working paper. https://bit.ly/3WOZeWQ

Carbon Credits, carboncredits.com A good place to keep an eye on what's happening in the carbon markets. Includes a live tracker on carbon pricing.

DEFRA, web.adas.co.uk/defra/regional.htm Regional maps of Nitrate Vulnerable Zones in England.

—, MAGIC Map (Multi-agency Geographical Information Centre), magic.defra.gov.uk An interactive mapping data website managed by Natural England.

—, *Reintroductions and Other Conservation Translocations: Code and guidance for England*, 2021, http://bit.ly/3Jwebdx

Ecosystems Knowledge Network, https://ecosystems knowledge.net/news/ A good place to keep up to date with the latest developments in private investment in nature restoration. See also Green Finance Institute (separate entry).

Endangered Landscapes Programme, https://www.endangeredlandscapes.org/project/understanding-voluntary-carbon-markets/ Project run by ELP to increase access to voluntary carbon markets and enable landscape restoration projects to understand the costs and benefits.

Environment Bank, environmentbank.com Establishing a network of 'habitat banks' with farmers and landowners in the UK so that developers can implement their biodiversity net gain obligations by investing in rewilding and nature restoration projects.

Farmer Clusters, https://www.farmerclusters.com/profiles/ Developed by the Game & Wildlife Conservation Trust and supported by Natural England, farmer clusters are designed to empower farmers to devise nature restoration projects collectively, at landscape scale. For existing groups and information on how to set up a farm group/cluster, and apply for funding, go to the website.

Finance for Biodiversity, www.financeforbiodiversity.org An initiative for restoring nature, driven largely by financial institutions and the business community.

Forest Trends, https://bit.ly/3WIuXJz Explains how funding can be stacked and bundled in a credible way.

Get Nature Positive, https://getnaturepositive.com/get-involved/ An initiative for restoring nature, driven largely by financial institutions and the business community.

Green Finance Institute, https://www.greenfinance institute.co.uk/gfihive/insights/welcome-to-gfi-hive/ See also Ecosystems Knowledge Network (separate entry).

—, https://www.greenfinanceinstitute.co.uk/gfihive/podcasts/ The 'Financing Nature' podcast showcases practical solutions to source funding for nature restoration.

—, https://www.greenfinanceinstitute.co.uk/gfihive/hive-toolkit/ GFI's Investment Readiness Project Toolkit explains how to develop a privately funded scheme.

Highlands Rewilding, *Second Natural Capital Report*, 2022. Groundbreaking analysis of the biodiversity and carbon value of two rewilded estates in Scotland, Benloit and Beldorney. https://www.highlandsrewilding.co.uk/blog/second-natural-capital-report

Hoffmann, Isabel, 'Mangroves, biodiversity and climate change', Interview, Sustainability Journey podcast, https://sustainabilityjourney.podbean.com/ See also https://open.spotify.com/episode/7dN8Lvuc-Z6lC1VFXtKxCuR Her interview on the excellent Sustainability Journey podcast provides a good summary of biodiversity and carbon credits.

IUCN UK Peatland Programme, https://www.iucn-uk-peatlandprogramme.org/peatland-code-0 Promotes peatland restoration in the UK. Its Peatland Code now has 100 projects registered under the UK Land Carbon Registry.

National Library of Scotland, https://maps.nls.uk/geo/explore/side-by-side An amazing tool for showing historic maps that can be overlaid over modern ones, including satellite data.

Natural England, *Valuing Ecosystem Services – case studies from lowland England*, 2012, http://publications.naturalengland.org.uk/publication/2319433

Project Drawdown, www.drawdown.org The world's leading resource for climate solutions.

Race to Zero, https://unfccc.int/climate-action/race-to-zero-campaign An initiative for restoring nature, driven largely by financial institutions and the business community.

RePLANET Wildlife, https://www.replanet.org.uk/wildlife/ In 2022 Charlie helped found rePLANET Wildlife to act as a broker between rewilding projects and the investment market. It was involved with the Wallacea Trust biodiversity credit working group that developed the Biodiversity Credit methodology, and works with Plan Vivo, Gold Standard and other accreditation organisations that are creating the protocols to certify biodiversity credits.

Rewilding Britain, *Rewilding and the Rural Economy,* 2021. https://www.rewildingbritain.org.uk/news-and-views/research-and-reports/rewilding-and-the-rural-economy
—, https://www.rewildingbritain.org.uk/explore-rewilding/reintroductions-key-species/reintroductions-and-bringing-back-species The page on species reintroductions.

SciHub, Shamefully, the majority of scientific nature conservation literature is closed access. This is the most commonly used online 'shadow' library providing free access to around 70% of all published academic articles without regard to copyright. SciHub is used by major institutions around the world. Its website link is often disconnected so if you're happy to go down this route you may need to search for the latest connection.

Taskforce on Nature-related Financial Disclosures, tnfd.global Advises companies and financial institutions on how to integrate nature into decision-making.

The Land App, thelandapp.com Provides some of the information given by the Magic App (see separate entry) but allows editing and enables you to build your project in as much detail as you like. Most land managers now use it. It is free, requiring payment only when printing or downloading.

Unpaywall, www.unpaywall.org An online shadow library that is both free and legal, though it has a smaller resource base than SciHub (see separate entry).

Wallacea Trust, https://www.opwall.com/biodiversity-credits/ A new methodology for creating biodiversity credits that is applicable to all 1,300 eco-regions across the world.

Wilder Carbon, www.wildercarbon.com High-quality conservation projects in the UK for carbon storage and biodiversity.

Wildlife Trusts, www.wildlifetrusts.org Find your local nature reserves so you can explore how your land relates to them, and tap into your local network of ecologists by joining your local Wildlife Trust.

Woodland Carbon Code, www.woodlandcarboncode.org.uk Backed by government, this is the quality assurance standard for woodland creation projects in the UK and generates independently verified carbon credits.

World Economic Forum, https://www.weforum.org/projects/nature-action-agenda Information about developments in the world of natural capital.

Get involved

Citizen Zoo, https://www.citizenzoo.org/get-involved/volunteer-with-us/ Community-focused rewilding organisation involved in reintroduction programmes for species such as water voles, beavers and large marsh grasshoppers. Much of its work is based on citizen science.

Derek Gow Consultancy, rewildingcoombeshead.co.uk and www.watervoles.com The Derek Gow Consultancy specialises in water vole reintroductions around the UK. Derek also runs courses on practical rewilding, regenerative agriculture and species introductions from his rewilded farm, Coombeshead Farm in Devon, where you can stay, camp and go on guided rewilding walks.

Vincent Wildlife Trust, www.vwt.org.uk Primarily focused on recovery programmes for bats and mustelids (such as polecats and pine martens) in the UK. It provides specialist advice and opportunities for volunteers.

CHAPTER 9: RECORDING, SURVEYING AND MONITORING

Books and articles

Henshaw, A. et al., *A Bird's-eye View of the Wildland: 20 years of vegetation regeneration at a pioneering rewilding project,* 2021. Research report of rewilding Knepp with LIDAR images, Queen Mary, University of London, http://bit.ly/3HDOR43

Sutherland, W. J. (ed.), *Transforming Conservation: A practical guide to evidence and decision making,* Open Book Publishers, 2022

Other sources

Biological Records Centre, https://www.brc.ac.uk/
home The iRecord system is a wonderful tool for a site
list. It is relatively easy to use, even for novice naturalists,
and a great way of drawing in family members, others
in your rewilding team and the wider community. Once
you have registered on iRecord you're able to upload
photos and records onto their website. Records are
checked by experts, and wildlife sightings for all non-
sensitive species are shared with other users and made
available to National Recording Schemes, Local Record
Centres and Vice-county Recorders, all of which contributes to research and decision-making at local and
national levels. iRecord can apply your data to maps and
graphs. You can also extrapolate specific information,
such as which birds were recorded in a certain year at a
certain location, or a plant list for the whole site, or all
the records over time for one species.

British Dragonfly Society, https://british-dragonflies.
org.uk/wp-content/uploads/2019/04/Survey-guidance.
pdf The protocol for setting up a dragonfly survey.

British Scarabs, https://britishscarabs.org/species/
dung-beetles/ Check the website for identification
of dung beetles.

British Trust for Ornithology (BTO), https://www.
bto.org/user/register You will need to register for a free
British Trust for Ornithology account to access some of
the pages below. The data collected by the BTO helps
inform critical national conservation efforts.

—, Breeding Bird Survey, https://www.bto.org/
our-science/projects/breeding-bird-survey

—, English Winter Bird Survey, https://www.bto.org/
our-science/projects/english-winter-bird-survey/
ewbs-survey-methods The same transects can be walked
as for the Breeding Bird Survey above.

—, Nest Record Scheme, https://www.bto.org/our-
science/projects/nest-record-scheme/develop-skills
Attend a nest-finding course, or find a mentor local to
you, to learn the field craft of finding nests.

—, Bird Ringing, https://www.bto.org/our-science/
projects/ringing Bird ringing can teach you a lot about
your site and contributes to an important national
database run by the BTO. Ringing must be carried out
by licensed ringers. You may be able to find a group or
individual near your project who can help.

—, BirdTrack app. This is an excellent way of keeping
your bird records, on either your mobile phone or your
computer. The data gets sent on to county bird clubs/
societies so the valuable information you're collecting
can be used locally and nationally.

Bumblebee Conservation Trust,
bumblebeeconservationteemill.com The Bumblebee
Conservation Trust's BeeWalk Survey Scheme is a
monthly survey. It doesn't need much prior knowledge
and can be carried out by volunteers, on a set route
from March to October. The data collected will feed
back to national conservation efforts and also inform
what is happening on a local level.

**Chartered Institute of Ecology & Environmental
Management (CIEEM),** https://cieem.net/i-need/
Represents ecologists and environmental managers in
the UK, Ireland and abroad. Their website links to their
Registered Practices Directory where you can find an
ecologist or environmental manager. It also provides
'A Householder's Guide to Engaging an Ecologist'.

Field Studies Council, https://www.field-studies-
council.org/fsc-natural-history-courses/ **or** https://
www.field-studies-council.org/biolinks-courses/
Runs a comprehensive suite of identification
and field skill courses (both online and in the field).

Joint Nature Conservation Committee,
hub.jncc.gov.uk For UK habitat mapping. The 'Phase
1' technique acts as the standard system for classifying
and mapping wildlife habitats survey across the UK,
including in urban areas. Phase 1 handbooks can be
downloaded from the website. Each habitat type/feature
is defined by way of a brief description and is allocated
a specific name, an alpha-numeric code and a unique
mapping colour.

Knepp Rewilding Project, https://knepp.co.uk/
rewilding/ We post all our wildlife records, survey
findings and university research based at Knepp on
our website, together with information about ecosystem
services provided by our project.

Local Environmental Records Centre,
www.alerc.org.uk In the UK, they will be able to put
you in contact with local biological recorders and specialist recording groups. To find the relevant LERC for
your project area, check the website. Many local natural
history groups run field trips and identification courses

too. From amphibians and reptiles to bats, mosses and grasshoppers, there are recording groups, nationally and locally, to promote recording and conserving them.

MapMate, www.mapmate.co.uk Software for recording, mapping and analysing your natural history sightings on your home computer. Data exchanges can be carried out with your Local Environmental Record Centre using this software. A small annual fee funds software and species dictionary updates, and email support.

National Biodiversity Network, https://nbn.org.uk/tools-and-resources/useful-websites/database-of-wildlife-surveys-and-recording-schemes/ Lists many national recording schemes and links to their websites, which often contain identification pages.

Pasture for Life Association, https://youtu.be/56BZiq3iK4I Dung Beetle Deep Dive with Sally-Ann Spence.

QGIS, www.qgis.org A free open-source Geographic Information System (GIS) which you can use to map your boundaries, habitats and species, and analyse and publish geospatial information.

Recorder 6, www.recorder6.info Software for recording, mapping and analysing your natural history sightings on your home computer. Data exchanges can be carried out with your Local Environmental Record Centre using this software. A small annual fee funds software and species dictionary updates, and email support.

Rewilding Britain, https://www.rewildingbritain.org.uk/start-rewilding/measuring-and-monitoring-rewilding A Monitoring Framework is being developed and there is a useful page on their website for getting started.

Soil Food Web, https://www.soilfoodweb.com/foundation-courses-2/ This foundation course, set up by the American queen of the soil, Elaine Ingham, offers a series of sixty lectures on soil biology, surveying and the science behind the soil food web.

Soilmentor, soils.vidacycle.com Aimed specifically at regenerative farmers, using simple methodologies and equipment to study the soil, it covers everything from topsoil depth and rain infiltration to soil pH and rhizosheaths (the coatings of soil particles that cling to healthy plant roots). It's designed to work in conjunction with their Soilmentor app, developed in collaboration with the Pasture for Life Association (PFLA) and independent UK soil health advisor Niels Corfield, who also runs training courses and workshops. You can learn a lot just

from their website, but the app gives you the ability to map your soil sample locations on GPS, compare the ecology of your soil in different areas, record worm counts, dung beetles and broader wildlife, measure your findings against UK benchmarks, and share data with your team.

Species identification apps, Try the following: iNaturalist (perfect for beginners and children), PlantNet, NatureID, PictureThis and Flora Incognita (for plant identification); BirdNET, Merlin Bird ID, Song Sleuth and ChirpOMatic (for birdsong).

UK Butterfly Monitoring Scheme, https://ukbms.org/methods More information on the methodology pros and cons of the Wider Countryside Butterfly Survey and the UK Butterfly Monitoring Scheme can be found on the website.

UK Habitat, ukhab.org Gives links to an app you can download and 'ecountability' to connect you with consultants who offer habitat mapping services.

CHAPTER 10: WHERE DO WE FIT IN?

Books and articles

Bond, M., *Wayfinding: The art and science of how we find and lose our way*, Picador, 2021. Fascinating insight on the influence of nature on behavioural development.

Linnell, J. D. C. et al., 'Framing the relationship between people and nature in the context of European conservation', *Conservation Biology*, 2015, vol. 29, issue 4, pp. 978–85

Painting, A., *Regeneration: The rescue of a wild land*, Birlinn, 2021. Ecologist Andrew Painting's account of landscape rewilding at Mar Lodge, an estate owned by the National Trust for Scotland, explores the dichotomy of encouraging public access to nature and the restoration of nature itself.

Tree, I., *Wilding: The return of nature to a British farm*, Picador, 2018. Much of our experience at Knepp is to do with challenging conventional attitudes about what our landscape should look like. Many people find 'injurious' weeds particularly affronting, though they are part of our native ecology and often important for insects and birds. Chapter 5 deals specifically with the prejudice against ragwort.

Wilson, E. O., *Biophilia: The human bond with other species*, Harvard University Press, 1984

Other sources

Buglife, https://www.buglife.org.uk/resources/policy-and-legislation-hub/ragwort/ An excellent page debunking the myths about ragwort and giving the case for its importance for wildlife.

Citizen Zoo, 'Wild Walking with Dogs', https://www.citizenzoo.org/wild-walking-and-dogs/ A helpful leaflet.

British Horse Society, https://www.bhs.org.uk/media/0k5pkioq/livestock-1219.pdf Advice on riding or driving through livestock in England and Wales.

Get involved

Council for the Protection of Rural England (CPRE), https://www.cpre.org.uk/get-involved/ Campaigns on a wide range of rural issues from sustainable transport, waste disposal and renewable energy to hedgerow restoration, dark skies and the protection of the countryside. For volunteering opportunities, see the website.

Country Trust, www.countrytrust.org.uk An educational charity in the UK dedicated to connecting children with the countryside.

Countryfile Magazine, https://www.countryfile.com/how-to/outdoor-skills/britains-growing-litter-problem-why-is-it-so-bad-and-how-to-take-action/ This online magazine advises how to take action on litter in the countryside.

Natural England, 'The Countryside Code', https://www.gov.uk/government/publications/the-countryside-code/the-countryside-code-advice-for-countryside-visitors The Countryside Code provides guidance on how to behave in the countryside including general courtesy, keeping to footpaths and protecting the environment. It is unhelpfully woolly, though, about dog control and wildlife disturbance.

CHAPTER 11: REWILDING YOUR GARDEN

There is an enormous number of books on nature-friendly and naturalistic gardening. These are some of our favourites.

Books and articles

Barkham, P., *Wild Child: Coming home to nature*, Granta, 2020

Bradbury, K., *Wildlife Gardening: For everyone and everything*, Bloomsbury Wildlife, 2019

—, *How to Create a Wildlife Pond: Plan, dig and enjoy a natural pond in your own back garden*, Dorling Kindersley, 2021

—, *The Tree in My Garden: Choose one tree, plant it, and change the world*, Dorling Kindersley, 2022

Butterfly Brothers, *Wild Your Garden*, Dorling Kindersley, 2020

Dunnett, N., *Naturalistic Planting Design, the Essential Guide: How to design high-impact, low-input gardens*, Filbert Press, 2019

Goulson, D., *The Garden Jungle: Or gardening to save the planet*, Jonathan Cape, 2019

—, *Gardening for Bumblebees: A practical guide to creating a paradise for pollinators*, Square Peg, 2021

Hitchmough, J., *Sowing Beauty: Designing flowering meadows from seed*, Timber Press, 2017

Howard, J., *The Wildlife Pond Book: Create your own pond paradise for wildlife*, Bloomsbury Wildlife, 2019

Jeffery, J., *Let's Wildflower the World – Save, swap and seedbomb to rewild our world*, Leaping Hare Press, 2022

Kingsbury, N., *Wild: The Naturalistic Garden*, Phaidon Press, 2022

Kircher, W., *How to Build a Natural Swimming Pool*, Filbert Press, 2016

Lewis-Stempel, J., *The Wildlife Garden*, How To Books, 2014

Littlewood, M., *Natural Swimming Pools*, Schiffer Publishing, 2004

Lloyd, C. and Garrett, F., *Meadows: At Great Dixter and Beyond*, Pimpernel Press, 2016

Maziarz, M. et al., 'Microclimate in tree cavities and nest-boxes – implications for hole-nesting birds', *Forest Ecology and Management*, 2017, vol. 389, pp. 306–13

Miller-Klein, J., *Gardening for Butterflies, Bees and Other Beneficial Insects*, Saith Ffynnon Books, 2010

Oudolf, P. and Gerritsen, H., *Dream Plants for the Natural Garden*, Frances Lincoln, 2013. Piet Oudolf is probably most famous for rewilding the New York High Line. Here, he and fellow Dutch garden designer Henk Gerritsen choose over 1,200 beautiful and robust plants for a natural-looking, low-input garden.

Pearson, D., *Tokachi Millennium Forest: Pioneering a new way of gardening with nature*, Filbert Press, 2020

Rich, C. and Longcore, T., *Ecological Consequences of Artificial Night Lighting*, Island Press, 2006

Robinson, W., *The Wild Garden*, Timber Press, 2009. First published in 1870, this groundbreaking book advocating a more natural style of gardening and the use of plants to suit local growing conditions is still hugely relevant today. This beautifully illustrated edition includes new chapters and photographs by US landscape consultant Rick Darke.

Rouse, D., *How to Attract Birds to Your Garden: Foods they like, plants they love, shelter they need*, Dorling Kindersley, 2020

Smith, L., *Tapestry Lawns: Freed from grass and full of flowers*, CRC Press, 2019

Stuart-Smith, S., *The Well Gardened Mind: Rediscovering nature in the modern world*, William Collins, 2020

Tophill, F., *Rewild Your Garden*, Greenfinch, 2020

Other sources

YouTube, Conventional swimming pools can be death traps for wildlife. There are some simple adaptations you can make to prevent creatures such as hedgehogs, voles, mice and frogs from drowning. See https://www.youtube.com/watch?v=sEoENXaWxhw&ab_channel =TechInsider, for example, for how to provide a float in the pool filter to give creatures a raft to climb onto, as this frog and mouse discovered: https://youtu.be/JTiST5Ux6gU (be sure to check the float regularly).

Get involved

Buglife, www.buglife.org.uk A UK charity that campaigns for insects. It leads campaigns against pesticides and light pollution, and runs numerous projects involving volunteers, children and schools, protecting and creating habitat for insects on green- and brownfield sites, and in gardens and small and marginal areas.

Froglife, www.froglife.org Does for amphibians and reptiles what Buglife does for insects. Provides information and advice on how to create habitat for frogs, toads, newts, snakes and lizards, and also welcomes volunteers.

Operation Future Hope, www.operationfuturehope.org A UK charity that aims to inspire young people to get involved in rewilding and nature restoration, and supports schools to create outdoor learning. It also publishes *The Rewilding Manual for Schools*.

CHAPTER 12: URBAN REWILDING

Books and articles

Beatley, T., *Biophilic Cities: Integrating nature into urban design and planning*, Island Press, 2011

Dunnett, N. and Clayden, A., *Rain Gardens: Managing water sustainably in the garden and designed landscape*, Timber Press, 2007

Dunnett, N. et al., *Small Green Roofs: Low-tech options for homeowners*, Timber Press, 2011

Dunnett, N. and Hitchmough, J. (eds), *The Dynamic Landscape: Design, ecology and management of naturalistic urban planting*, Taylor & Francis, 2008

Dunnett, N. and Kingsbury, N., *Planting Green Roofs and Living Walls*, Timber Press, 2008

Goode, D., *Nature in Towns and Cities*, Collins New Naturalist Library, William Collins, 2014

Green, T., 'Urban solar panels', *Arb*, 2021, issue 195. Tree specialist Ted Green presents the incredible benefits of open-grown trees in cities.

Hoare, B., *Wild City*, Macmillan, 2020. A beautifully illustrated children's book, touring cities of the world where unusual creatures from penguins to opossums share the city space with humans.

Hodges, S. and Hodges, S., *Wild London: Urban escapes in and around the city*, Square Peg, 2019

Hughes, J., Taylor, E. and Jupiter, T., *Living Cities: Towards ecological urbanism*, Scottish Wildlife, 2018

London Wildlife Trust, *London in the Wild: Exploring nature in the city*, Kyle Books Trust, 2022

Malpas, L., *The Rewilding Manual for Schools – A guide to restoring nature in your school grounds*, Kindle Direct Publishing, 2023

Pettorelli, N. et al., *Rewilding Our Cities*, Zoological Society of London report, 2022

Poizner, S., *Growing Urban Orchards: How to care for fruit trees in the city and beyond*, Book Publishing Co, 2017. Susan Poizner also runs workshops, courses and webinars. https://orchardpeople.com/

Reynolds, R., *On Guerrilla Gardening: A handbook for gardening without boundaries*, Bloomsbury, 2008. This lively, impassioned call to arms covers the illicit greening of neglected public spaces around the world, from Berlin to Montreal, and provides practical advice for fellow activists.

Roe, J. and McCay, L., *Restorative Cities: Urban design for mental health and wellbeing*, Bloomsbury, 2021

Straka, T. M. et al., 'Urban cemeteries as shared habitats for people and nature', *Land*, 2022, vol. 11, p. 1,237

Stuart, K. and Thompson-Fawcett, M. (eds), *Indigenous Knowledge and Sustainable Urban Design*, New Zealand Centre for Sustainable Cities, University of Otago, Wellington, New Zealand, 2010

Wilkinson, F., *Wild City: Encounters with urban wildlife*, Orion Spring, 2022

Wilson, B., *Urban Jungle: Wilding the City*, Jonathan Cape, 2023

Wood, P., *London Tree Walks*, Safe Haven Books, 2020

—, *London is a Forest*, Quadrille, 2022

Other sources

Beatley, Tim, *The Nature of Cities*, 2010. This film follows Tim Beatley, Professor of Sustainable Communities at the University of Virginia, as he explores urban nature projects around the world. https://www.filmsforaction.org/watch/the-nature-of-cities/

Biophilic Cities, www.biophiliccities.org This organisation partners with cities, scholars and advocates from around the world to promote the creation of nature in cities. It produces short documentary films showcasing the greening of cities around the world, and its bi-annual Biophilic Cities Journal includes projects, academic research and book reviews from design and planning practitioners.

C40 Cities, www.c40.org A network of mayors around the world taking action to confront the climate crisis and create greener, more sustainable cities.

Citizen Zoo, https://www.citizenzoo.org/CZ/urban-rewilding/ A dynamic rewilding organisation in the UK with particular interest in supporting community-led projects in cities.

City Trees, This tree database and identification app displays the number of trees and their respective species for over 20 cities around the world.

Ecocities Emerging, ecocitiesemerging.org Curated by Ecocity Builders, a non-profit company in Oakland, California, this website carries interesting articles on ecocity design and innovation, and interviews with the movers and shakers.

Edible Schoolyard Project, edibleschoolyard.org Pioneered by chef Alice Waters in a state school in Berkeley, California, it has inspired school projects around the world. Teaching through food, this amazing project brings growing, cooking, ecology, social science and economics into the classroom. See also Waters, A. and Liittschwager, D., *Edible Schoolyard*, Chronicle Books, 2008.

Greater London Authority, The Green Infrastructure maps and tools page of the GLA's website provides maps of trees, green cover, green infrastructure, green roofs and sustainable drainage systems (SuDS) in the capital. It aims to help organisations and individuals improve London's green infrastructure. http://bit.ly/3HGxFuy

Hedgehog Street, www.hedgehogstreet.org A child-friendly website championing the protection of urban hedgehogs in the UK.

Smart Surfaces Coalition, smartsurfacescoalition.org Made up of 40 international organisations promoting the use of smart surfaces in cities around the world to reduce heat and flooding and improve public health.

Susdrain Grey to Green scheme, The Grey to Green scheme in Sheffield, begun in 2014, is the UK's largest retro-fit SuDS (Sustainable Drainage) project, and also the UK's largest inner city 'Green Street'. It was designed by Nigel Dunnett, Professor of Planting Design and Urban Horticulture at the University of Sheffield. https://bit.ly/3Y2D0le

Trees for Cities, www.treesforcities.org Works with volunteers, schools and local communities in the UK and abroad, and has planted more than 1 million trees since 1993. It specialises in working in socially deprived urban areas.

Trees in Cities Challenge, treesincities.unece.org A global campaign of mayors implementing pledges to make their cities greener, more sustainable and resilient.

Get involved

Caring For God's Acre, www.caringforgodsacre.org.uk This organisation provides advice on protecting and restoring churchyards in the UK as wildlife havens and runs a Churchyard Task Team of volunteers. In June each year there is a national Love Your Burial Ground campaign, encouraging people to celebrate and care for their local churchyards and cemeteries.

Community Rewilding, www.communityrewilding.org.uk Community rewilding in Scotland, with projects for volunteers in Glasgow, Clydebank and Yoker.

Grass Roof Company, www.grassroofcompany.co.uk John Little, who manages the company, specialises in brownfield landscapes, green roofs and urban green space using unusual substrates and reclaimed structures for wildlife habitat. A great website full of practical ideas and information. He runs master classes, too.

Guerrilla Gardening, www.guerrillagardening.org This began in 2004 as a blog run by Richard Reynolds as a record of his illicit cultivation of areas around London. It now supports a growing community of would-be guerrilla gardeners aiming to reclaim neglected public spaces for nature and society.

London Wildlife Trust, www.wildlondon.org.uk Events, youth programmes, volunteering and campaigning, and wildlife sightings from the public.

Nature of Cities, www.thenatureofcities.com An international platform sharing ideas about how to transform cities into sustainable ecosystems for people and nature.

Operation Future Hope, www.operationfuture hope.org Based in the UK, it provides online environmental and ecological education for students and teachers, and the skills to rewild their school grounds.

Orchard Project, www.theorchardproject.org.uk The Big Rewild campaign encourages wildlife-friendly gardens and orchards in London, Manchester, Swansea and Glasgow. Also community events and volunteering.

Pollinating London Together, www.pollinating londontogether.com Run by the livery companies and guilds of London, it aims to increase pollinator-friendly planting and habitats in and around the City of London. Also offers non-members advice on habitat creation and how to conduct basic pollinator surveys.

Rewild My Street, www.rewildmystreet.org Provides advice for people wanting to adapt their homes, gardens and streets to encourage wildlife. Its 'habitats and activities' pages provide useful information for residents of most cities.

Thames Marine Mammal Survey, Zoological Society of London, https://sites.zsl.org/inthethames/ For public sightings of harbour seals, grey seals, harbour porpoises, dolphins and whales in the Thames and Thames estuary. To report a sighting, visit the website.

Thames 21, www.thames21.org.uk Works with communities across Greater London to improve rivers, canals, ponds and lakes for people and wildlife. It runs projects involving thousands of volunteers every year.

Glossary

abiotic Physical rather than biological; not derived from living organisms, or biota.

additionality Something new that happens as the result of an intervention. In the case of carbon offsetting, this would be the additional carbon storage shown to have been achieved by an intervention such as rewilding or tree planting, in comparison to an established baseline.

agroforestry Agriculture incorporating the cultivation of trees, generally to improve biodiversity, productivity and/or conditions for pastured livestock, and sometimes as a 'two-tier' system to produce an additional crop such as nuts or fruit.

Anthropocene A name (not yet formally adopted) given to the epoch that dates from when human impact on the planet's climate and ecosystems became significant. Most popularly this is estimated as the start of the Industrial Revolution in the 1800s; others believe it should be 1945 when the first atomic bomb was tested, or the 1950s when the Great Acceleration – a dramatic increase in human population and activity – took off. Paleoclimatologists propose a much earlier date, when the invention of agriculture 8,000 years ago increased atmospheric CO_2; or even earlier, in the late Pleistocene, when mass extinctions of megafauna caused by humans resulted in huge changes to the planetary ecosystem.

apex predator A predator at the top of a food chain that has no natural predators of its own.

arable (of land or soil) Suitable for growing crops such as maize, barley and wheat (as opposed to keeping livestock or growing fruit and vegetables).

aurochs (**Bos primigenius**) An extinct species of European wild ox, ancestor of modern domestic cattle.

biodiversity The variety of plant and animal life in the world in general, or in a particular habitat. Generally, a good measure of the health of an ecosystem.

biodiversity net gain (BNG) An approach to development that aims to leave biodiversity on and/or off site in a measurably better state than it was before.

biota The plant and animal life of a region or habitat.

brownfield An area in a town or city that has previously been built on or developed, usually for industrial purposes, that is no longer in use (as opposed to greenfield, which has never been built on).

browse (verb, of an animal) To eat leaves, shoots, twigs and other high-growing vegetation, as opposed to graze (to eat grass). 'Browse' (noun) is the range of available food on shrubs and trees.

carbon sequestration The process of capturing carbon dioxide gas from the atmosphere and storing it in solid or liquid form in the oceans, soils, vegetation (especially trees) and geologic formations.

carbon sink A natural system that absorbs more carbon dioxide than it releases, such as the sea, plants and soil.

catastrophic shift The sudden tipping of an ecosystem from one stable state to a more depleted, less functional stable state, from which it may be difficult to return. Examples are the eutrophication of shallow lakes and the desertification of drylands. In some cases, catastrophic shifts can trigger land abandonment.

catchment The area of land through which water drains and eventually collects in a body of water such as a river, lake or stream.

clear-felling A practice in which all the trees in an

area of plantation or forest are cut down at the same time.

continuous cover forestry The practice of felling trees individually, so that overall tree cover is maintained.

coppicing The traditional practice of cutting back trees or shrubs such as hazel, sweet chestnut, ash and lime to near ground level in order to stimulate the growth of multiple new shoots. These shoots can then be used for fencing, building, tool handles, bean poles and so on.

corvid A bird of the crow family; includes jackdaws, jays, magpies, rooks, choughs and ravens.

cover crop A close-growing crop planted primarily to protect and enrich the soil, increase water availability, smother weeds and help control pests and diseases, though it can sometimes also be harvested for profit.

DEFRA (The Department for Environment, Food and Rural Affairs) The UK government office responsible for agriculture, food supplies, environmental protection and rural communities.

dressed carcass The carcass of an animal once the internal organs, hide and, usually, head and feet, have been removed.

ecosystem services The goods and services provided by nature that directly or indirectly benefit humans, such as water, clean air, food, timber (and other raw materials), flood, storm and drought protection, climate regulation, pollinating insects for crops, pest and disease control, and mental and physical health. Also known as 'public goods'.

ELM The Environmental Land Management schemes, to be rolled out in various stages by the UK government from 2022, replacing the previous farm subsidies under the EU's Common Agricultural Policy. The schemes, which are divided into three tiers (Sustainable Farming Incentive, Local Nature Recovery and Landscape Recovery) aim to encourage ecologically sensitive land management, through which farmers and land managers can be paid for ecosystem services.

eutrophication The nutrient enrichment of an environment, generally a water body, often caused by agricultural fertilisers, which causes harmful algal blooms, depleting oxygen and suffocating life.

food web All the interconnected food chains that contribute to the movement of energy and nutrients through the ecosystem.

habitat The natural home or environment of an animal, plant or other organism.

heathland An open natural habitat on acidic, free-draining, low-nutrient, generally sandy soil or peat, often characterised by heather, bracken, gorse and heathland grasses, with scattered open-grown trees.

herbivory The eating of plants by animals.

Holocene The official term for the current geological epoch, covering the last 11,700 years of Earth's history since the last glacial period.

hydrology The study of the parts of the water cycle that take place on land.

keystone species A plant or animal which has a dis-proportionately large effect on its natural environment and therefore plays a critical role in maintaining an ecosystem and the organisms which depend on it.

ley Arable land put down to grass (generally of an intensive commercial variety) for hay, silage or grazing for a single season or limited number of years; as opposed to permanent pasture.

megafauna Large animals, generally referring to vertebrates with an adult body weight of more than 44 kilograms.

mob grazing A way of managing livestock, allowing a large number of animals to graze a small area of pasture intensively for a short duration, and moving them on regularly (every one to three days), giving the pasture a chance to rest and recover behind them.

mycorrhizae The tiny fungal threads, or hyphae, that connect a fungus with the roots of vascular plants, through which nutrients are exchanged. The 'mycorrhizal network' is the underground net-like web of all these complex interconnections that also allows plants to 'talk' to each other via chemical signals.

natural capital A way of thinking about nature as a stock that provides a flow of benefits to people and the economy, thereby putting nature on the balance sheet so it can be valued and protected.

pannage The traditional practice of allowing pigs to roam in woodland areas so they can feed on fallen beechmast, chestnuts, acorns and other nuts.

peat hagg A pit or hollow in peatland with steep, exposed edges, caused by erosion, fire, over-grazing and/or digging up peat for fuel or garden compost.

Pleistocene The geological epoch before the current Holocene. It ran from approximately 2,500,000 years ago to 11,700 years ago, and was characterised by glacial cycles.

poaching (with reference to animal impact on the land) The excessive trampling of fields or pasture, particularly in wet weather, which breaks up the sward, resulting in loss of grass and other plants.

polder A level, low-lying area of land reclaimed from the sea or a river or a lake, often protected by dykes or embankments, especially in the Netherlands.

pollarding The practice of cutting the top and branches off a tree, leaving club-headed stems that grow new branches (often used as fodder for livestock or for construction). Pollarding can be used to keep a tree trimmed to a certain height and generally extends its life.

process-led (with reference to rewilding projects) Where human management takes a backseat and nature is allowed to evolve in a spontaneous and dynamic way; also described as 'self-willed'.

regenerative agriculture A holistic approach to farming that focuses on improving soil health, increasing biodiversity and protecting water resources.

riffle The shallower, faster moving sections of a river or stream where rocks break the water surface, agitating the flow.

ring-barking Removing a band of bark around the circumference of a tree trunk or branch, resulting in the death of the tissue above the cut. It can be done as a deliberate tree-management practice, or naturally by animals.

riparian Relating to or situated on the banks of a river or stream.

riverine Relating to a river; of a species, one that lives in or next to rivers.

sallow The common name used in the UK for native broad-leafed species of willow – great or goat willow (*Salix caprea*), white willow (*Salix alba*) and grey or common sallow (*Salix cinerea*), which naturally hybridise in the wild.

saproxylic Relating to dead or dying wood; saproxylic insects, for example, depend on dead wood.

scrub A habitat consisting of low woody shrubs, often described as 'successional habitat', meaning that it is in transition to becoming woodland.

shifting baseline The phenomenon by which people's accepted standards for environmental conditions are continually being lowered, with each generation unwittingly accepting a depleted state of nature as the new 'normal'.

silage A type of animal fodder made from grass that has been fermented to preserve its nutrients.

SSSI (Site of Special Scientific Interest) A conservation designation given to defined areas of land that feature species or habitats of particular biological or geological interest, and which aims to protect these features.

substrate In biology, the material in which an organism grows. In construction, an underlying layer that supports another layer on top, typically crushed rock and soil.

sward An area of ground featuring a mat of grass and grass roots or turf.

tarpan (**Equus ferus ferus**) The European wild horse, now extinct, considered likely to be the ancestor of modern horses. It once ranged from southern France and Spain to central Russia.

taxa Plural of *taxon*, a unit used in the science of classification or taxonomy, ranging from kingdom to subspecies.

top (verb) To mow or scythe an area of grass, meadow or hedgerow.

transect A line across a habitat or part of a habitat used in ecological surveys. The number and/or diversity of species along a transect can be observed and recorded at regular intervals to estimate changes over time.

trophic cascade A set of interactions between species that cause a series of knock-on effects for other species, and affect an entire ecosystem.

understorey A layer of vegetation lower than the main canopy of a wooded area.

ungulates A group of large herbivorous mammals with hooves, including cattle, deer, camels, sheep and pigs.

vascular plant A land plant with specialised tissue for transporting water and nutrients through its stem, leaves and roots. Most land plants, including ferns and flowering plants, are vascular.

vegetation succession The natural process by which communities of plants change and replace each other over time. In the absence of natural disturbance (such as animals, fire, disease, windblow and drought) vegetation evolves from pioneer species to a 'climax community' – such as a mature forest.

virtual fencing A system that keeps animals within a bounded area without using physical fencing. A collar-mounted GPS device emits a sound or mild electric shock if the animal crosses a geographical boundary that has been set by the farmer or rancher using a remote device. Also known as 'No Fence'.

windthrow The uprooting and toppling of trees by strong winds.

wood pasture An area of land grazed by free-roaming animals and/or livestock, characterised by scattered open-grown trees.

Index

Figures in *italics* refer to illustrations, captions and tables.

545

Y

Z

Acknowledgements

We are deeply indebted to an enormous number of people for helping us put this book together. Countless experts have responded to what must have seemed, at times, rather random questions with enormous generosity and patience. Thank you to Pamela Abbott, Graham Bathe, Craig Bennett, Henri Brocklebank, Jill Butler, Nik Cole, Roy Dennis, Charlotte Faulkner, Fergus Garrett, Charles Godfray, Ronald Goderie, Ben Goldsmith, Derek Gow, Ted Green, Neil Hulme, Paul Jepson, Carl Jones, Tony Juniper, Andrew Lanson, John Lawton, Leo Linnartz, Tim Mackrill, Darren Mann, James Moss, Ross Mounce, Matthew Oates, Nancy Ockendon, Jenny Phelps, Christopher Price, Matt Rees-Warren, Alison Richard, Patrick Roper, Chris Sandom, Lee Schofield, Matt Somerville, Tom Tew, Edwin Third, David Tudor, Joep van der Vlasakker, Andrew Wallace, Peter Welsh, Tony Whitbread, Angus Winchester, Rob Wise, Hayley Wiswell, Kristiina Yang and Philine zu Ermgassen.

Thank you, too, to Martin Janes and Pascale Nicolet for their invaluable contributions to the Rewilding Water chapter; to Graeme Lyons for helping with the Surveying and Monitoring chapter; to Tim Coles, Sarah Darrah, Alicia Gibson, Ben Hart, David Hill and Isabel Hoffmann for help on the funding sections of the Your Rewilding Project chapter; Mick Crawley, Dave Goulson, James Hitchmough and Tom Stuart-Smith for their inspired vision on rewilding gardens; Elliot Newton, Lily Ginsberg-Keig, Lucas Ruzo and Ben Stockwell for their expertise on Urban Rewilding; and to our own team at Knepp for their considerable input to various aspects of the book – Russ Carrington, Charlie Harpur, Ivan de Klee, Jason Emrich, Penny Green, Ian Mepham, Rina Quinlan and Patrick Toe.

To the many others, too many to mention, who have helped with research and fact-checking along the way, we are eternally grateful.

Several extremely kind friends, experts in their own right, spared time in busy lives to read the manuscript, or considerable parts of it, and provided hugely helpful notes. Thank you, Alastair Driver, David Hetherington, Jonathan Spencer, Bill Sutherland and, above all, Lisbet Rausing, consummate rewilder and communicator, whose clarity of thinking encouraged in us zero tolerance for jargon and science-speak, and a fresh eye on pretty much everything.

It is tremendously complicated compiling a book of this size. Any omissions or errors that have crept in over various incarnations are our responsibility alone.

It has been such fun working with Jeroen Helmer on the illustrations for the book. There can be few natural history artists with such a holistic understanding of ecology. We've learnt so much from every picture. And we'd like to give special thanks to Benedict Dempsey for his ingenious quiz which shows us that, wherever we are as individuals on the rewilding spectrum, we have a vital contribution to make to the recovery of our planet.

Thanks, as always, to literary agent extraordinaire David Godwin; and to the amazing team at Bloomsbury – Laura Brodie, Ros Ellis, Jonny Coward, Faye Robinson, Genista Tate-Alexander, Lauren Whybrow and Rowan Yapp – along with picture researcher Jo Carlill, editor Laura Gladwin and designer Anna Green, for pushing this mammophant of a book over the line, always with such enthusiasm, perfectionism and good humour.

Finally, we would like to thank Frans Vera, for the inspiration.

Illustrators

Cover illustrations: Lucy Boydell

Lucy is a fine artist based in Norfolk, UK, who specialises in life-size or larger works of animals in charcoal, chalk and pastel on paper. The brown hare on the front cover is a non-native species, introduced during the Iron Age, that has become an important part of the UK's ecology and now has its own Biodiversity Action Plan. The red stag on the back cover, often associated with remote highlands, may well once have been a species of floodplains and lowland wetlands before humans pushed them out of their habitat. www.lucyboydell.com

Interior illustrations: Jeroen Helmer

Jeroen is a nature illustrator based in the Netherlands who works with ARK Rewilding Netherlands, an organisation that pioneers rewilding. He specialises in visualising the unfolding of natural processes and the ecological role of keystone species. His work is an invaluable tool in communicating the principles of rewilding. www.ark.eu/en/projects/artist-impressions

Picture Credits

Key: t: top, b: below, m: middle, l: left, r: right

Agricarbon UK: 338; **Alamy**: /A.D.Fletcher 412, /Andrew Smith 508, /Ashley Cooper pics 176, /dbimages 485, Gabriela Insuratelu 256, /Goran Šafarek 428, /Jack Barr 108, /Joe Giddens/PA 475, /Katya Tsvetkova 424, /Marcin Rogozinski 401, /Nature Picture Library 197, /RooM the Agency 513, /Stefan Rousseau/PA 478, /Steve Joynson Photography 152; **ANS Global/ANS Living Wall System**: 497; © **Boeri Studio**: /project credit Stefano Boeri Architetti 499r, /photo Dimitar Harizanov 499l; **Bridgeman Images**: /© 2022 Museum of Fine Arts, Boston 163t; **Charlie Burrell**: 4, 13, 14–15, 56–7, 94–5, 138–9, 161, 163b, 169br, 172, 173, 174, 182, 188–9, 226–7, 231bl, 231br, 232l, 232r, 248, 249, 250, 292, 300l, 300r, 304, 357, 358–9, 382–3, 392, 397, 402–3, 449, 466; **CC BY-SA 3.0**: /https://commons.wikimedia.org/wiki/File:Elephasantiquus. jpg. Apotea, http://www.quagga.cat/en/reconstructions 31; **David Hetherington**: 141r; **dustygedgephotography**: 502l, 502r; **ESRI World Imagery**: Content is the intellectual property of Esri and is used herein with permission. Copyright © 2022 Esri and its licensors. All rights reserved 320tl; **Filip Trnka**: /www.naturabohemica.cz 378; **Foster + Partners**: /Nigel Young 470; **Frankfurt Zoological Society**: /© Daniel Rosengren 105; **Garden Exposures Photo Library**: /© Andrea Jones 427; **Getty Images**: /Bruce Yuanyue Bi 484, /Fine Art Images/Heritage Images 231tl, /Jason Hawkes 405, /Pictures From History/Universal Images Group 231tr, /Richard Newstead 506, /Universal History Archive 184; ©**Google Earth Imagery**: /©2001 169tl, /©2005 169ml, /©2012 169bl, /©2015 169tr, /©2018 169mr, 318, /©2022 CNES / Airbus, Getmapping plc, Infoterra Ltd & Bluesky, Maxar Technologies, The GeoInformation Group, Map Data ©2022 435t, 435b; **Jack Hague jbhague.co.uk**: 364; **Jak Wonderly**: 445; **Jason Emrich**: 308–9; **Jason Ingram**: 457; **Jeroen Helmer / ARK Rewilding Netherlands**: 29, 64, 66, 69, 72, 73, 75, 76, 99, 102, 109, 111, 116, 122, 131, 132, 144, 145, 147, 158, 164, 166, 203, 204, 205, 230, 234, 237, 247, 252, 257, 258, 261, 286, 408t, 408b, 409, 415, 440; / From the report: Natuurdroom 2050 Noord Holland, ARK Natuurontwikkeling 2018 462; / Onder Het Maaiveld / IUCN NL 376; / Rewilding Britain: 77, 87, 91; / Rewilding Europe 21; **Leo Linnartz / ARK Rewilding Netherlands**: 458–9; **Lia Brazier**: 380; **Loren McClenachan**: /Center for Marine Biodiversity & Conservation, La Jolla, CA 49br; **Lucy Boydell, www.lucyboydell. com**: "The Trickster", original drawing in charcoal and chalk, 107 × 107cms plus frame FRONT COVER, 515; "Watchful Stag", original drawing in charcoal and chalk on paper, 108 × 102cms plus frame BACK COVER; **Megan Hughes, South Downs National Trust Volunteers**: @meganhughesinsta www.southdownsntv.org.uk 288–9; **Mike Langman (rspb-images.com)**: 322; **Monroe County Public Library, Florida Keys**: /courtesy 49tl, 49tr, 49bl; **Murdo MacLeod**: 54, 177; **National**

Data Credits

pages 18 and 481 Fernández, N. et al. (2020). *Boosting Ecological Restoration for a Wilder Europe*. iDiv & Rewilding Europe. DOI: https://dx.doi.org/10.978.39817938/57 / **page 25** Bunney, K., Bond, W. J. and Henley, M. (2017). 'Seed dispersal kernel of the largest surviving megaherbivore – the African savannah elephant'. *Biotropica*, 49(3), pp. 395–401. DOI: https://doi.org/10.1111/btp.12423 / **page 34** Calculated by Our World in Data based on information from the UN Food and Agriculture Organization (FAO). Licensed under CC BY (Creative Commons licence) / **page 35** Our World in Data. Data source from UN Food and Agriculture Organization (FAO). Licensed under CC BY by the authors Hannah Ritchie and Max Roser in 2019 / **page 40** Adapted from an original diagram by John Lawton / **page 42** Dr Tony Whitbread and Knepp / **page 53** Jepson, P. and Blythe, C. (2020). 'Table 2: Rewilding and restoration ecology compared' in *Rewilding – the radical new science of ecological recovery* / **page 99** Water Cycle v1.11 (2016) by Ehud Tal. Licensed under CC BY-SA / **page 103** Westcountry Rivers Trust / **pages 131–3** Freshwater Habitats Trust / **page 150** Attributes of British and Irish Plants: Status, Size, Life History, Geography and Habitats. PLANTATT, 2004. © UK Centre of Ecology and Hydrology (UKCEH). Reproduced with permission from UKCEH / **page 258** Licenced under CC BY/ **page 297** Data from *British Wildlife* 32.4, February 2021, p. 256 / **page 305** Dr Tony Whitbread and Knepp / **page 313** Land App / **page 315** © WENP (West of England Nature Partnership). Flood Zone 2: © UKCEH. © Crown Copyright and database rights 2018 Ordnance Survey / **page 317** Weald to Waves, licensed under Crown Copyright / **page 328** The Natural Capital Protocol Framework. Natural Capital Coalition (2016). 'Natural Capital Protocol'. Available online: www.naturalcapitalcoalition.org/protocol, CC BY-NC-ND 4.0 / **pages 332–3** Alicia Gibson and Sarah Darrah of Finance Earth / **page 440** Adapted from an illustration by The Master Gardeners of

Otero County, New Mexico / **page 452** Greenspace Information for Greater London CIC, 2022 / **page 468** Gloucester City Council / **page 469** Matt Brown and Londonist / **page 476** Creative Commons BY 4.0: A dataset from a structured project (structured dataset), consisting of abundance data from the standardised Berlin breeding bird survey (BBS) in 2017, coordinated by the German ornithological association (Dachverband Deutscher Avifaunisten – DDA, www.dda-web.de) and the Berlin Senate Department for Environment, Transport and Climate Protection (SenUVK).

A note on the credits

Isabella Tree is an award-winning journalist and author, and lives with her husband, the conservationist Charlie Burrell, in the middle of a pioneering rewilding project in West Sussex. She is author of five non-fiction books. Her book *Wilding*, the story of the ambitious journey she and Charlie undertook to rewild their farm, has sold a quarter of a million copies worldwide and been translated into eight languages. It won the Richard Jefferies prize for nature writing, was shortlisted for the Wainwright prize and was one of the Smithsonian's top ten science books for 2018. In 2022, Isabella served on the Mayor of London's Rewilding London Task Force.

Charlie Burrell is chair of Foundation Conservation Carpathia, the biggest rewilding project in Europe, and the White Stork Project. He is a co-founder of Rewilding Britain, and sits on the board of the Arcadia Fund and the oversight committee for the Endangered Landscapes Programme. He also chairs rePlanet Wildlife and Nattergal Ltd, sister companies involved in nature restoration.

BLOOMSBURY PUBLISHING
Bloomsbury Publishing Plc
50 Bedford Square, London, WC1B 3DP, UK
29 Earlsfort Terrace, Dublin 2, Ireland

BLOOMSBURY, BLOOMSBURY PUBLISHING and the
Diana logo are trademarks of Bloomsbury Publishing Plc

First published in Great Britain 2023

A catalogue record for this book is available from the British Library

Library of Congress Cataloguing-in-Publication data has been applied for

ISBN: HB: 978-1-5266-5929-3; eBook: 978-1-5266-5930-9; ePDF: 978-1-5266-5931-6

2 4 6 8 10 9 7 5 3

Commissioning Editor: Rowan Yapp
Project Editor: Laura Gladwin
Text design and layout: Anna Green

Printed and bound in Germany by Mohn Media

FSC
www.fsc.org
MIX
Paper | Supporting
responsible forestry
FSC® C011124

To find out more about our authors and books visit www.bloomsbury.com and sign up for our newsletters